The Chloroplast: Basics and Applications

The Chloroplast: Basics and Applications

Editor: Eldon Dixon

MURPHY & MOORE
www.murphy-moorepublishing.com

www.murphy-moorepublishing.com

ⓂMURPHY & MOORE

Cataloging-in-Publication Data

The chloroplast : basics and applications / edited by Eldon Dixon.
    p. cm.
Includes bibliographical references and index.
ISBN 978-1-63987-735-5
1. Chloroplasts. 2. Chloroplast DNA. 3. Biochemistry. I. Dixon, Eldon.
QK898.C52 C45 2023
581.192 7--dc23

Murphy & Moore Publishing
1 Rockefeller Plaza,
New York City,
NY 10020, USA

ISBN 978-1-63987-735-5

Contents

Preface

Plants are the foundation of all life that exists on the planet Earth. Plant cells contain plastids, which are double-membrane cell organelles. Plastids perform the function of manufacturing and storing food. These cell organelles can be classified into three types, namely, chloroplasts, chromoplasts, and leucoplasts. Chloroplasts are the organelles responsible for food production in plants. Stroma, inner membrane, outer membrane, thylakoid membrane, and intermembrane space are specific parts of chloroplasts. There are various functions of chloroplasts, including photosynthesis, pathogen defense, amino acid synthesis, carbon fixation and G3P synthesis. Photosynthesis is the main function of chloroplasts, which involves the conversion of light into chemical energy in order to produce food in the form of sugars. This book outlines the basic concepts and applications of chloroplast in detail. It consists of contributions made by international experts. This book aims to serve as a resource guide for students and experts.

This book is the end result of constructive efforts and intensive research done by experts in this field. The aim of this book is to enlighten the readers with recent information in this area of research. The information provided in this profound book would serve as a valuable reference to students and researchers in this field.

At the end, I would like to thank all the authors for devoting their precious time and providing their valuable contribution to this book. I would also like to express my gratitude to my fellow colleagues who encouraged me throughout the process.

Editor

Comparative Analyses of the Chloroplast Genomes of Patchouli Plants and their Relatives in *Pogostemon* (Lamiaceae)

Cai-Yun Zhang [1,†]**, Tong-Jian Liu** [2,*,†]**, Xiao-Lu Mo** [1]**, Hui-Run Huang** [2,3]**, Gang Yao** [4]**, Jian-Rong Li** [2]**, Xue-Jun Ge** [2,3] **and Hai-Fei Yan** [2,5]

[1]   Guangdong Food and Drug Vocational College, Guangzhou 510520, China; zhangcy@gdyzy.edu.cn (C.-Y.Z.); moxl@gdyzy.edu.cn (X.-L.M.)

[2]   Key Laboratory of Plant Resources Conservation and Sustainable Utilization, South China Botanical Garden, Chinese Academy of Sciences, Guangzhou 510650, China; huirun.huang@scbg.ac.cn (H.-R.H.); lijianrong@scbg.ac.cn (J.-R.L.); xjge@scbg.ac.cn (X.-J.G.); yanhaifei@scbg.ac.cn (H.-F.Y.)

[3]   Center of Conservation Biology, Core Botanical Gardens, Chinese Academy of Sciences, Guangzhou 510650, China

[4]   South China Limestone Plants Research Centre, College of Forestry and Landscape Architecture, South China Agricultural University, Guangzhou 510642, China; gyao@scau.edu.cn

[5]   Center of Plant Ecology, Core Botanical Gardens, Chinese Academy of Sciences, Guangzhou 510650, China

*   Correspondence: liutongjian@scbg.ac.cn

†   These authors contributed equally to this work.

Abstract: *Pogostemon* Desf., the largest genus of the tribe Pogostemoneae (Lamiaceae), consists of ca. 80 species distributed mainly from South and Southeast Asia to China. The genus contains many patchouli plants, which are of great economic importance but taxonomically difficult. Therefore, it is necessary to characterize more chloroplast (cp) genomes for infrageneric phylogeny analyses and species identification of *Pogostemon*, especially for patchouli plants. In this study, we newly generated four cp genomes for three patchouli plants (i.e., *Pogostemon plectranthoides* Desf., *P. septentrionalis* C. Y. Wu et Y. C. Huang, and two cultivars of *P. cablin* (Blanoco) Benth.). Comparison of all samples (including online available cp genomes of *P. yatabeanus* (Makino) Press and *P. stellatus* (Lour.) Kuntze) suggested that *Pogostemon* cp genomes are highly conserved in terms of genome size and gene content, with a typical quadripartite circle structure. Interspecific divergence of cp genomes has been maintained at a relatively low level, though seven divergence hotspot regions were identified by stepwise window analysis. The nucleotide diversity (*Pi*) value was correlated significantly with gap proportion (indels), but significantly negative with GC content. Our phylogenetic analyses based on 80 protein-coding genes yielded high-resolution backbone topologies for the Lamiaceae and *Pogostemon*. For the overall mean substitution rates, the synonymous (d_S) and nonsynonymous (d_N) substitution rate values of protein-coding genes varied approximately threefold, while the d_N values among different functional gene groups showed a wider variation range. Overall, the cp genomes of *Pogostemon* will be useful for phylogenetic reconstruction, species delimitation and identification in the future.

Keywords: Pogostemoneae; nucleotide diversity; phylogeny; genome skimming

1. Introduction

Pogostemon Desf. is the largest genus of the tribe Pogostemoneae (Lamiaceae), consisting of about 80 species distributed mainly from South and Southeast Asia to China [1]. According to the recent infrageneric classifications [2,3], this genus has been separated into three subgenera, which are subg. *Pogostemon*, subg. *Allopogostemon* Bhatti & Ingr., and subg. *Dysophyllus* Bhatti & Ingr. The former two (*Pogostemon* sensu stricto) consist of terrestrial herb and subshrubs, whereas the latter is made up of aquatic and marshland plants [2]. However, the most recent molecular phylogenetic study showed that none of the three morphologically defined subgenera of *Pogostemon* were supported as monophyletic, and that *Pogostemon* should be classified into two subgenera (i.e., subg. *Pogostemon* and subg. *Dysophyllus*) [4].

The newly circumscribed subgenus *Pogostemon* contains 27 aromatic species [2,4], which are of great economic importance as the sources of patchouli oil (an essential oil extracted from the dried tops of aromatic *Pogostemon* plants) [5]. *Pogostemon cablin* (Blanoco) Benth. (patchouli) is a well-known aromatic species in the world, since it is one of the top 20 essential oil-yielding plants traded on the world market [5]. Like *P. cablin*, other *Pogostemon* taxa in subg. *Pogostemon*, such as *P. plectranthoides* Desf., *P. heyneanus* Benth., and *P. benghalensis* (Burm.f.) Kuntze, have also been cultivated for their essential oils (also known as patchouli oil) [2]. Nonetheless, these oils are of inferior and often variable quality [6]. Patterns of morphological variation among patchouli plants are complicated, since different methods of cultivation and diverse local climatic conditions (especially when transferred between countries) can result in substantial morphological changes in patchouli species [6]. Patchouli species have therefore been considered to be a taxonomically difficult group [2]. Thus, it is essential to generate new data to solve the species delimitation and identification issues in the group.

Chloroplast (cp) genomes have been considered as "ultra-barcode" for species/cultivar identification [7–9] and phylogenomic analyses [10–12]. Recently, obtaining hundreds to thousands of organelle loci has become routine because of rapid improvements in high-throughput sequencing (HTS) technology in the past decade [13,14]. The number of whole cp genomes has rocketed in recent years. However, there is still a lack of cp genome data for the genus *Pogostemon* (especially for patchouli species). So far, only nine cp genome sequences derived from three *Pogostemon* species are publicly available, of which only the *P. cablin* cp genomes have been well studied [9]. It is necessary to sequence and characterize more cp genomes for analyzing the infrageneric phylogeny and species identification of *Pogostemon*, especially for the economically important patchouli plants. In this study, genome-skimming data based on a high-throughput sequencing platform was generated for *P. plectranthoides*, *P. septentrionalis* C. Y. Wu & Y. C. Huang, and two cultivars of *P. cablin* ("Indonesia" and "Vietnam"). We then assembled whole chloroplast genomes for these taxa and comparatively analyzed them (including the sequences of subg. *Dysophyllus* available in GenBank). Specifically, we aimed to (1) characterize the cp genomes of *Pogostemon* and identify high-divergence hotspots, (2) characterize nucleotide substitution patterns across *Pogostemon* cp genomes, and (3) revisit the infrageneric relationships of *Pogostemon* and the backbone phylogeny of Lamiaceae based on cp genomes.

2. Results

2.1. Characteristics of Chloroplast Genomes

The detailed information of *Pogostemon* cp genomes (including publicly available data in GenBank) is listed in Table 1. We obtained 10,041,957 and 10,334,686 paired-end reads from genome skimming sequencing for *P. plectranthoides* and *P. septentrionalis*, respectively. The average sequencing depth of the cp genome was 550.4× for *P. plectranthoides* and 844.6× for *P. septentrionalis*. The length of the cp genome is 152,430 bp for *P. plectranthoides* and 152,514 bp for *P. septentrionalis*. These two cp genomes have the typical quadripartite structure, consisting of a pair of identical inverted repeats regions (IRs; 25,666–25,665 bp), separated by a large single-copy region (LSC; 83,614–83,514 bp) and a small single-copy region (SSC; 17,584–17,570 bp). The overall GC content of the cp genome is 38.3% for *P. plectranthoides* and 38.2% for *P. septentrionalis*. The newly generated cp genomes of the two *P. cablin* cultivars ("Indonesia" and "Vietnam") were effectively identical to those of other *P. cablin* cultivars, with only one base difference [9]. We therefore arbitrarily chose the cp genome of the cultivar "Gaoyao" (GenBank no. MF445415) as a representative of *P. cablin* for further comparative analyses.

Table 1. Characteristics of chloroplast genomes from five species in the genus *Pogostemon*.

Genome Feature	*P. plectranthoides*	*P. septentrionalis*	*P. yatabeanus*	*P. stellatus*	*P. cablin* "Gaoyao"
Genome size	152,430	152,514	152,707	151,824	152,461
LSC length	83,514	83,614	83,791	83,012	83,553
IR length	25,666	25,665	25,674	25,644	25,662
SSC length	17,584	17,570	17,568	17,524	17,584
Total coding length	91,424	91,445	91,132	91,135	91,442
Protein-coding length	79,500	79,521	79,275	79,278	79,518
rRNA-coding length	9064	9064	9064	9064	9064
tRNA-coding length	2860	2860	2793	2793	2860
Total GC content (%)	38.3	38.2	38.2	38.2	38.2
LSC GC content (%)	36.4	36.4	36.2	36.3	36.4
IR GC content (%)	43.4	43.4	43.4	43.4	43.4
SSC GC content (%)	32.1	32.1	32	32.1	32.1
Total number of genes (total/different)	132/114	132/114	132/114	132/114	132/114
Number of duplicated genes in IR	18	18	18	18	18
Number of genes with introns (with 3 exons)	18(2)	18(2)	18(2)	18(2)	18(2)
Number of protein-coding genes (total/in IR)	80/7	80/7	80/7	80/7	80/7
Number of tRNA genes (total/in IR)	30/7	30/7	30/7	30/7	30/7
Number of rRNA genes (total/in IR)	4/4	4/4	4/4	4/4	4/4

LSC—large single-copy region; SSC—small single-copy region; IR—inverted repeats region.

We identified a total of 114 unique genes in the *P. plectranthoides* and *P. septentrionalis* cp genomes, including 80 protein-coding genes (PCGs), 30 tRNA genes, and four rRNA genes, consistent with the numbers identified in three other *Pogostemon* species. The gene content was conserved in these genomes, each of which has seven coding genes, seven tRNA genes, and four rRNA genes duplicated in IR regions. Besides, twelve PCGs and six tRNA genes were interrupted by introns; sixteen of these genes contained a single intron each and two genes (*clpP* and *ycf3*) possessed two introns (Figure 1).

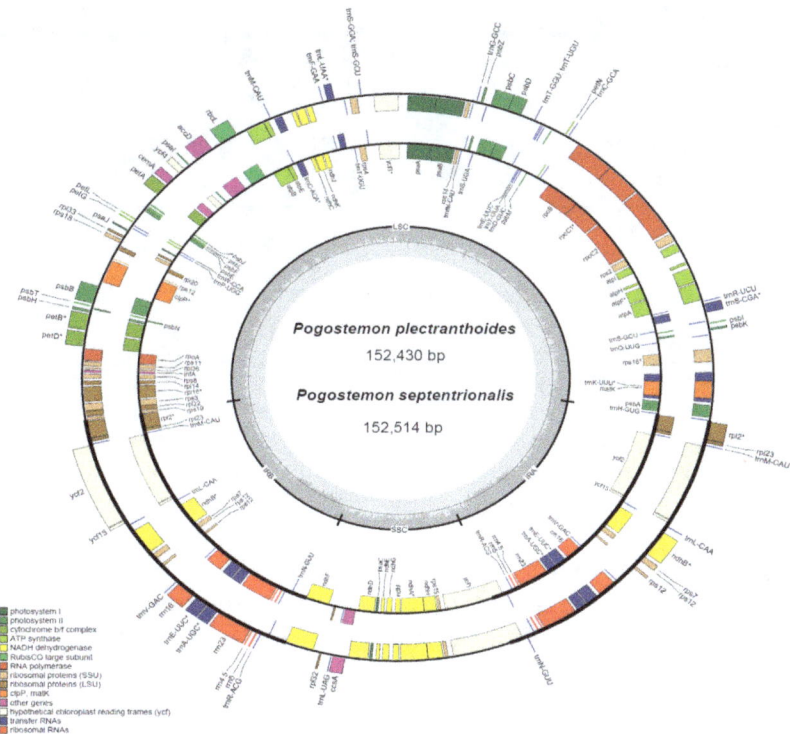

Figure 1. Physical map of two chloroplast genomes from *Pogostemon plectranthoides* (outer circle) and *Pogostemon septentrionalis* (inner circle). Genes inside the circle are transcribed clockwise and those outside are transcribed counterclockwise. The light gray inner circle corresponds to the AT content and the dark gray circle to the GC content. Genes belonging to different functional groups are shown in different colors. Asterisks indicate intron-containing genes.

In addition, *P. yatabeanus* (Makino) Press has the largest cp genome (152,707 bp), followed by *P. septentrionalis* (152,514 bp) (Table 1). The cp genome structures of *Pogostemon* are highly conserved, since no large genome rearrangement was detected among the five cp genomes (Figure S1). A limited variation occurred in the junction between IRs and the SSC (Figure S2). We found that the largest difference in genome size was mainly due to the indels in intergenic spacers (e.g., *trnC–petN* has 431 bp, and *trnF–ndhJ* contains 258 bp).

2.2. Divergence Hotspots in Chloroplast Genomes

Intergenic spacers (IGS) were the most highly variable regions, followed by introns, while coding regions were the most conserved regions (Figure 2). Specifically, the five most variable non-coding loci ($Pi > 0.03$) identified in this study were *rps16–trnQ*, *petA–psbJ*, the *rpl16* intron, *ndhF–rpl32*, and *rpl32–trnL* (Figure 2). *Ycf1* was the only coding region with high sequence divergence ($Pi > 0.03$, Figure 2). In a comparison of PCGs using phylogenetically informative characters (PICs), *ycf1* was still the most variable region, contributing 146 PICs from an alignment length of 5583 bp (2.6%) (Figure 3A, Table S1). The other coding genes with high PICs were *rpoC2*, *ndhF*, *matK*, and *rpoB*, having 58 (1.39%), 47 (2.10%), 44 (2.86%), and 30 (0.93%) PICs, respectively (Table S1). In addition, these divergence hotspots (coding and non-coding regions) were all located in the single-copy regions (LSC and SSC).

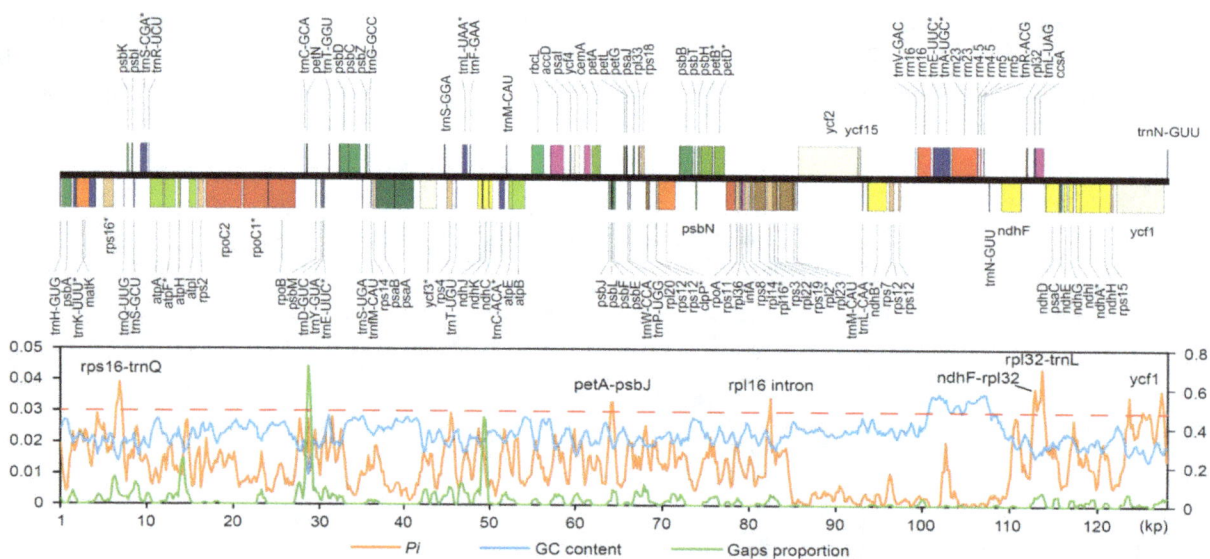

Figure 2. Sliding window analysis showing the nucleotide diversity (*Pi*), GC content, and gap proportion in the five *Pogostemon* chloroplast genome sequences, using 600 bp windows and a 200 bp step size.

We identified a significant negative correlation between *Pi* and GC content (Figure 4A), while the relationship between *Pi* and gap (indel) ratio was a positive correlation (Figure 4B), suggesting a close association between sequence variation and nucleotide composition. Our results seem to indicate that indels have a potential impact on sequence divergence in the *Pogostemon* chloroplast genome. Furthermore, large indels (see above) resulted in length polymorphism in these chloroplast genomes.

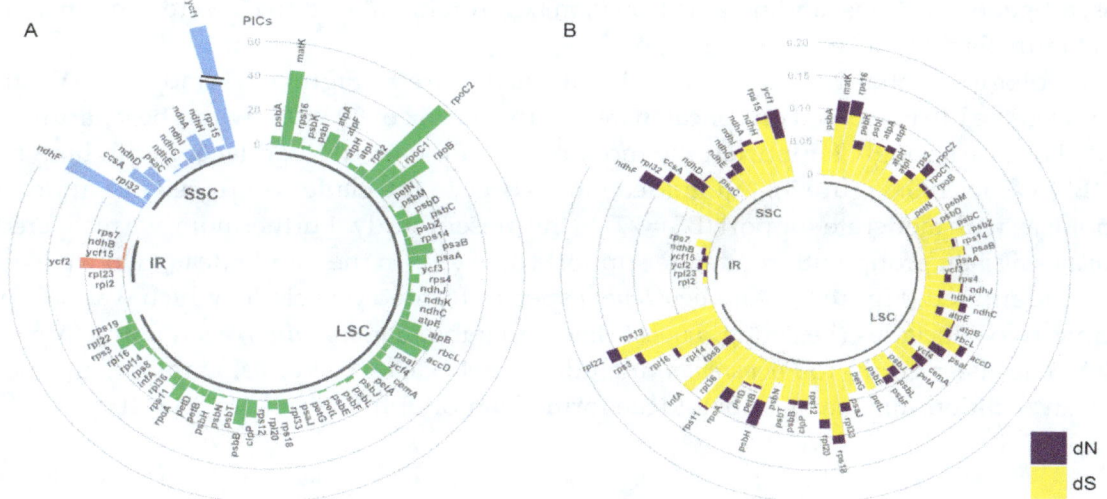

Figure 3. Summary of variation in phylogenetically informative characters (PICs) and substitution rates in *Pogostemon*. (**A**) Phylogenetically informative character variation in protein-coding genes in the five *Pogostemon* chloroplast genomes indicated by PICs. (**B**) Nonsynonymous (d_N) and synonymous (d_S) substitution rate for each protein-coding gene in the five *Pogostemon* chloroplast genomes.

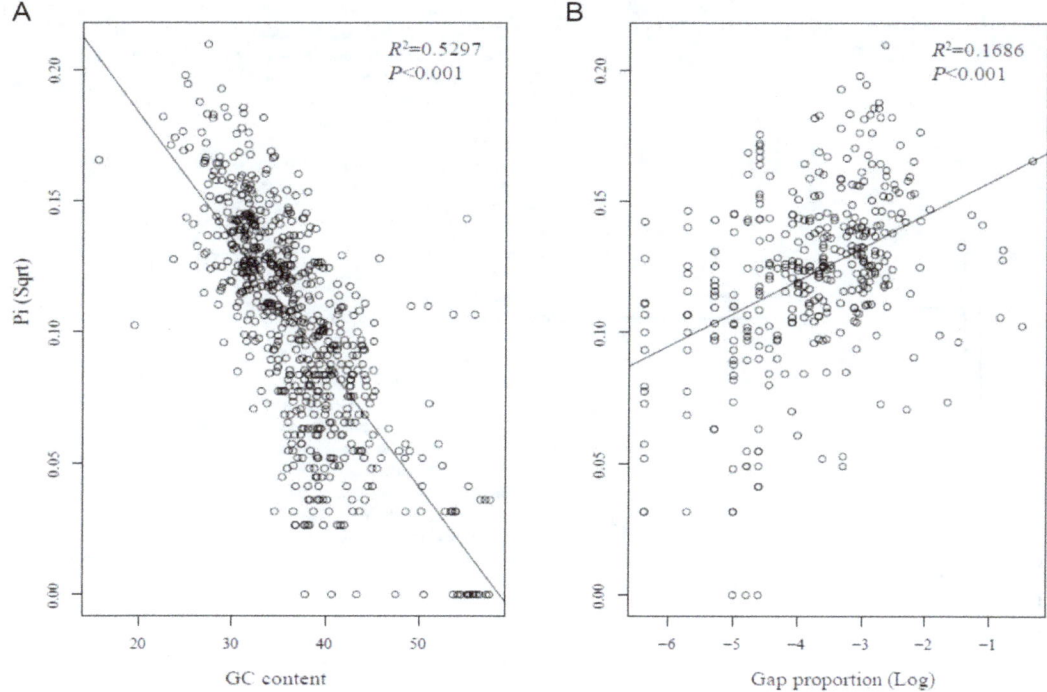

Figure 4. GC content and gap proportion are correlated with *Pi*. (**A**) Scatter plot showing a significant negative correlation ($p < 0.001$) between GC content and *Pi* (Sqrt). (**B**) Scatter plot showing a significant positive correlation ($p < 0.001$) between gap (indels) proportion and *Pi* (Sqrt) in the five *Pogostemon* chloroplast genomes. All three features were measured using 600 bp windows and a 200 bp step size.

2.3. Phylogenetic Findings

The chloroplast dataset comprised 80 concatenated PCGs with a total aligned length of 68,514 bp, after stripping out the sites containing more than 20% gaps. For large matrices of this type, the best practice for accommodating rate heterogeneity is grouping genes and codon positions that fitted the same best models into different partitions. Using three partitioning strategies (the best inferred gene–codon partitions, all-gene partitions, and no partitions), maximum likelihood trees were constructed and are shown in Figure 5 and Figures S3 and S4.

The topologies of the three trees were identical (Figure 5, Figures S3 and S4). We, therefore, mainly showed the ML tree based on concatenated codon matrix of 80 PCGs with a best partition scheme (Figure 5). The tree strongly supported the monophyly of the family Lamiaceae with high bootstrap support (BS) value (100; Figure 5). Ajugoideae was sister to the clade comprised of Lamioideae and Scutellarioideae with moderate support (BS = 77) in the present study. Furthermore, we recovered *Premna* and *Tectona* as a sister group with moderate support (BS = 73). In the Lamioideae, two well-supported clades were identified (Figure 5). All *Pogostemon* species formed one clade, which was subsequently divided into two subclades (i.e., subg. *Pogostemon* and subg. *Dysophyllus*), each with strong support (BS = 100). The sister group to *Pogostemon* included Stachydeae, *Galeopsis*, Lamieae, and Leonureae (Figure 5). In addition, all taxa of Nepetoideae formed a monophyletic clade (BS = 100).

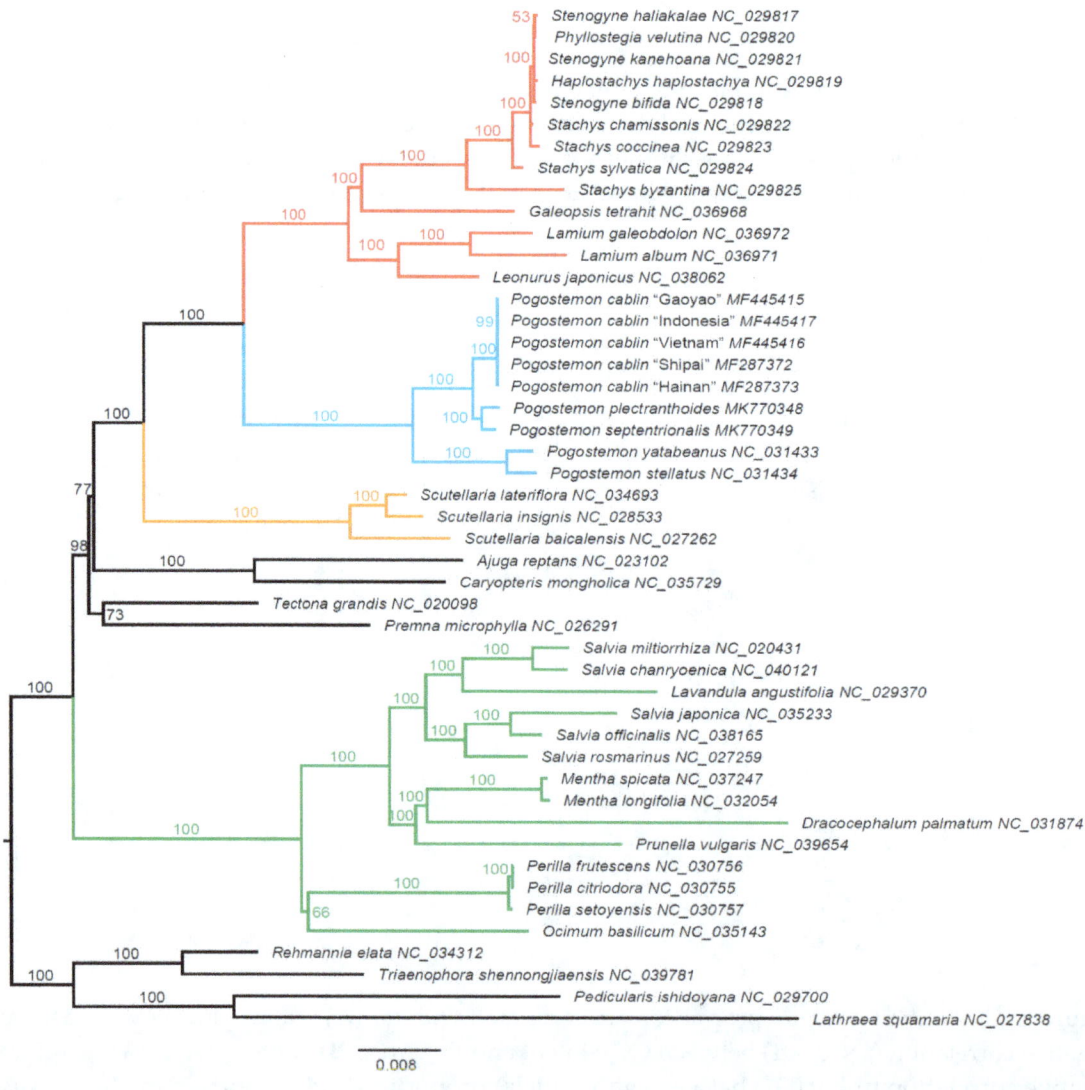

Figure 5. Maximum likelihood tree of 43 Lamiaceae accessions and four outgroups inferred from a concatenated codon supermatrix of 80 chloroplast protein-coding genes with a best partition scheme. Red clade: Stachydeae + *Galeopsis* + Lamieae + Leonureae, blue clade: *Pogostemon*, yellow clade: Scutellarioideae, and green clade: Nepetoideae. The group including red and blue colors is the Lamioideae. Numbers on the branches are bootstrap support values.

2.4. Estimation of Evolutionary Rates among Protein-Coding Genes

Substitution rates varied considerably among the chloroplast genes across the five *Pogostemon* species (Figure 3B). For example, eighteen genes had only synonymous substitutions, whereas three contained solely nonsynonymous substitutions (Table S1). In addition, five genes had no variation representing either type of substitution event (Table S1). The mean substitution rate for all chloroplast genes was 0.008 ± 0.0095 for nonsynonymous substitution rates (d_N) and 0.0467 ± 0.0303 for synonymous substitution rates (d_S) (Table S1).

Among gene groups, *rpo* genes had the highest mean nonsynonymous substitution rate (0.0127 ± 0.0072), followed by *ycf* genes (0.0125 ± 0.0176), whereas the lowest nonsynonymous substitution rate occurred in *pet* genes (0.0018 ± 0.0020) (Table S2). In contrast, mean synonymous substitution rates (d_S) among gene groups ranged from 0.0197 to 0.0630 (Table S2).

3. Discussion

3.1. Divergence Hotspots in Cp Genomes

Genome-wide comparisons have examined sequence divergence to decide which genes or regions to utilize for phylogenetic studies within angiosperms [15–17] and chloroplast biotechnology [18]. The poor phylogenetic relationships within *Pogostemon* revealed in previous studies partly resulted from the sequences without specific highly variable chloroplast regions and adequately phylogenetically informative sites [4].

In this study, we detected six highly variable loci in *Pogostemon* cp genomes, all of which are located within the eight top-ranking regions identified in 25 angiosperm lineages by Shaw et al. [17], consistent with previous findings in *Pogostemon* [9]. The phylogenetic utility of protein-coding sequences in *Pogostemon* was also evaluated by a comparison of the number of PICs. Among the protein-coding genes, *ycf1* has the highest PICs, followed by *rpoC2*, *ndhF*, and *matK*. This result is similar to the findings of Walker et al. [19]. Of the most commonly used DNA barcodes, Walker et al. [19] found that gene *rpoC2* is the top-performing gene and *matK* greatly outperformed *rbcL*. They excluded *ycf1* because of its long alignment length. However, the region *ycf1* has been recommended as a powerful barcode for use within angiosperms [20]. In this study, at least three hotspots were identified in the *ycf1* region (Figure 2), which will facilitate primer design and sequencing by the Sanger method as suggested by Dong et al. [20].

Identifying divergence hotspots has long been a central issue in chloroplast genome comparisons. In general, sequence divergence in chloroplast genomes results from several common types of variation, including substitution, insertion/deletion, duplication, and rearrangement [21]. However, few studies have taken indels or inversions into consideration when comparing sequences along chloroplast genomes [22], since a high level of indels and inversions may complicate the choice of phylogenetic markers in highly variable genomes. Given the conserved structure of cp genomes in *Pogostemon*, nucleotide diversity per site (*Pi*), gap proportion, and GC content were calculated separately in each 600 bp window. A positive relationship between nucleotide substitutions and genomic rearrangements (including indels) has been detected in previous studies of angiosperms [23]. In this study, nucleotide diversity showed a significant positive correlation with gap proportion ($p < 0.001$, Figure 4B), confirming the above finding by Jansen et al. [23]. One hypothesis to explain this is that aberrant DNA repair mechanisms accelerate substitution rates [24,25]. Moreover, we found that nucleotide diversity was negatively associated with GC content (Figure 4A). Previous studies indicated that base composition often plays an important role in chloroplast DNA sequence evolution, resulting in mutations that are spatially biased across the genome [23,26].

3.2. Phylogenetic Positions of Pogostemon and Its Related Taxa

In recent years, complete or nearly complete organelle genomes have become increasingly accessible, providing a powerful tool for phylogenetic studies. Our molecular phylogenetic analyses based on 80 PCGs produced a high-resolution backbone topology for the Lamiaceae. All three trees strongly supported the monophyly of the family Lamiaceae (BS = 100; Figure 5, Figures S3 and S4). Our results recognized several subfamilies as presented by Li et al. [27], i.e., the Lamioideae, Scutellarioideae, and Nepetoideae, which were fully supported by bootstrap analysis. In the Lamioideae, we confirmed the robust sister relationship between *Pogostemon* and the remaining taxa within the subfamily, a finding which supports the phylogeny of Bendiksby et al. [28] based on four chloroplast molecular markers. All *Pogostemon* species formed one group, which could be subsequently divided into two subclades with strong support (BS = 100), in agreement with the result of Yao et al. [4]. The contentious relationships among the subfamilies Ajugoideae, Viticoideae, and the genus *Tectona* were still not fully resolved, but we considerably improved the resolution of these relationships using chloroplast phylogenomics. For example, our result moderately supported the Ajugoideae as sister to the clade comprising the Lamioideae and Scutellarioideae (BS = 77). We also recovered *Premna* and *Tectona* as a sister group with moderate support (BS = 73), which agreed with the phylogenetic results of Bendiksby et al. [28] but differed from those in a recent study by Li et al. [27].

3.3. Substitution Rate Variation in Cp Genomes

Substitution rate heterogeneity has been observed in chloroplast genomes across different lineages of plants and among different classes of chloroplast genes [29,30]. This lineage-specific or locus-specific variation in nucleotide substitution may have played a major role in adaptive evolution in different kinds of plants. In particular, the acceleration of substitution rates for specific gene classes has been found repeatedly among many plant groups [24,31]. Drouin et al. [32] showed that the d_N and d_S rates for different genes within the cp genome are very similar, not varying by more than two- to four-fold. However, their viewpoint was based on only a few cp genes (i.e., *atpB*, *matK*, *psaA*, *psbB*, and *rbcL*), which probably resulted in an underestimation of the variation in cp genes. In the present study, the mean d_S values for different cp genes varied by about threefold (0.0197 to 0.0630), in agreement with the results of Drouin et al. [32]. However, the d_N values among different functional gene groups showed greater variation than d_S values. For example, the d_N values for *rpo* (RNA polymerase genes) and *ycf* (unknown function) genes had elevated nonsynonymous substitution rates compared with other genes, to the extent that they were seven times higher than the lowest d_N value (*pet* genes) (Table S2). The pattern of genes with accelerated d_N has been found in *Pelargonium* and *Sileneae* [33,34] and has also been reported in a comparative study on cp genomes of flowering plants [24].

The molecular mechanism underlying extremely high variation in nonsynonymous substitution rates in most plants is still elusive. It is often inferred that the variation in nonsynonymous substitution rates was more likely to result from strong coevolution with synonymous substitutions, instead of consequence from the relaxation of purifying selection [35]. However, our data did not significantly support this hypothesis, since several genes (i.e., *ycf1*, *psbH*, and *matK*) showed high nonsynonymous rates but low synonymous substitutions rates, which led to relatively higher d_N/d_S (ω) values (0.7186 for *ycf1*, 0.6482 for *psbH*, and 0.4317 for *matK*) (Table S1). We inferred that positive diversifying selection acting on the above genes would be expected to result in higher nonsynonymous substitution rates.

4. Materials and Methods

In this study, we newly sequenced two patchouli species (*P. plectranthoides* and *P. septentrionalis*) and two cultivars of *P. cablin* ("Indonesia" and "Vietnam") using genome-skimming technology. These species were cultivated and collected in a greenhouse at South China Botanical Garden. No specific permissions were required for the relevant locations/activities. Voucher specimens were deposited in the Herbarium of South China Botanical Garden. In order to carry out a comprehensive analysis of *Pogostemon*, publicly available chloroplast genomes sequences (*P. cablin*, *P. yatabeanus*, and *P. stellatus* (Lour.) Kuntze) from *Pogostemon* in GenBank were used in this study (Table S3).

4.1. DNA Extraction and Sequencing

Total genomic DNA was isolated from fresh healthy leaves using a modified cetyltrimethylammonium bromide (CTAB) method [36]. The DNA concentration was estimated using a Qubit Fluorometer with a Qubit dsDNA HS Assay Kit (ThermoFisher, Waltham, MA, USA). Short-insert (ca. 500 bp) paired-end libraries were prepared using a TruePrepTM DNA Library Prep Kit V2 for Illumina (Vazyme Biotech Co., Ltd., Nanjing, China), following the manufacturer's protocol. Genome skimming sequencing with a 2 × 150 bp chemistry reaction system was performed on an Illumina HiSeq X Ten platform at the Beijing Genomics Institute (Shenzhen, China).

4.2. Assembly and Annotation

The raw reads generated from Illumina paired-end sequencing were assessed by FastQC 0.11.5 [37] and quality control was carried out using Trimmomatic 0.35 [38] to remove adapters and low-quality nucleotide bases.

Here, we used two strategies to assemble the chloroplast genome. The first was the "seed-and-extend" approach in NOVOPlasty 2.5.9 [39], which assembles adapter-free reads into rough scaffolds. The whole chloroplast genome and *rbcL* protein-coding sequence of *P. cablin* "Shipai" (GenBank no. MF287372) were used separately as seed to trigger the assembly. To assess the quality and to correct errors that occurred in the initial assembly, we mapped cleaned reads back to the resulting scaffolds in Bowtie2 v.2.3.4 [40]. After checking and refining the reads mapping graph, a consensus sequence was generated with a 90% matching threshold. The circular and ungapped consensus sequence was considered to represent the complete chloroplast genome.

In the other approach, a "de novo assembly" strategy was employed using SPAdes assembler 3.13.0 [41], with parameters "-k 21,33,55,77 –careful –cov-cutoff 10". We used Bandage 0.8.1 [42] to visualize and manipulate the assembly graph. Each node (or contig) in the graph was used to blast against the reference cp genome (*P. cablin* "Shipai", MF287372). Nodes without blast hits were filtered out. For unresolved loops (two or more alternative nodes with the same blast hit) in the assembly graph, we dropped the nodes with lower read coverage and retained the most reliable one. For each shared node on two intersectional paths with different blast hits, we duplicated this node as repeat sequences and placed it in two distinct paths. Finally, complete chloroplast genomes were obtained from the refined graphs.

The chloroplast genomes were annotated using Dual Organellar Genome Annotator [43]. We checked and adjusted the annotations of protein-coding genes by comparison with homologous genes from the above well-annotated reference cp genomes (*P. cablin* "Shipai", MF287372, and *P. yatabeanus*, KP718618). The tRNA genes were annotated using ARAGORN v1.2.38 [44]. Circular genome maps were drawn on the OrganellarGenomeDRAW (OGDRAW) online service following by manual modification [45]. Finally, well-annotated cp genomes were submitted to GenBank (Table S3). Raw sequencing data in this study are archived at the National Center for Biotechnology Information under BioProject PRJNA671793.

4.3. Sequence and Structure Divergence

The gene order and structures were analyzed in the pairwise comparisons across the five *Pogostemon* cp genomes using Geneious R11.1.5 [46]. Three single-copy (SC) and inverted repeat (IR) region (SC–IR) junctions of all *Pogostemon* cp genomes were manually examined to investigate expansion and contraction in the IR. Multiple sequence alignments of all five *Pogostemon* cp genomes were performed in MAFFT v7 [47] under default parameters and manually adjusted in Geneious. The mVISTA program in Shuffle-LAGAN mode [48] was used to visualize the overall similarities among different cp genomes in *Pogostemon*.

To identify divergence hotspots in chloroplast genomes, sliding window analysis (the window length was 600 bp with a 200 bp step size) was employed to calculate nucleotide diversity (*Pi*),

GC content, and gap proportion for each window in a *Pogostemon* cp genome alignment, with the DendroPy library and a custom Python script [49]. *Pi* indicates the average number of nucleotide differences per site between two sequences from all possible pairs in an alignment. Gap proportion shows the level of missing data caused by indels. GC content gives fundamental information about the stability of nucleotide sequences, which usually corresponds to highly conserved regions constrained by mRNA secondary structures [50]. The correlation between *Pi* and GC content or gap proportion was tested in R 3.6.1 [51]. Before fitting to a linear regression model, *Pi* and gap proportion were transformed to fit the normal distribution.

4.4. Phylogenetic Analyses

In order to determine the phylogenetic position of *Pogostemon* and relationships among the major groups of Lamiaceae, we collected 47 chloroplast genomes, including nine accessions from the genus *Pogostemon*, 34 other accessions in the Lamiaceae, and four outgroup accessions from the Orobanchaceae (Table S3).

Nucleotide sequences of protein-coding genes (PCGs) were extracted from the 47 annotated chloroplast genomes from the Lamiaceae. Protein-coding sequences were first translated into amino acid sequences and aligned using MAFFT v7 under the L-INS-I method. These protein alignments were then back-translated to codon alignments in PAL2NAL [52]. A matrix was generated by concatenating the codon alignments of 80 PCGs and removing sites with >20% of gaps. Given the gene and codon position partitions, the best-fitting model of nucleotide substitution for the entire dataset was determined by the corrected Akaike information criterion (AICc) in PartitionFinder 2.1.1 [53]. Maximum likelihood (ML) analyses were performed in RAxML-HPC v8.2.12 [54] based on three partitioning strategies (one partition, gene-partitioning, and best-fit partitioning schemes). All three ML analyses were evaluated with 1000 rapid bootstrap replicates.

4.5. Estimation of Evolutionary Rates

The nucleotide sequences of 80 protein-coding genes were extracted from five *Pogostemon* cp genomes. Each gene was codon-aligned using the L-INS-I method in MAFFT v7. Phylogenetically informative characters (PICs) were counted for each gene using a Python script. Given both codon and related protein alignments of each gene, average nonsynonymous (d_N) and synonymous (d_S) substitution rates were estimated using the maximum likelihood method [55] with the F3 × 4 model implemented in the codeml program in PAML v4.9 [56]. In addition, protein-coding genes were assigned to nine functional groups according to the conventional classification. The genes within a functional group were concatenated for the above tests as well. The best ML tree based on PCGs was used as a constraint tree.

Supplementary Materials:
Figure S1: mVISTA plot showing the percent identity of plastid genomes between each of four *Pogostemon* species and the reference *Pogostemon cablin* based on pairwise global sequence alignments, Figure S2: IRa-SSC and SSC-IRb boundaries in five *Pogostemon* chloroplast genomes, Figure S3: Maximum likelihood tree of 43 Lamiaceae accessions and four outgroups inferred from a concatenated codon matrix of 80 plastid protein-coding genes with gene partitions, Figure S4: Maximum likelihood tree of 43 Lamiaceae accessions and four outgroups inferred from a concatenated codon supermatrix of 80 plastid protein-coding genes without partitions, Table S1: Number of phylogenetically informative characters (PICs), and d_N, d_S, and ω values for each protein-coding gene, Table S2: Mean d_N, d_S, and ω values for each group of protein-coding genes, Table S3: Samples used for chloroplast phylogenomic analyses in this study.

Author Contributions: Conceptualization, T.-J.L., C.-Y.Z., and H.F.Y.; methodology, T.-J.L.; formal analysis, C.-Y.Z. and T.-J.L.; investigation, T.-J.L. and C.-Y.Z.; resources, X.-L.M., J.-R.L., and X.-J.G; data curation, T.-J.L., C.-Y.Z., and H.-F.Y.; writing—original draft preparation, T.-J.L. and C.-Y.Z.; writing—review and editing, T.-J.L., C.-Y.Z., H.-R.H., G.Y., X.-J.G., and H.F.Y.; visualization, T.-J.L. and C.-Y.Z.; supervision, X.-J.G. All authors have read and agreed to the published version of the manuscript.

Acknowledgments: We thank the technician Yu-Ying Zhou from the Molecular Ecology lab in South China Botanical Garden for preparing DNA samples.

References

1. Harley, R.M.; Atkins, S.; Budantsev, A.L.; Cantino, P.D.; Conn, B.J.; Grayer, R.; Harley, M.M.; de Kok, R.; Krestovskaja, T.; Morales, R.; et al. Labiatae. In *Flowering Plants Dicotyledons: Lamiales (Except Acanthaceae Including Avicenniaceae)*; Kadereit, J.W., Ed.; Springer: Berlin/Heidelberg, Germany, 2004; pp. 167–275.
2. Bhatti, G.R.; Ingrouille, M. Systematics of Pogostemon (Labiatae). *Bull. Nat. Hist. Mus.* **1997**, *27*, 77–147.
3. Yao, G.; Deng, Y.F.; Ge, X.J. A taxonomic revision of Pogostemon (*Lamiaceae*) from China. *Phytotaxa* **2015**, *200*, 1–67. [CrossRef]
4. Yao, G.; Drew, B.T.; Yi, T.S.; Yan, H.F.; Yuan, Y.M.; Ge, X.J. Phylogenetic relationships, character evolution and biogeographic diversification of Pogostemon s.l. (*Lamiaceae*). *Mol. Phylogen. Evol.* **2016**, *98*, 184–200. [CrossRef] [PubMed]
5. Swamy, M.K.; Sinniah, U.R. Patchouli (Pogostemon cablin Benth.): Botany, agrotechnology and biotechnological aspects. *Ind. Crops Prod.* **2016**, *87*, 161–176. [CrossRef]
6. Weiss, E.A. *Essential Oil Crops*; CAB International: Wallingford, UK, 1997.
7. Kane, N.; Sveinsson, S.; Dempewolf, H.; Yang, J.Y.; Zhang, D.; Engels, J.M.M.; Cronk, Q. Ultra-barcoding in cacao (*Theobroma* spp.; *Malvaceae*) using whole chloroplast genomes and nuclear ribosomal DNA. *Am. J. Bot.* **2012**, *99*, 320–329. [CrossRef]
8. Nock, C.J.; Waters, D.L.E.; Edwards, M.A.; Bowen, S.G.; Rice, N.; Cordeiro, G.M.; Henry, R.J. Chloroplast genome sequences from total DNA for plant identification. *Plant Biotechnol. J.* **2011**, *9*, 328–333. [CrossRef] [PubMed]
9. Zhang, C.Y.; Liu, T.J.; Yuan, X.; Huang, H.R.; Yao, G.; Mo, X.L.; Xue, X.; Yan, H.F. The plastid genome and its implications in barcoding specific-chemotypes of the medicinal herb Pogostemon cablin in China. *PLoS ONE* **2019**, *14*, e0215512. [CrossRef] [PubMed]
10. Bock, D.G.; Kane, N.C.; Ebert, D.P.; Rieseberg, L.H. Genome skimming reveals the origin of the Jerusalem Artichoke tuber crop species: Neither from Jerusalem nor an artichoke. *New Phytol.* **2014**, *201*, 1021–1030. [CrossRef] [PubMed]
11. Gitzendanner, M.A.; Soltis, P.S.; Yi, T.S.; Li, D.Z.; Soltis, D.E. Plastome phylogenetics: 30 years of inferences into plant evolution. *Adv. Bot. Res.* **2018**, *85*, 293–313.
12. Zhang, M.Y.; Fritsch, P.W.; Ma, P.F.; Wang, H.; Lu, L.; Li, D.Z. Plastid phylogenomics and adaptive evolution of Gaultheria series Trichophyllae (Ericaceae), a clade from sky islands of the Himalaya-Hengduan Mountains. *Mol. Phylogen. Evol.* **2017**, *110*, 7–18. [CrossRef]
13. Zimmer, E.A.; Wen, J. Using nuclear gene data for plant phylogenetics: Progress and prospects II. Next-gen approaches. *J. Syst. Evol.* **2015**, *53*, 371–379. [CrossRef]
14. McKain, M.R.; Johnson, M.G.; Uribe-Convers, S.; Eaton, D.; Yang, Y. Practical considerations for plant phylogenomics. *Appl. Plant Sci.* **2018**, *6*, e1038. [CrossRef]
15. Shaw, J.; Lickey, E.B.; Beck, J.T.; Farmer, S.B.; Liu, W.; Miller, J.; Siripun, K.C.; Winder, C.T.; Schilling, E.E.; Small, R.L. The tortoise and the hare II: Relative utility of 21 noncoding chloroplast DNA sequences for phylogenetic analysis. *Am. J. Bot.* **2005**, *92*, 142–166. [CrossRef] [PubMed]
16. Shaw, J.; Lickey, E.B.; Schilling, E.E.; Small, R.L. Comparison of whole chloroplast genome sequences to choose noncoding regions for phylogenetic studies in angiosperms: The tortoise and the hare III. *Am. J. Bot.* **2007**, *94*, 275–288. [CrossRef]
17. Shaw, J.; Shafer, H.L.; Leonard, O.R.; Kovach, M.J.; Schorr, M.; Morris, A.B. Chloroplast DNA sequence utility for the lowest phylogenetic and phylogeographic inferences in angiosperms: The tortoise and the hare IV. *Am. J. Bot.* **2014**, *101*, 1987–2004. [CrossRef] [PubMed]
18. Ruhlman, T.; Verma, D.; Samson, N.; Daniell, H. The role of heterologous chloroplast sequence elements in transgene integration and expression. *Plant Physiol.* **2010**, *152*, 2088–2104. [CrossRef] [PubMed]
19. Walker, J.F.; Stull, G.W.; Walker-Hale, N.; Vargas, O.M.; Larson, D.A. Characterizing gene tree conflict in plastome-inferred phylogenies. *bioRxiv* **2019**. [CrossRef]

20. Dong, W.P.; Xu, C.; Li, C.H.; Sun, J.H.; Zuo, Y.J.; Shi, S.; Cheng, T.; Guo, J.J.; Zhou, S.L. ycf1, the most promising plastid DNA barcode of land plants. *Sci. Rep.* **2015**, *5*, 8348. [CrossRef]

21. Jansen, R.K.; Ruhlman, T.A. Plastid genomes of seed plants. In *Genomics of Chloroplasts and Mitochondria*; Bock, R., Knoop, V., Eds.; Springer: Dordrecht, The Netherlands, 2012; pp. 103–126.

22. Dong, W.P.; Liu, J.; Yu, J.; Wang, L.; Zhou, S.L. Highly variable chloroplast markers for evaluating plant phylogeny at low taxonomic levels and for DNA barcoding. *PLoS ONE* **2012**, *7*. [CrossRef] [PubMed]

23. Jansen, R.K.; Cai, Z.; Raubeson, L.A.; Daniell, H.; Depamphilis, C.W.; Leebens-Mack, J.; Muller, K.F.; Guisinger-Bellian, M.; Haberle, R.C.; Hansen, A.K.; et al. Analysis of 81 genes from 64 plastid genomes resolves relationships in angiosperms and identifies genome-scale evolutionary patterns. *Proc. Natl. Acad. Sci. USA* **2007**, *104*, 19369–19374. [CrossRef]

24. Guisinger, M.M.; Kuehl, J.N.V.; Boore, J.L.; Jansen, R.K. Genome-wide analyses of Geraniaceae plastid DNA reveal unprecedented patterns of increased nucleotide substitutions. *Proc. Natl. Acad. Sci. USA* **2008**, *105*, 18424–18429. [CrossRef]

25. Guisinger, M.M.; Kuehl, J.V.; Boore, J.L.; Jansen, R.K. Extreme reconfiguration of plastid genomes in the angiosperm family *Geraniaceae*: Rearrangements, repeats, and codon usage. *Mol. Biol. Evol.* **2011**, *28*, 583–600. [CrossRef]

26. Decker-Walters, D.S.; Chung, S.M.; Staub, J.E. Plastid sequence evolution: A new pattern of nucleotide substitutions in the Cucurbitaceae. *J. Mol. Evol.* **2004**, *58*, 606–614. [CrossRef]

27. Li, B.; Cantino, P.D.; Olmstead, R.G.; Bramley, G.L.C.; Xiang, C.L.; Ma, Z.H.; Tan, Y.H.; Zhang, D.X. A large-scale chloroplast phylogeny of the *Lamiaceae* sheds new light on its subfamilial classification. *Sci. Rep.* **2016**, *6*, 34343. [CrossRef]

28. Bendiksby, M.; Thorbek, L.; Scheen, A.C.; Lindqvist, C.; Ryding, O. An updated phylogeny and classification of *Lamiaceae* subfamily *Lamioideae*. *Taxon* **2011**, *60*, 471–484. [CrossRef]

29. Gaut, B.; Yang, L.; Takuno, S.; Eguiarte, L.E. The patterns and causes of variation in plant nucleotide substitution rates. *Annu. Rev. Ecol. Evol. Syst.* **2011**, *42*, 245–266. [CrossRef]

30. Wicke, S.; Schneeweiss, G. Next-generation organellar genomics: Potentials and pitfalls of high-throughput technologies for molecular evolutionary studies and plant systematics. In *Next Generation Sequencing in Plant Systematics*; Hörandl, E., Appelhans, M., Eds.; Koeltz Scientific Books: Konigstein, Germany, 2015.

31. Bock, D.G.; Andrew, R.L.; Rieseberg, L.H. On the adaptive value of cytoplasmic genomes in plants. *Mol. Ecol.* **2014**, *23*, 4899–4911. [CrossRef]

32. Drouin, G.; Daoud, H.; Xia, J. Relative rates of synonymous substitutions in the mitochondrial, chloroplast and nuclear genomes of seed plants. *Mol. Phylogen. Evol.* **2008**, *49*, 827–831. [CrossRef]

33. Park, S.; Ruhlman, T.A.; Weng, M.L.; Hajrah, N.H.; Sabir, J.S.M.; Jansen, R.K. Contrasting patterns of nucleotide substitution rates provide insight into dynamic evolution of plastid and mitochondrial genomes of Geranium. *Genome Biol. Evol.* **2017**, *9*, 1766–1780. [CrossRef]

34. Sloan, D.B.; Triant, D.A.; Forrester, N.J.; Bergner, L.M.; Wu, M.; Taylor, D.R. A recurring syndrome of accelerated plastid genome evolution in the angiosperm tribe Sileneae (*Caryophyllaceae*). *Mol. Phylogen. Evol.* **2014**, *72*, 82–89. [CrossRef]

35. Wicke, S.; Naumann, J. Molecular evolution of plastid genomes in parasitic flowering plants. In *Advances in Botanical Research*; Chaw, S.-M., Jansen, R.K., Eds.; Academic Press: London, UK, 2018; Volume 85, pp. 315–347.

36. Porebski, S.; Bailey, L.G.; Baum, B.R. Modification of a CTAB DNA extraction protocol for plants containing high polysaccharide and polyphenol components. *Plant Mol. Biol. Rep.* **1997**, *15*, 8–15. [CrossRef]

37. Andrews, S.; FASTQC. A Quality Control Tool for High Throughput Sequence Data. Available online: http://www.bioinformatics.babraham.ac.uk/projects/fastqc/ (accessed on 20 September 2019).

38. Bolger, A.M.; Lohse, M.; Usadel, B. Trimmomatic: A flexible trimmer for Illumina sequence data. *Bioinformatics* **2014**, *30*, 2114–2120. [CrossRef]

39. Dierckxsens, N.; Mardulyn, P.; Smits, G. NOVOPlasty: De novo assembly of organelle genomes from whole genome data. *Nucleic Acids Res.* **2017**, *45*, e18.

40. Langmead, B.; Salzberg, S.L. Fast gapped-read alignment with Bowtie 2. *Nat. Methods* **2012**, *9*, 357–359. [CrossRef]

41. Bankevich, A.; Nurk, S.; Antipov, D.; Gurevich, A.A.; Dvorkin, M.; Kulikov, A.S.; Lesin, V.M.; Nikolenko, S.I.; Pham, S.; Prjibelski, A.D.; et al. SPAdes: A new genome assembly algorithm and its applications to single-cell sequencing. *J. Comput. Biol.* **2012**, *19*, 455–477. [CrossRef] [PubMed]

42. Wick, R.R.; Schultz, M.B.; Zobel, J.; Holt, K.E. Bandage: Interactive visualization of de novo genome assemblies. *Bioinformatics* **2015**, *31*, 3350–3352. [CrossRef]

43. Wyman, S.K.; Jansen, R.K.; Boore, J.L. Automatic annotation of organellar genomes with DOGMA. *Bioinformatics* **2004**, *20*, 3252–3255. [CrossRef] [PubMed]

44. Laslett, D.; Canback, B. ARAGORN, a program to detect tRNA genes and tmRNA genes in nucleotide sequences. *Nucleic Acids Res.* **2004**, *32*, 11–16. [CrossRef]

45. Lohse, M.; Drechsel, O.; Kahlau, S.; Bock, R. OrganellarGenomeDRAW-a suite of tools for generating physical maps of plastid and mitochondrial genomes and visualizing expression data sets. *Nucleic Acids Res.* **2013**, *41*, W575–W581. [CrossRef]

46. Kearse, M.; Moir, R.; Wilson, A.; Stones-Havas, S.; Cheung, M.; Sturrock, S.; Buxton, S.; Cooper, A.; Markowitz, S.; Duran, C.; et al. Geneious Basic: An integrated and extendable desktop software platform for the organization and analysis of sequence data. *Bioinformatics* **2012**, *28*, 1647–1649. [CrossRef]

47. Nakamura, T.; Yamada, K.D.; Tomii, K.; Katoh, K. Parallelization of MAFFT for large-scale multiple sequence alignments. *Bioinformatics* **2018**, *34*, 2490–2492. [CrossRef] [PubMed]

48. Brudno, M.; Malde, S.; Poliakov, A.; Do, C.B.; Couronne, O.; Dubchak, I.; Batzoglou, S. Glocal alignment: Finding rearrangements during alignment. *Bioinformatics* **2003**, *19*, i54–i62. [CrossRef]

49. Sukumaran, J.; Holder, M.T. DendroPy: A python library for phylogenetic computing. *Bioinformatics* **2010**, *26*, 1569–1571. [CrossRef]

50. Šmarda, P.; Bureš, P. The Variation of Base Composition in Plant Genomes. In *Plant Genomes, Their Residents, and Their Evolutionary Dynamics*; Wendel, J.F., Greilhuber, J., Dolezel, J., Leitch, I.J., Eds.; Springer: Vienna, Austria, 2012; pp. 209–235. [CrossRef]

51. Team, R.D.C. *R: A Language and Environment for Statistical Computing*; R Foundation for Statistical Computing: Vienna, Austria, 2019.

52. Suyama, M.; Torrents, D.; Bork, P. PAL2NAL: Robust conversion of protein sequence alignments into the corresponding codon alignments. *Nucleic Acids Res.* **2006**, *34*, W609–W612. [CrossRef]

53. Lanfear, R.; Frandsen, P.B.; Wright, A.M.; Senfeld, T.; Calcott, B. PartitionFinder 2: New methods for selecting partitioned models of evolution for molecular and morphological phylogenetic analyses. *Mol. Biol. Evol.* **2017**, *34*, 772–773. [CrossRef]

54. Stamatakis, A. RAxML version 8: A tool for phylogenetic analysis and post-analysis of large phylogenies. *Bioinformatics* **2014**, *30*, 1312–1313. [CrossRef]

55. Goldman, N.; Yang, Z. A codon-based model of nucleotide substitution for protein-coding DNA sequences. *Mol. Biol. Evol.* **1994**, *11*, 725–736. [CrossRef]

56. Yang, Z.H. PAML 4: Phylogenetic analysis by maximum likelihood. *Mol. Biol. Evol.* **2007**, *24*, 1586–1591. [CrossRef] [PubMed]

2

Chloroplast Distribution in the Stems of 23 Eucalypt Species

Geoffrey E. Burrows *⦿ and Celia Connor

School of Agricultural and Wine Sciences, Charles Sturt University, Locked Bag 588, Wagga Wagga, NSW 2678, Australia; cconnor@csu.edu.au

* Correspondence: gburrows@csu.edu.au

Abstract: Small diameter branchlets and smooth barked stems and branches of most woody plants have chloroplasts. While the stems of several eucalypt species have been shown to photosynthesise, the distribution of chloroplasts has not been investigated in detail. The distribution of chloroplasts in branchlets (23 species) and larger diameter stems and branches with smooth bark (14 species) was investigated in a wide range of eucalypts (species of *Angophora*, *Corymbia* and *Eucalyptus*) using fresh hand sections and a combination of bright field and fluorescence microscopy. All species had abundant stem chloroplasts. In both small and large diameter stems, the greatest concentration of chloroplasts was in a narrow band (usually 100–300 μm thick) immediately beneath the epidermis or phellem. Deeper chloroplasts were present but at a lower density due to abundant fibres and sclereids. In general, chloroplasts were found at greater depths in small diameter stems, often being present in the secondary xylem rays and the pith. The cells of the chlorenchyma band were small, rounded and densely packed, and unlike leaf mesophyll. A high density of chloroplasts was found just beneath the phellem of large diameter stems. These trees gave no external indication that green tissues were present just below the phellem. In these species, a thick phellem was not present to protect the inner living bark. Along with the chlorenchyma, the outer bark also had a high density of fibres and sclereids. These sclerenchyma cells probably disrupted a greater abundance and a more organised arrangement of the cells containing chloroplasts. This shows a possible trade-off between photosynthesis and the typical bark functions of protection and mechanical strength.

Keywords: *Angophora*; bark; corticular photosynthesis; *Corymbia*; *Eucalyptus*; fluorescence; phellem; stem photosynthesis; wood photosynthesis

1. Introduction

Leaves, specifically the lamina or blade, are the primary site of photosynthesis for most species. For some plants, modified petioles (phyllodes) or modified stems (cladodes) are the primary photosynthetic organs. Nonetheless, a wide range of parts of seedlings and mature plants are green and photosynthetic, e.g., hypocotyl, cotyledons, flowers, fruits, and roots [1]. In addition, small and large diameter stems, although not obviously green, can have numerous chloroplasts in the living bark cells as long as the phellem or rhytidome are relatively thin. Nearly all woody plants carry out photosynthesis in their twigs, but far fewer do so in larger diameters [2,3].

Several ways of classifying photosynthesis in stems have been proposed. There are structural or anatomical classifications (e.g., [4–7]) (see also tables in [2,8]), such as:

1. CAM plants (stem succulents, often with cladodes).
2. Stem photosynthesis—green stems with epidermis (no or delayed periderm development) anda high stomatal density [9], with some species having a well-developed palisade layer

(e.g., [10]). They are mainly herbaceous plants but not exclusively and include desert shrubs [10] and early successional legumes (references in [11]).

3. Corticular or bark photosynthesis in woody plants after periderm development. Ávila et al. [9] indicate that this type of structure should be termed 'cortical photosynthesis'. In stems of trees and shrubs, the zone between the periderm and the secondary phloem is often referred to as 'cortex'. However, the cortex can be defined as a primary ground tissue between the epidermis and vascular bundles in a stem [12]. Thus, 'secondary cortex' [13] and 'pseudo-cortex' [14] have also been used for this zone. Rosell [15,16] considers that these cells are initially derived from the shoot apical meristem but then divide radially, thus are primary in origin (not derived from the vascular or cork cambia). Anatomical aspects of corticular photosynthesis have been more commonly investigated in small diameter/young stems (Table 1). Chloroplasts have been recorded in current year shoots to trunks 60 cm in diameter. In the few studies where both small and large diameters were examined, chloroplasts were found at greater depths in the smaller diameters (e.g., [17]). This may be related to light being conducted in the axial system of vessels, tracheids, and fibres (e.g., [18]). As could be expected, the density of chloroplasts decreases as phellem thickness increases (e.g., [3]).

4. Wood or woody tissue photosynthesis usually occurs in the xylem ray parenchyma cells and can also extend into the pith in small diameter stems.

Ávila et al. [9] proposed a physiological/anatomical classification: stem net photosynthesis (SNP) where CO_2 from the atmosphere enters the stem via stomata (equivalent to stem photosynthesis) and stem recycling photosynthesis (SRP) where the chloroplasts fix CO_2 released from respiration of living stem cells (equivalent to corticular and woody tissue photosynthesis). In the latter, a periderm is present and this has high resistance to gas diffusion and is a barrier to light [9]. Internal transport of CO_2 from roots to stems in the xylem transpiration stream can also occur [19].

In species with SNP, cells can be maximised for photosynthesis (i.e., presence of palisade, [10]). Bark has multiple functions including protection, storage, mechanical support, translocation of photosynthates and photosynthesis [3,16]. It is probable, especially in plants with well-developed leaves, that cell types and arrangement in the bark would be maximised for protection and mechanical strength rather than photosynthesis. As noted, in species with SRP, the periderm blocks CO_2 from entering and leaving the stem and thus the chloroplasts refix the CO_2 released by living cells in the bark. SRP can make a significant contribution to the overall carbon balance, water use efficiency [20], and drought tolerance [21]. Cernusak and Cheesman [11] note that with thin phellem, there is a possible trade-off between protection of stems from fire and high rates of SRP.

There are over 800 species of eucalypts (species of *Angophora*, *Corymbia* and *Eucalyptus*) [22]. They have what is probably the greatest diversity of bark types in a single group of trees. Bark types vary from entirely smooth (gum), smooth with scribbles, ironbark, stringybark, bloodwood, box, minniritchi, tessellated, as well as half-barks (half-butts) that have a rough bark on the lower half of the tree and smooth bark on the upper half. All eucalypt species have green or reddish branchlets covered by an epidermis and cuticle. They then develop relatively smooth bark on their small diameter (e.g., 0.5–2 cm) stems, before developing the more distinctive bark types as stem diameter increases. Many eucalypt species are smooth barked throughout their ontogeny. The bark of about 50% of eucalypt species is wholly smooth over the main stem and major branches and in 75% of species it is smooth over canopy branches [20]. Gillison ([23] p. 41) notes that species with green stem or cortical photosynthesis are evident in holarctic and boreal regions but most species occur in the tropics "in particular among the gum-barked eucalypts of Australia". Along with a large percentage of species with smooth bark, most eucalypt species have relatively narrow (lanceolate) leaves that hang vertically, thus relatively high light intensities are intercepted by the trunk and branches. In addition, there is a high proportion of half-butt or half-bark (where the basal thicker dead bark can provide fire protection while the upper thin phellem might allow sufficient light to enter a branch for photosynthesis) eucalypt species, especially in the more arid areas of Australia. SRP is associated with increased whole plant water-use efficiency and reduced drought effects as CO_2 is refixed with minimal water loss [11]. This is important for this widely distributed group of three genera in an arid continent being affected by climate change and longer fire periods. As noted, plants with SNP (long lived green stems) can have stem stomatal densities similar to those found in leaves (e.g., [10]). Young stems of tree species usually have very low stomatal densities (e.g., [6]). As recently formed eucalypt shoots are greenish (or reddish), it was of interest to quantify stomatal density as this would give some indication of whether SNP or SRP (or a combination of both) occurred.

Physiological studies of SRP have been made in eucalypt species such as *C. citriodora* [24], probably *E. dunnii* (although the species name is given as *EuCahetus dunnii*) [25], *E. globulus* [26,27], *E. grandis* × *E. urophylla* [24,28], *E. miniata* [20], *E. nitens* [29], and *E. saligna* [30,31]. All these species were smooth barked (gums), except for *E. miniata* (half-butt), where photosynthesis research centered on the smooth upper half of the trees.

Few studies of chloroplast distribution in eucalypt stems have been made (e.g., [3,20,24,28,32–37]). Mostly these studies only recorded that greenish stem tissues were observed. Chen et al. [24] studied four tree species with SRP, two eucalypt species and two other species not in the Myrtaceae. They noted the eucalypts' cortical green tissue was about 2 mm thick, while in the other species, this tissue was only about 0.5 mm thick. The only anatomical study was that of Mishra et al. [37] where the autofluorescence of chloroplasts in secondary xylem ray parenchyma of two-year-old plants of *E. bosistoana* was recorded.

Table 1. Anatomical studies of stem structure and chloroplast distribution relevant to stem recycling photosynthesis in woody plants. Studies are arranged in chronological order.

Author	Date	Species	Materials Examined
Scott [38]	1907	30 woody species	Young shoots and inflorescence axis
Sokolov [33]	1953	100 angiosperm spp., 16 gymnosperm spp.	
Pearson & Lawrence [39]	1958	*Populus tremuloides*	19 yr old, 16 cm diameter at 2 m above ground level
Gómez-Vazquez & Engleman [40]	1984	*Bursera longipes, B. copallifera*	Trees 8 and 5 m high, respectively (no diameter given)
Kauppi [41]	1991	*Betula pendula, B. pubescens*	0, 1, 3, 4, 5, 10 and 20 yr old branches
van Cleve et al. [42]	1993	*Populus* × *canadensis*	2–3 yr old twigs, 1–2 cm diameter
Pilarski [43]	1999	*Syringa vulgaris*	Current year, 1 and 3 yr old stems
Pilarski & Tokarz [17]	2006	*Fagus sylvatica*	2 and 6 yr old stems and trunk (3.4, 6.9 and 600 mm diameter, respectively)
Dima et al. [34]	2006	20 species	2 yr old twigs
Berveiller et al. [44]	2007	9 tree species	Current year stems, 2–5 mm diameter
Filippou et al. [5]	2007	*Olea europaea*	1–30 yr old stems
Rentzou & Psaras [6]	2008	3 Mediterranean species	1–2 yr old and 2–5 yr old twigs
Kotina et al. [45]	2012	14 species Apiaceae	Branch tips to thicker stems with mature bark
Schmitz et al. [46]	2012	13 Australian mangrove species	Mainly 6 mm diameter (range 2.8–14 mm diameter)
Cocoletzi et al. [47]	2013	*Myriocarpa longipes, Urera glabriuscula*	10 yr old trees
Kocurek et al. [48]	2015	*Clusia* multiflora, *C. rosea*	2.0–2.5 cm diameter at 1 m above ground level
Wittmann & Pfanz [49]	2016	5 woody species	Current year shoots, 3–5 mm diameter
Kotina et al. [14]	2017	*Adansonia digitata*	Young twigs to stems 12–17 cm diameter
Blagitz et al. [50]	2019	*Monquinastrum polymorphumZanithoxylum rhoifolium*	Large diameter stems

Given that there have been several physiological studies of SRP in eucalypts, but only one detailed study of chloroplast distribution in the stems of a single eucalypt species, this study concentrates on the anatomical aspects of SRP in this important group of trees. When green tissues are found just beneath the phellem of stems, it is usually assumed that this colouration is from chloroplasts (e.g., Rosell [3]), that these chloroplasts will be able to photosynthesise and this photosynthesis would be SNP. We also make this assumption.

It is clear that SRP is a key aspect of eucalypt carbon balance (e.g., [20]) and this can impact drought tolerance. To better understand these features, it is important to survey the presence of chloroplasts in the stems of a wide taxonomic diversity of eucalypt species, in a wide range of stem diameters, and to quantify the location of the chloroplasts in those stems. In addition, it has been suggested that there may be a trade-off between fire protection and photosynthesis in larger diameter trunks and branches. An anatomical study of chloroplast distribution is essential to evaluate any trade-off. We studied chloroplast distribution in stems from a wide range of eucalypt species and hypothesised that:

i) given the findings of the studies in Table 1, chloroplasts would be found in all the small diameter branches as well as large diameter branches and trunks of smooth barked stems,
ii) given the wide taxonomic diversity of species examined, different chloroplast arrangements would be recorded,
iii) given that light appears to be conducted more effectively in small diameter stems, chloroplasts would be found at greater depth in smaller diameter stems than larger diameter ones,
iv) the youngest stems (no periderm) would have stomata but they would be at a low density compared to adjacent leaves, and
v) given that the species examined had thin phellem (provides little protection), the bark inside this layer would be optimised for defence and mechanical support rather than photosynthesis.

2. Results

2.1. External Morphology

As the phellem of rough barks such as iron and stringy barks can be centimetres thick, thus totally excluding light from the living bark tissues, this study was of smooth barked species (Figure 1). Smooth bark was maintained by shedding of the periderm, either on an annual basis (Figure 1c) or in sections over time (Figure 1a,c). Although the phellem was generally very thin in almost all species with periderm formation, the surface of the bark was not green or greenish. The only exception was the sections of the minniritchi bark of *Eucalyptus caesia* that had recently peeled (Figure 1b). In the studied species, a gentle scape of the phellem cells revealed the bright green tissues of the outer bark (Figure 1f).

2.2. Small Diameter Stems Without Periderm

Transverse sections showed that almost all regions of the stems, except the epidermis (i.e., cortex, secondary phloem, secondary xylem, pith) have chloroplasts (Figures 2 and 3a,c). In the secondary xylem, chloroplasts were restricted to the ray parenchyma cells (along with starch grains) (Figure 2a,b,g,h and Figure 3a,c), thus with fluorescence microscopy, radiating lines of red chloroplasts were visible. The highest density of chloroplasts was in a 100–400 μm wide cortical band just under the epidermis (Figure 2a,b and Figure 3a,c). The chloroplasts were in compact, rounded cortical cells with little intercellular airspace (Figure 2a–d), not in elongated palisade-type cells. Chloroplasts in the inner secondary xylem and pith could be more than 1500 μm in from the epidermis (Figure 3a,c).

Figure 1. Bark surfaces of various eucalypt species showing the wide diversity of bark morphologies and the delayed bark formation that occurs in some species. (**a**) Surface of the main trunk of a *Eucalyptus blakelyi* tree showing that the phellem gets darker and slightly thicker as it ages and is shed in irregularly shaped patches. (**b**) Minniritchi bark on a main stem of *Eucalyptus caesia*. Note the alternating strips of green and brownish red bark. Scale 5 cm. (**c**) Trunk (40 cm diameter) of *Corymbia citriodora* in early summer during the bark shedding phase. Note that when the bark was initially formed (right hand side), it was almost white, but when shed a year later, it was much darker. Scale 10 cm. (**d**) Young branch of *Eucalyptus caesia* showing the highly reflective glaucous layer that has been partially rubbed away to reveal the epidermis and greenish stem. Scale 2 cm. (**e**) Part of the trunk of a large *Eucalyptus cladocalyx* tree that had formed rapidly growing epicormic shoots after the tree had been extensively pruned. Note the smooth gum bark on the large diameter bole and the bright green epicormic shoots. The largest shoot still had an epidermis and stomata (the minute white spots) although it was over 2 cm diameter. Scale 5 cm. (**f**) Bark of *Corymbia citriodora* with some of the thin (<100 µm thick) phellem removed with a scalpel, revealing the dark green tissues directly below the phellem. Scale in mm.

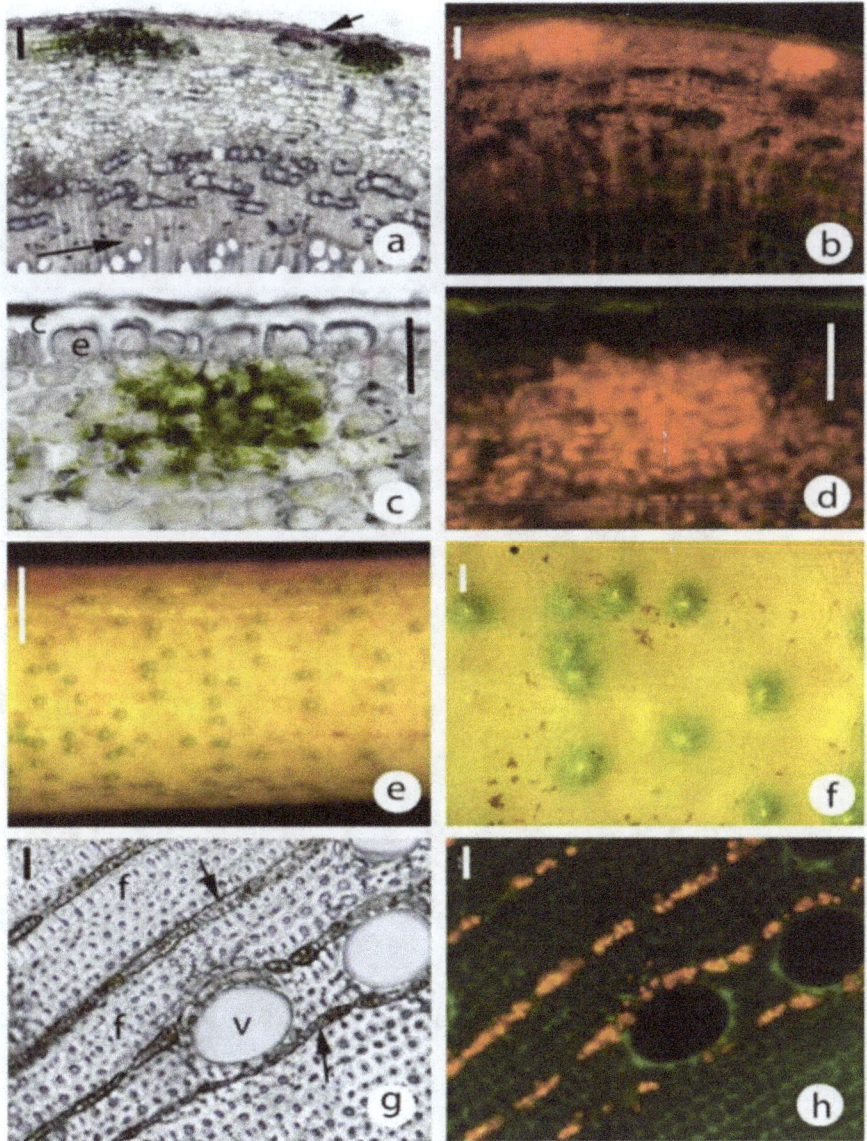

Figure 2. Chloroplast distribution in small diameter eucalypt stems. Note in (**a–f**), the much greater chloroplast density in the vicinity of the low density stomata. All sections (**a–d,g,h**) are transverse. (**a**) *Corymbia ficifolia* stem 3.5 mm diameter, bright field illumination, showing a green band between the epidermis (short arrow) and the secondary phloem. Note the two darker green regions just beneath the epidermis that were probably associated with stomata. vascular cambium, long arrow. Scale 100 μm. (**b**) as per (**a**) but showing chloroplast auto-fluorescence. Scale 100 μm. (**c**) *Eucalyptus sideroxylon*, 3 mm diameter shoot, showing detail of clustered chloroplasts associated with a stoma (bright field illumination). c, cuticle; e, epidermis. Scale 50 μm. (**d**) as per (**c**) but showing chloroplast auto-fluorescence. Scale 50 μm. (**e**) Segment of small diameter *Eucalyptus sideroxylon* branch showing the distribution of stomata and associated areas of higher chloroplast density. Note the wide variation in stomatal density. This stem had, on average, about 5 stomata/mm^2. Scale 1 mm. (**f**) detail of a stem like that in (**e**). Note the outline of the typical epidermal cells and the guard cells. Scale 100 μm. (**g**) Transverse section of the secondary xylem in a *Eucalyptus cladocalyx* stem 13 mm diameter. Note the presence of starch grains and chloroplasts that were almost entirely restricted to the ray parenchyma cells. f, fibres; v, vessels; arrow, xylem ray. Scale 20 μm. (**h**) as per (**g**) but showing chloroplast auto-fluorescence. Scale 20 μm.

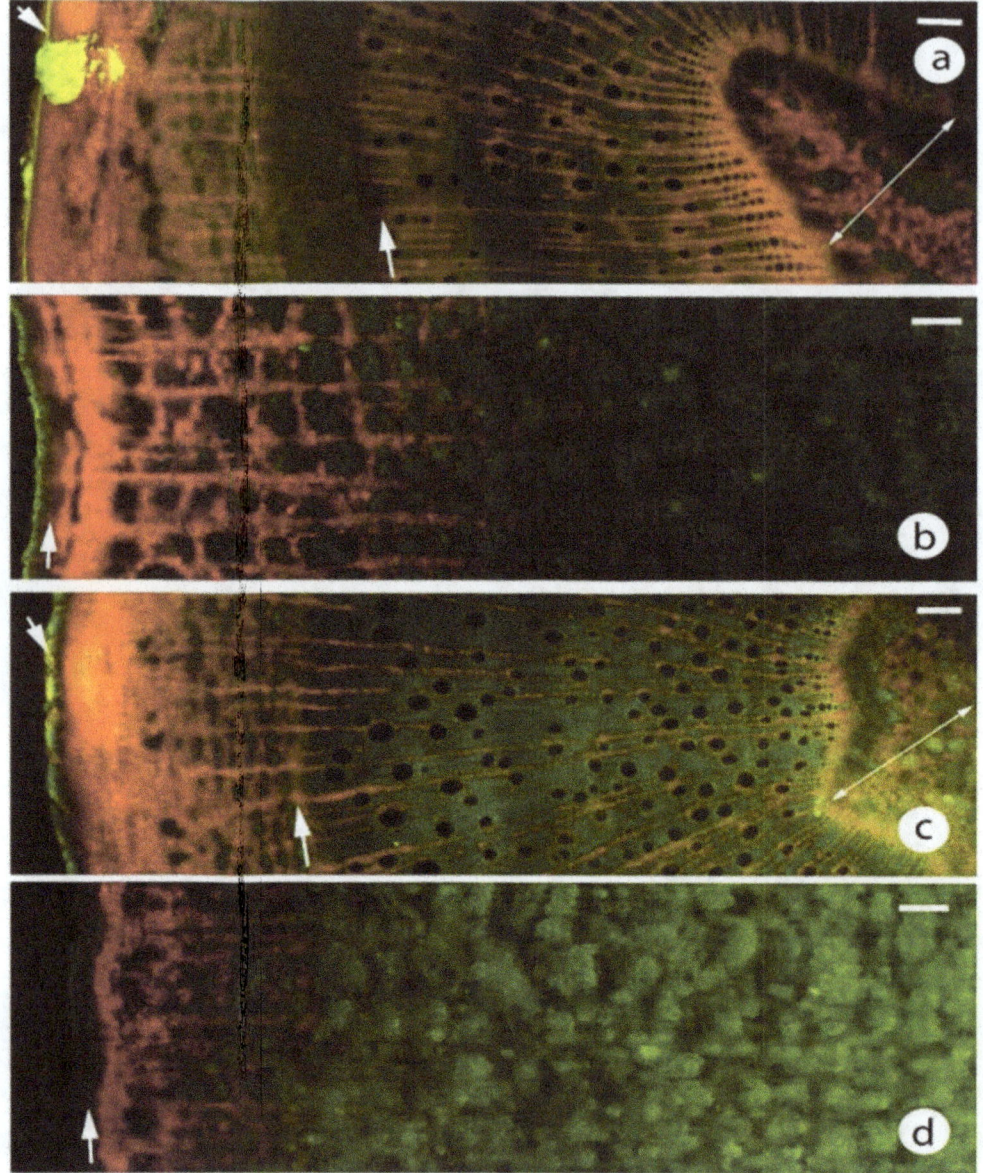

Figure 3. Differences in depth of chloroplasts in small and large diameter eucalypt stems. Red and green indicate the location of the chloroplasts and sclerenchyma, respectively. Note that in the small diameter stems chloroplasts were present from immediately beneath the epidermis to the pith, while in the large diameter stems they were only present in the outer bark. (**a,b**) *Eucalyptus globulus*, (**c**) and (**d**) *Eucalyptus cladocalyx*. All scale bars 100 μm. (**a**) 6 mm diameter. The bright yellow area near the epidermis was possibly the contents of a resin duct. short arrow, epidermis; long arrow, vascular cambium; line, pith. (**b**) 50 cm diameter. arrow, phellogen. (**c**) 4 mm diameter. Labelling as per (**a**). (**d**) 30 cm diameter. arrow, phellogen.

In most species, small circular to oval areas were observed in the outer cortex that were a brighter green in bright field microscopy (Figure 2a,c) and much brighter red in fluorescence microscopy (Figure 2b,d) than the general distribution of chloroplasts. Examination of the stem surface with a stereomicroscope revealed the presence of scattered small darker green areas (Figure 2e), but their conspicuousness depended on the species and the area of the stem examined. High power stereomicroscopy and epidermal replicas indicated that the green areas were below stomata (Figure 2f). Stomatal density was much lower on stems than leaves (Table 2) and much more variable in density

(Figure 2e) (relative standard deviation 17% leaves, 59% stems). These darker green areas were a small volume of the cortex (Figure 2e,f).

Table 2. Leaf and small diameter stem average stomatal densities (number/mm^2 ± SD) for six eucalypt species.

Species	Leaf (Adaxial)	Leaf (Abaxial)	Stem
Corymbia citriodora	267 (± 28)	320 (±39)	2.3 (± 1.2)
Corymbia ficifolia	0 (± 0)	209 (±39)	1.6 (± 0.8)
Eucalyptus cladocalyx	0 (± 0)	279 (±27)	1.9 (± 0.8)
Eucalyptus melliodora	184 (± 31)	194 (±60)	4.1 (± 2.1)
Eucalyptus sideroxylon	134 (± 17)	169 (±22)	1.3 (± 1.1)
Eucalyptus torquata	105 (± 24)	124 (±26)	2.2 (± 1.6)

The small diameter stems of some species (*E. caesia*, *E. kruseana*, *E. macrocarpa*) were covered by a dense glaucous wax layer. These stems were almost pure white (Figure 1d) but still had substantial chloroplast development.

2.3. Large Diameter Stems

In the 13 species studied, the phellem was thin (≤100 µm) (Table 3) and in *C. citriodora* was very thin (<50 µm) in all stem diameters studied (2–45 cm) (Figure 1f). The greatest density of chloroplasts was usually in a narrow band (80–400 µm) (Table 3) immediately beneath the phellem (Figure 3b,d). Less abundant chloroplasts were found 600–1800 µm further into the bark, usually in parenchyma cells of the secondary phloem or secondary cortex (Figure 3b,d or Figure 4b,c). The outer region of the living part of the bark (the phellogen and outer secondary phloem or secondary cortex) had a large proportion of thick walled, lignified cells (fibres and sclereids) (Figure 3b,d or Figure 4e). The sclereids usually had relatively thin walls. Both cell types were usually devoid of chloroplasts in their lumens. Some sclerenchyma cells were present in the chlorenchyma band but were far more abundant in the remainder of the bark Figure 3b,d or Figure 4e).

In *C. citriodora* and the closely related *C. maculata*, a well defined dark green band was present directly beneath the phellem (Figure 4a,b) and under a stereo microscope this band had a palisade-like appearance. Sectioning showed these cells were probably phelloderm, about 10–15 cells deep (Figure 4c–e). Most cells of this band contained chloroplasts, but relatively small and relatively thin-walled sclereids were also present (Figures 4e and 5). Tangential longitudinal sections of the band (paradermal-type sections) showed that the chlorenchyma and sclereids were usually segregated into different blocks of tissues rather than being a homogeneous mix (Figure 5). The chlorenchyma cells were thin walled, with little intercellular air space development. Generally, a distinct differentiation was present between the two cell types but some sclereids with chloroplasts were observed.

As noted, *C. citriodora* sheds its outermost bark in early summer (Figure 1c). Bark examined in late spring had a band of chlorenchyma directly below the phellem but the subsequent phellogen was already formed (800–1500 µm in from the outer older phellogen) and its phelloderm chlorenchyma layer had already formed (Figure 6). Thus two concentric bands of chlorenchyma were present and the tree would be ready to photosynthesise as soon as the outer bark was shed. When moist, the bark layer that will be shed (i.e., still living, measured immediately after being peeled from the trunk) transmitted 6% of incident radiation. When dry (e.g., brown bark on the left-hand side of Figure 1c), the bark transmitted 0.4% of incident radiation.

Table 3. Species studied and various anatomical measurements. Subgenera and sections are from Nicolle, D. (2019) Classification of the eucalypts (*Angophora, Corymbia* and *Eucalyptus*) Version 4 http://www.dn.com.au/ Bark types and species distribution (states or territories) are from an online version of the EUCLID Eucalypts of Australia 4th Edition (2015). A '-' in 'phellem thickness' indicates a periderm was not present (epidermis and cuticle still present). Stem diameter: small (S) < 8 mm, medium (M) 8–40 mm, large (L) > 4 cm.

Genus/Species	State/Territory	Subgenus–Section	Bark Type	Stem Diameter	Phellem Thickness (µm)	Chlorenchyma Bright Band (µm)	Chlorenchyma Maximum Depth (µm)
Angophora							
A. hispida (Sm.) Blaxell	NSW		Rough to small branches, fibrous	S (2 mm)	-	100–200	600–1000
				S (6–8 mm)	30–70	150–200	800–1000
				M (13–15 mm)	30–60	200	1700–1800
Corymbia							
C. citriodora (Hook.) K.D. Hill & L.A.S. Johnson	Qld	*Blakella*	Smooth throughout—gum	S (1.5–3 mm)	-	150–200	1000–1400
				S/M (5–11 mm)	50–70	100–150	500–1500
				L (10–40 cm)	20–90	100–330	700–1300
C. maculata (Hook.) K.D. Hill & L.A.S. Johnson	NSW, Qld, Vic	*Blakella*	Smooth throughout	S (3–6 mm)	0–60	150–200	1000
				M (11–14 mm)	30–40	120–150	1400–1500
				L (30–50 cm)	30–50	120–200	900–1800
C. ficifolia (F. Muell.) K.D. Hill & L.A.S. Johnson	WA	*Corymbia*	Rough to small branches, fibrous	S (3–6 mm)	-	220–400	1200–2000
				M (14–15 mm)	50–100	250–300	1800–2100
				M (26 mm)	50–80	300	1800–2100
C. eximia (Schauer) K.D. Hill & L.A.S. Johnson	NSW	*Blakella*	Rough to small branches; tessellated, flaky	S (2–7 mm)	-	200–400	1100–2000
				M (10–30 mm)	50–100	100–400	1000–3000
				L (7–10 cm)	100	100–120	1100–1400
Eucalyptus							
E. erythrocorys F. Muell.	WA	*Eudesmia*	Smooth; can have rough bark on lower trunk	L (4–5 cm)	50–80	80–100	2300–3000
				L (15–20 cm)	60–100	100	2500–2800
E. macrorrhyncha F. Muell. ex Benth.	NSW, Vic, SA	*Eucalyptus*	Rough to small branches (stringybark)	S (2 mm)	-	120–150	900–1000
				M (8–18 mm)	20–100	120–220	1000–3000
E. rossii R.T. Baker & H. G. Sm.	NSW	*Eucalyptus*	Smooth (scribbly gum)	S (2 mm)	-	120–150	900–1000
				M (11–20 mm)	30–80	200–400	1500–2000
				L (10–20 cm)	30–100	150–280	1000–1300
E. leucoxylon F. Muell.	SA, Vic	*Symphyomyrtus—Adnataria*	Smooth throughout, sometimes with rough box-type at base	S (1.5–4.5 mm)	-	170–200	800–2000
				S (7 mm)	20	150	1300
				L (17 cm)	80–90	150–300	700–1100
E. melliodora A. Cunn. ex Schauer	NSW, Vic, Qld	*Symphyomyrtus—Adnataria*	Rough box-type at base, smooth higher	S (2.5–3 mm)	-	150–200	1100–1500
				S (6–8 mm)	30–50	150–200	700–1400
				L (50–80 cm)	50–100	100–150	700–800

Table 3. *Cont.*

Genus/Species	State/Territory	Subgenus-Section	Bark Type	Stem Diameter	Phellem Thickness (μm)	Chlorenchyma Bright Band (μm)	Chlorenchyma Maximum Depth (μm)
E. sideroxylon A. Cunn. ex Woolls	NSW, Vic, Qld	Symphyomyrtus—Adnataria	Rough throughout (ironbark)	S (3–8 mm) / M (10–12 mm) / M (23–26 mm)	– / 30–70 / 40–60	200 / 150–250 / 100–200	1000–1900 / 1800–2200 / 800–2300
E. albens Benth.	NSW, Vic, Qld, SA	Symphyomyrtus—Adnataria	Rough on trunk and large branches (box-type)	S (2–4 mm) / M (12–28 mm)	– / 50–80	250–400 / 150–350	700–1900 / 900–1500
E. caesia Benth.	WA	Symphyomyrtus—Bisectae	Minnirichi on trunk and branches	*1			
E. macrocarpa Hook.	WA	Symphyomyrtus—Bisectae	Smooth	S/M (3–15 mm) / L (8–10 cm)	– / 40–50	200–500 / 100–200	2000–3200 / 800–1100
E. kruseana F. Muell.	WA	Symphyomyrtus—Glandulosae	Rough but thin on trunk base, smooth above	S (3–8 mm) / M (15–40 mm)	– / 100	200–300 / 50–100	1000–2600 / 500–1500
E. torquata Luehm.	WA	Symphyomyrtus—Dumaria	Rough, hard, shortly fibrous to almost tessellated	S (3–8 mm) / M (15–30 mm)	– / 100–300	250–400 / 100–300	1000–3500 / 500–1400
E. blakelyi Maiden	NSW, Vic, Qld	Symphyomyrtus—Exsertaria	Smooth throughout (gum)	S (3–7 mm) / M (10–40 mm) / L (20–37 cm)	– / 20–60 / 30–100	200–300 / 100–300 / 100–200	1200–1600 / 1500–2000 / 300–700
E. camaldulensis Dehnh.	Most states	Symphyomyrtus—Exsertaria	smooth	S (3–6 mm) / M (8–25 mm) / L (6–80 cm)	– / 30–50 / 30–70	150–200 / 100–200 / 100–200	900–1800 / 500–800 / 500–800
E. globulus Labill.	Tas, Vic	Symphyomyrtus—Maidenaria	Smooth apart from the base with persistent slabs (gum)	S (3–8 mm) / M (18–20 mm) / L (50–60 cm)	– / 30–40 / 30–80	200–300 / 150–200 / 150–300	1300–2800 / 1000–1800 / 800–1200
E. cinerea F. Muell. ex Benth.	NSW	Symphyomyrtus—Maidenaria	Thick, furrowed, fibrous to small branches	S (2–6 mm) / M (10–20 mm)	– / 10–80	150–200 / 100–250	900–2700 / 1500–3500
E. nicholii Maiden & Blakely	NSW	Symphyomyrtus—Maidenaria	Thick, fibrous to small branches (peppermint)	S (3–4 mm) / M (8–18 mm)	– / 20–50	150–200 / 200–350	1300–1900 / 1700–2300
E. scoparia Maiden	Qld	Symphyomyrtus—Maidenaria	Smooth–gum	S (3–8 mm) / L (4–26 cm)	– / 30–50	150–300 / 150–400	1400–2800 / 900–1300
E. cladocalyx F. Muell.	SA	Symphyomyrtus—Sejunctae	Smooth–gum	S (2–8 mm) / L (24–60 cm)	0–60 / 40–70	180–300 / 80–150	800–3900 / 600–800

*1 see Section 2 for detail.

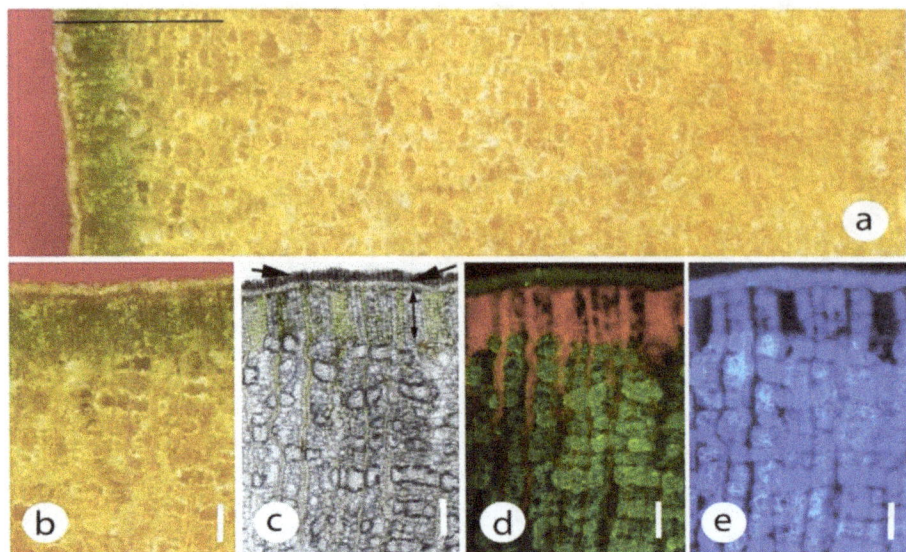

Figure 4. Various images, either transverse section or transverse plane, of the bark from a 40 cm diameter trunk (bark thickness 9–10 mm) of *Corymbia citriodora*. Note the greater density of chloroplasts in the phelloderm. Chloroplasts occurred deeper (mainly in the ray-like parenchyma cells) but overall density was much lower. All images were of the same small block of bark. (**a**) Outer 5 mm bark thickness observed with a stereo microscope. Note the thin phellem and green layer directly beneath the phellem. Scale 1000 μm. (**b**) Area of (**a**) at higher magnification. Note that the cells in the green layer were in files of cells perpendicular to the surface. Scale 200 μm. (**c**) Hand cut section of outer bark observed with a compound microscope and bright field illumination. short arrow, phellem; long arrow, phellogen; line, phelloderm. Scale 200 μm. (**d**) as per (**c**) but fluorescence microscopy. Note that from the red auto-fluorescence, most of the chloroplasts were in a 300 μm deep band immediately beneath the phellem, although deeper chloroplasts were found in ray-like structures. Scale 200 μm. (**e**) As per (**d**), except UV light shows the high density of fibres and sclereids (blue fluorescence) especially below the chlorenchyma band. Scale 200 μm.

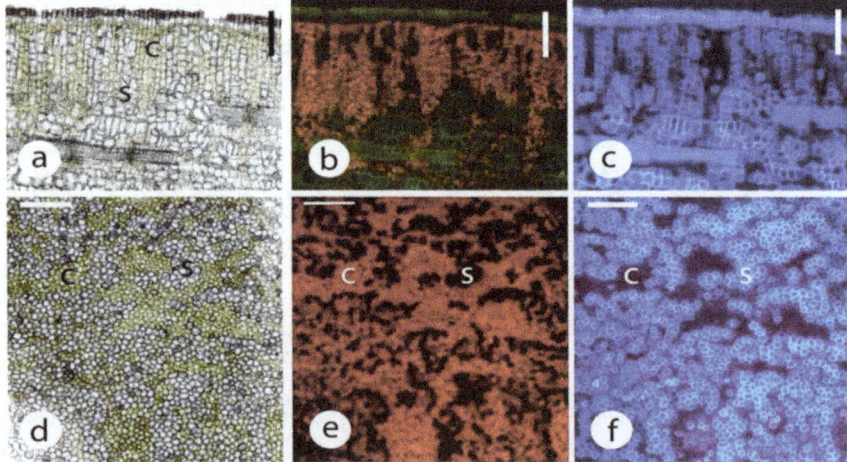

Figure 5. Bark from the main stem of *Corymbia citriodora*, transverse (**a–c**) and tangential longitudinally sectioned (**d–f**). c, chlorenchyma; s, sclereids. Sections were imaged with white (**a,d**), blue (**b,e**) and UV (**c,f**) light. Note that the phellem was very thin and the greatest concentration of chloroplasts was in the phelloderm cells. Note also that while the phelloderm cells are exposed to relatively high light intensities, many of these cells have differentiated into sclereids. The tangential longitudinal sections show the chlorenchyma and sclerenchyma were apparently arranged in random, but relatively discrete, tissue groupings. Scales 200 μm.

Figure 6. Outer bark of 40 cm diameter tree of *Corymbia citriodora* collected in November (late spring) shortly before the tree had its annual bark shedding (December). Note that a deeper, newly initiated phelloderm with a high density of chloroplasts had already been formed although outer bark abscission was still 2–4 weeks away. (**a**) Transverse plane observed using a stereo microscope with white light. Note the green band of chlorenchyma directly beneath the thin phellem. Note also an additional inner phellogen had already been initiated and an inner band of chlorenchyma had formed even though it was under 1500 μm of bark. Scale 1000 μm. (**b**) as per (**a**) but a hand-cut section showing chloroplast auto-fluorescence. short arrow, older phellogen; long arrow, recently initiated phellogen; line, layer of bark to be shed. Scale 1000 μm.

The bright green *E. cladocalyx* epicormic shoots still possessed an epidermis with stomata on relatively large diameter stems (Figure 1e). Stereo microscope observation of these stems cut in a transverse plane showed the usual intense green band immediately below the epidermis (Figure 7a) but a greenish tinge was also present across the secondary xylem and a deeper green in the innermost xylem and pith (Figure 7a). Fluorescence microscopy confirmed the visual observation that chloroplasts were present all the way across the secondary xylem and were at relatively high density in the middle of the epicormic shoot (Figure 7b). Chloroplasts were recorded to a maximum depth of 10,000 μm.

Smaller diameter branches of *Eucalyptus caesia* were covered in a white waxy bloom which, when rubbed away, revealed a reddish or deep green stem (Figure 1d). These stems had not developed a periderm and, in the trees examined, were up to 6 cm in diameter. The stems, while almost a pure white with a high reflectance, still had a high chloroplast density (Figure 7c). On larger diameter stems, a minniritchi bark develops (Figure 1b). Numerous chloroplasts were present in green bark (Figure 7e) and also below the brown/red bark (dead and about 400 μm thick) (Figure 7d).

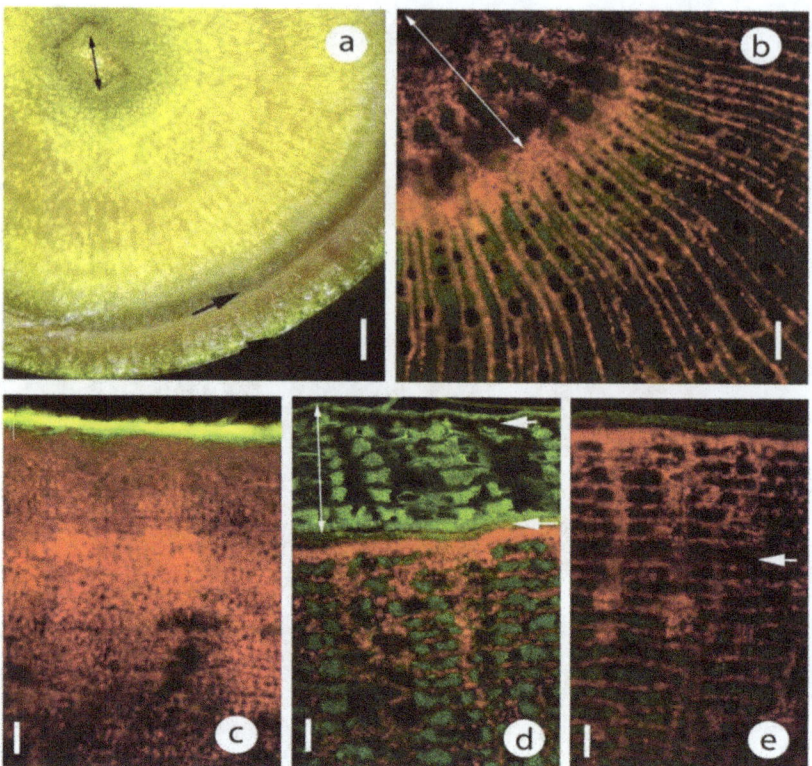

Figure 7. (**a,b**) Chloroplast distribution in an epicormic shoot of *Eucalyptus cladocalyx*, similar to that shown in Figure 1d. (**a**) transverse plane of shoot about 15 mm diameter. Note the deep green layer immediately below the epidermis, the low density of chloroplasts in the most recent secondary phloem, the greenish tint of the secondary xylem and the relatively bright green in and around the pith. short arrow, epidermis; long arrow, vascular cambium; line, pith. Scale 1000 μm. (**b**) transverse section of an 18 mm diameter shoot showing strong chlorophyll auto-fluorescence in the innermost secondary xylem and the pith. line, pith. Scale 100 μm. (**c–e**) *Eucalyptus caesia* stems sectioned in transverse plane, fluorescence microscopy. Scales 100 μm. (**c**) 3 cm diameter stem, similar to Figure 1d. Note that while the stem has a relatively large diameter, no periderm was present. Yellow surface layer is auto-fluorescence of a combination of the cuticle and glaucous wax layer. Note also that while the surface was highly reflective abundant chloroplasts were found to 700 μm below the cuticle. (**d**), (**e**) Similar to Figure 1b, stem 15 cm diameter. (**d**) Note that abundant chloroplasts were present below the 400 μm thick, dark brown outer bark tissues. short arrow, non-functional phellogen, long arrow, active phellogen; line, bark to be shed. (**e**) Very thin, greenish coloured bark. Note that the position of the next phellogen was apparent about 400 μm below the surface. arrow, location where next phellogen will be initiated.

3. Discussion

3.1. General Distribution of Chloroplasts

Chloroplasts were found in the small diameter stems of all species examined and in the large diameter stems of all the smooth ('gum') barked species. For all species and all stem diameters, the highest concentration of chloroplasts was in a relatively narrow (usually 100–300 μm) band just inside the epidermis or phellem. While several studies (see listing in the Introduction) have noted the presence of greenish tissues in eucalypt stems, the only anatomical study with images is that of Mishra et al. [37]. They studied autofluorescence in the sapwood and heartwood of 2 and 11-year-old plants of *E. bosistoana*. Only the sapwood of the younger plants showed chlorophyll autofluorescence. The small diameter stems in the present study were obviously green, except when glaucous. Only in a couple of species (e.g., *E. caesia*) were the larger diameter stems obviously green. In the other species examined the thin phellem very effectively obscured the green tissues beneath.

3.2. Differences in Chloroplast Distribution Between Species

Relatively little variation was observed between species when comparing similar diameter stems. Chattaway [32] noted that the structure of twigs and young stems of eucalypts was uniform. The similarities of the outer bark of the examined species (very thin phellem, relatively thin phelloderm with numerous chloroplasts, numerous sclereids and fibres) were more notable than any minor species differences, especially when compared to the photosynthetic bark of a wide range of other Australian trees and shrubs (G. Burrows unpublished data).

One distinctive feature in the larger diameter material was that initial stereoscopic views of *C. citriodora* and *C. maculata* bark cut in transverse section indicated that these species might have had palisade-type tissue. Sectioning showed that the 'palisade' was columns of phelloderm cells. While this well-defined band had a relatively high percentage of chloroplast containing cells, it also had numerous thin-walled sclereids and little air space development and was thus completely unlike typical palisade tissue. For trees and shrubs, Angyalossy et al. [51] note that a thin phellogen (1–3 cell layers) is most commonly observed. At 10–15 cell layers thick, the phellogen of these two species would appear to be relatively deep.

In SNP, the stems remain green and periderm development is delayed or does not occur. In these stems, the development of palisade cells in the outer cortex has been recorded (e.g., [4,10,52–54]). In none of the eucalypt species examined, at none of the stem diameters sectioned, were palisade-like cells observed and none appear to have been recorded in the SRP literature.

3.3. Chloroplast Distribution in Small vs Large Diameter Stems

In the gum barked species, where comparisons could be made between small and large diameter stems, chloroplasts were generally found at greater depths in small diameter stems than in large diameter stems. A similar finding has been recorded in the SRP literature. For example, chloroplasts have been recorded in the secondary xylem rays or pith of small diameter stems at depths of 1500–5000 μm [6,34, 42–44,46,48,49]. Far fewer studies have been made of larger diameter material (Table 1), with most studies indicating that the chloroplasts were in a thin band, usually 100–350 μm in depth [14,39,41,45,50]. Pilarski and Tokarz [17] investigated chloroplast distribution in both small (7 mm) and large (60 cm) diameter stems of *Fagus sylvatica*. In the former, chloroplasts were recorded to 4000 μm deep and in the latter chloroplasts were mostly to 200 μm deep, with a few scattered to 500 μm deep. This is similar to several eucalypt species in the present study with relatively deep chloroplasts in smaller diameters, and only a narrow peripheral band in larger diameters (Figure 3).

Phellem can drastically reduce light penetration into a stem. As this layer can increase in depth as a stem increases in diameter, the light transmission to the chloroplasts often reduces with increasing age of a stem. In the present study those eucalypt species with smooth bark on large diameter branches had very thin phellem thickness (almost all measurements ≤ 100 μm). In *C. citriodora*, annual bark shedding resulted in a very thin (30–50 μm) phellem on all branch diameters assessed (stems 2.5–45 cm diameter). Tausz et al. [29] reported that for *E. nitens*, for the trunk of 25 cm diameter trees, the periderm transmitted 57% of incident photosynthetically active radiation. They indicated this was "higher than normally reported" (p. 418 [29]) and the chlorenchyma had a photosynthetic pigment composition like that of sun leaves. For *C. citriodora*, the layer of bark that is shed annually, when alive, transmitted 6% of light even though it was around 800 μm thick, thus the thin phellem would probably transmit relatively high intensities into the chlorenchyma band. Rosell [15] examined outer (dead) and inner (living) bark thickness in 640 species, including 12 eucalypt species (see their Figure 3). Only about 15 species (2%) had an outer bark thickness of less than 50 μm and most of these values were for small stem diameters (e.g., <2 cm). Thus, smooth barked eucalypts in general, and *C. citriodora* in particular, would appear to have very thin phellem for trees. Eucalypts generally have pendulous leaves thus the bark receives relatively high light intensities. It may be that SRP is of greater relative importance in smooth barked eucalypts than in many other tree groups due to relatively high light intensities in the outer bark.

In the present study, the outermost layer of bark that *C. citriodora* sheds annually had a much higher light transmission when living (moist) than when it was dead (dry) and ready to be shed. It seems probable that SRP in *C. citriodora* would be reduced for a couple of weeks in early summer while this outer bark is dead but not yet shed (e.g., Figure 1c). The transverse sections of *C. citriodora* bark made in the weeks before annual bark shedding (Figure 6) indicate more than just light transmittance was involved in the distribution and/or differentiation of chloroplasts. A high density of chloroplasts was already present in the inner phelloderm layer, even though in November it still had an overlying c. 1000 μm thick layer of bark and the inner region of this overlying bark layer had few chloroplasts. These deeper phelloderm cells would appear to be predisposed to develop numerous chloroplasts, even when in a relatively low light intensity environment.

In rapidly growing epicormic shoots of *E. cladocalyx* (bright green, no periderm) (Figure 1e), relatively abundant chloroplasts were found in and near the pith of stems up to 20 mm diameter (Figure 7). How deep have chloroplasts been recorded in stems? Pfanz and Aschan [2] noted a few studies where green 'halos' were observed around the stem pith of woody plants. Schmitz et al. [46] found that in two *Ceriops* species with relatively thick bark, chloroplast fluorescence was only found near the pith. Van Cleve et al. [42] and Pilarski and Tokarz [17] both illustrated chloroplasts in pith cells 4000–5000 μm in from the stem surface. Thus, the 8000-10000 μm chloroplast depths found in this study are deeper but comparable to previous studies.

How do these deeply located chloroplasts receive light? Sun et al. [18], who studied the stems (0.5–3.0 cm diameter) of 21 species of woody plants, found that vessels, tracheids, and fibres conducted light efficiently in an axial direction. In vessels, the light was conducted in the lumina and in fibres and tracheids in the cell walls. They found that red/far red wavelengths were conducted most efficiently. Karabourniotis et al. [55] found that sclereids in sclerophyllous leaves of *Olea europea* act as optical fibres and may improve the light microenvironment within these relatively thick leaves. In bean seedlings, Kakuszi and Böddi [56] and Kakuszi et al. [57] recorded chlorophyll formation to 4–5 cm below the soil surface due to light piping by the hypocotyl. The information in these references does not explain why chloroplasts formed at such depths in *E. cladocalyx* but it does show that the internal light environment of stems is influenced by a complicated range of factors. The presence of chloroplasts does not necessarily mean cells are photosynthetic; chlorophyll formation can occur at light intensities below that needed for photosynthesis [58]. Deeper chloroplasts can have a starch storage function [43,49]. Further research is needed to ascertain whether the deeply located eucalypt chloroplasts are photosynthetic.

3.4. Stem and Leaf Stomatal Densities

In SNP species (long-lived green stems), high stem stomatal densities (50–190 stomata/mm^2) have been recorded [4,10,59]. Young tree stems generally have either no (e.g., [6]) or low stomatal densities (e.g., [58]). A similar finding was made in the present study (average 1–4 stomata/mm^2 with a very uneven distribution in the six species assessed). A much more intense chlorophyll fluorescence was observed in a radius of about 100 μm of the stomata. The CO_2 concentration has been reported to be 500–800 times higher in stems or branches than ambient air [60,61]. Thus, stomata should not make any difference to internal CO_2 concentration. However, the high concentrations referred to above are for larger diameter stems. In very young stems, there are less respiring cells, chloroplasts are present in all tissues, and thus there may be a CO_2 shortage during the day. Higher CO_2 concentrations have been associated with increased chloroplast density [62,63]. Most eucalypt stem chloroplasts would contribute to SRP, although a small amount of SNP probably occurs in small diameter stems in the immediate vicinity of the low density stomata.

Lenticels permit gas exchange through the otherwise impervious periderm. For *Olea europaea* and *Cercis siliquastrum*, the parenchyma cells beneath the lenticels showed a brighter red auto-fluorescence than those elsewhere [64], similar to that described for stomata in the young eucalypt branchlets. It has also been suggested that lenticels may also permit more light into the stem than the general periderm but this was not substantiated in a study of 10 species that examined one-year-old twigs [65]. Eucalypt lenticels are apparently rare. They were not observed on any of the investigated species and are rarely referred to in the literature.

3.5. Bark Functional Trade-Offs

Bark is a complex tissue with numerous functions (protection, storage, mechanical support, photosynthesis, translocation of photosynthates, xylem embolism repair, wound closure) [3,16]. This leads to trade-off relationships [16], e.g., protection from fire vs SRP and strength vs SRP. For eucalypts that shed bark a trade-off exists between protection from fire and maintenance of thin phellem for SRP [11]. In smooth barked eucalypts, many of the cells in prime position for photosynthesis (i.e., directly in from the phellogen, i.e., phelloderm and outermost secondary phloem or secondary cortex) would need to be fibres and sclereids for protection, e.g., in *C. citriodora*, the phelloderm had many sclereids which reduced the volume of photosynthetic tissue. In savanna ecosystems (grass, not crown fires), eucalypts with rough basal and smooth upper bark (half-barks) may have a particularly effective combination of bark types.

It would be useful to compare the bark density in the inner living bark of eucalypts with thin smooth phellem with those with thick phellem. Bark with a high proportion of thick walled sclerenchyma should have a relatively high density. We hypothesise that inner bark from species with thin phellem would be denser than inner bark that is protected by thick phellem.

3.6. Phellem Thickness and Chloroplast Development

Rosell et al. [3] examined several bark traits, including photosynthesis, in twigs and main stems of 85 shrub and tree species (including eight eucalypt species) in six sites from Australia and Mexico. The presence of photosynthetic bark was assessed by scraping off the rhytidome or phellem. Almost all (94%) of the twigs had photosynthetic bark, while about half (45%) of the main stems did. Their Figure 3 investigated the relationship between outer bark (i.e., dead bark/phellem) thickness and the probability of photosynthetic bark occurring in main stems. A trend existed of thin dead bark being associated with photosynthetic bark and vice versa. However, there did not appear to be a specific phellem thickness that was associated with the presence/absence of photosynthetic bark. As an example, their data indicates that about 10 species had photosynthetic bark with a dead bark thickness of >1 mm, while about 13 species had non-photosynthetic bark with an outer bark thickness of <1 mm. Other authors have noted chlorenchyma below a 1000 μm thick phellem [45] and below a 10 mm thick rhytidome [66]. In a study of relatively small diameter stems (6–13 mm) of 13 mangrove species, Schmitz et al. [46] found that xylary chlorophyll fluorescence and bark thickness (full bark thickness, not just phellem thickness) were not "strictly related" (p. 42 [46]). They also indicated that branch diameter and bark colour were not related to the distribution of xylary chloroplasts and this was found both between and within species. While the influence of phellem thickness on chloroplast depth was not directly addressed in the present study, in *E. blakelyi* (Figure 1a) and *E. caesia* (Figure 1b), chloroplasts were present beneath thicker and darker phellem.

3.7. Conclusions

From comparison with data in Rosell [15] it appears, on a worldwide basis, that smooth barked eucalypts may have very thin phellem. Combined with open canopies (pendulous leaves), this could mean relatively high light intensities in the chlorenchyma band in stems of all diameters and perhaps a greater importance for SRP in this group. In large diameter stems, this trade-off between maximal SRP and protection has resulted in a relatively similar outer bark anatomy in the wide diversity of species studied of: (i) thin phellem, (ii) an outer narrow band with most of the chloroplasts, and (iii) inner bark with a very high proportion of fibres and sclereids.

4. Materials and Methods

4.1. Materials

Material was sampled from 23 eucalypt species, including species of *Angophora*, *Corymbia*, and *Eucalyptus* and each of the three major subgenera (*Eudesmia*, *Eucalyptus*, *Symphomyrtus*) within *Eucalyptus* (Table 3). The trees sampled were mainly a mix of specimen trees and remnant vegetation on the Charles Sturt University, Wagga Wagga campus (35.06° S, 147.36° E). Thirteen of the species had smooth bark on at least the major branches and nine species had rough bark on the main stem and major branches. For species with rough bark, samples were taken from green stems where the epidermis was still present, to where the phellem had developed but was still relatively thin. For species with smooth bark throughout samples were excised from the trunk and/or main branches as well small diameter branchlets (Table 3). At least two trees per species were sampled, with at least three or four diameter sizes sampled per tree. For most of the 13 species with smooth bark on both small and large diameter stems, four to five trees were sampled and sectioned. For these species, only a single phellogen was present for most of the year. To maintain a thin layer of dead bark cells, when a new deeper phellogen was initiated, the tissues external to this meristem were soon shed.

4.2. Microscopy

All sections were hand cut with double-edge razor blades, then mounted in water and coverslipped (usually less than 2 h between sectioning and microscope observation). Small diameter (<8 mm) stems were sectioned whole (from epidermis to pith), medium diameter (0.8–4 cm) stems the bark, with a thin slice of secondary xylem, was sectioned and for large diameter stems (>4 cm) with >1 cm bark thickness the outer 4–5 mm of bark was examined. Most stem samples were cut in transverse section. These three diameter divisions (small, medium, large) do not always precisely correlate with the diameter bands in Table 3, especially for small and medium. The bands used in Table 3 depended on the available material and the growth stage of that material. The bark of *C. citriodora* and *E. globulus* was also sectioned in radial longitudinal and tangential longitudinal planes to further investigate the distribution of chloroplasts, fibres, and sclereids.

Chloroplast distribution was visualised with bright field and epifluorescence microscopy (Nikon Eclipse Ni microscope) using fluorescence optics (Nikon Intensilight C-HGFI) and a blue excitation filter (Nikon B-2A). Chloroplasts were identified by their bright red fluorescence. A UV filter (UV-2A) was used to examine lignified tissue (bright blue fluorescence) distribution. The depth of phellem was measured as was the depth of the band of tissue with the greatest concentration of chloroplasts. This was based on the brightest red fluorescence. In all species and all diameters, there was usually a distinctly brighter band near the stem surface. For example, in Figure 3 the bright band (greatest chloroplast density) was the following depth in these images: (a) 200–250 μm, (b) 100–150 μm, (c) 200 μm and (d) 100–150 μm. The approximate maximum depth of chloroplasts were also measured. For example, in Figure 3 the approximate maximum depths were: (a) 1800–2000 μm, (b) 1000 μm, (c) 2000 μm and (d) 700–800 μm. These measurements were estimations as discrete delimitations were not present.

4.3. Light Transmission

In *C. citriodora*, complete shedding of the outer bark (periderm plus some tissues beneath this) occurs over a 2–3 weeks in December in Wagga Wagga (Figure 1c). As this layer (when both dead and alive) can be easily collected in relatively large sheets, this provided a way to quantify the influence of the outermost bark on light intensity reaching inner regions of the stem. Transmission of light through bark that was soon to be shed but still alive (moist) was compared to bark that had already been shed (on the ground below trees, naturally dried). A Li-Cor quantum sensor, connected to a Li-Cor LI-185B, was placed perpendicular to the sun's rays (1600 μmol m^{-2} s^{-1}), and sheets of the two bark types were placed over the sensor and light transmission measured.

4.4. Maximum Chloroplast Depth

Most eucalypts begin to develop periderm on relatively small diameter stems (e.g., 4–10 mm diameter). Some eucalypt species (e.g., *E. caesia*, *E. cladocalyx*) had relatively large diameter stems (up to 3 cm) that had no periderm development (Figure 1d,e). For *E. cladocalyx*, this development was best expressed on vigorous epicormic shoots that formed after tree damage (storms, severe pruning) (Figure 1e). When these epicormic shoots were cut with secateurs in a transverse plane, the exposed surface appeared greenish from the epidermis to the pith. These shoots were sectioned to determine if chloroplasts were present from immediately below the epidermis to the pith. For *E. caesia*, the development of relatively large diameter shoots that had not commenced periderm development was a normal part of stem development. These shoots were usually distinctly glaucous (Figure 1d). These stems were sectioned to determine if the thick waxy and highly reflective layer reduced chloroplast development.

4.5. Stomatal Density in Stems and Leaves

Small green areas (e.g., Figure 2f) were observed with a stereo microscope on the small diameter stems of most species. They appeared to be associated with stomata and thus stomatal density (stomata/mm^2) was measured on small diameter stems that had not begun bark formation and also for leaves from adjoining nodes for six species (Table 2). For leaves, clear nail polish was applied to abaxial and adaxial surfaces, allowed to harden, peeled off, mounted on glass slides, photographed under a compound microscope, then stomata were counted (3 positions × 3 leaves × 2 plants). Stems were observed with a stereo microscope (Nikon SMZ25), images taken (e.g., Figure 2e) and number of stomata counted in 1 × 3 mm areas (3 positions × 2 stems × 2–3 plants).

Author Contributions: G.E.B. planned and designed the research. G.E.B. and C.C. performed the experiments. G.E.B. analysed the data and wrote the text. All authors have read and agreed to the published version of the manuscript.

Acknowledgments: We thank Alison Pound for assistance with some of the sectioning.

References

1.   Simkin, A.J.; Faralli, M.; Ramamoorthy, S.; Lawson, T. Photosynthesis in non-foliar tissues: Implications for yield. *Plant J.* **2020**, *101*, 1001–1015. [CrossRef] [PubMed]

2.   Pfanz, H.; Aschan, G. The existence of bark and stem photosynthesis in woody plants and its significance for the overall carbon gain. An eco-physiological and ecological approach. *Prog. Bot.* **2001**, *62*, 477–510. [CrossRef]

3.   Rosell, J.A.; Castorena, M.; Laws, C.A.; Westoby, M. Bark ecology of twigs vs. main stems: Functional traits across eighty-five species of angiosperms. *Oecologia* **2015**, *178*, 1033–1043. [CrossRef] [PubMed]

4.   Yiotis, C.; Psaras, G.K. *Dianthus caryophyllus* stems and *Zantedeschia aethiopica* petioles/pedicels show anatomical features indicating efficient photosynthesis. *Flora* **2011**, *206*, 360–364. [CrossRef]

5.   Filippou, M.; Fasseas, C.; Karabourniotis, G. Photosynthetic characteristics of olive tree (*Olea europaea*) bark. *Tree Physiol.* **2007**, *27*, 977–984. [CrossRef]

6.   Rentzou, A.; Psaras, G.K. Green plastids, maximal PSII photochemical efficiency and starch content of inner stem tissues of three Mediterranean woody species during the year. *Flora* **2008**, *203*, 350–357. [CrossRef]

7.   Saveyn, A.; Steppe, K.; Ubierna, N.; Dawson, T.E. Woody tissue photosynthesis and its contribution to trunk growth and bud development in young plants. *Plant Cell Environ.* **2010**, *33*, 1949–1958. [CrossRef]

8.   Nilsen, E.T. Stem photosynthesis: Extent, patterns, and role in plant carbon economy. In *Plant Stems: Physiology and Functional Morphology*; Gartner, B.L., Ed.; Academic Press: San Diego, CA, USA, 1995; pp. 223–240.

9.   Ávila, E.; Herrera, A.; Tezara, W. Contribution of stem CO_2 fixation to whole-plant carbon balance in nonsucculent species. *Photosynthetica* **2014**, *52*, 3–15. [CrossRef]

10. Gibson, A.C. Anatomy of photosynthetic old stems of nonsucculent dicotyledons from North American deserts. *Bot. Gaz.* **1983**, *144*, 347–362. [CrossRef]

11. Cernusak, L.A.; Cheesman, A.W. The benefits of recycling: How photosynthetic bark can increase drought tolerance. *New Phytol.* **2015**, *208*, 995–997. [CrossRef]

12. Esau, K. *Anatomy of Seed Plants*; John Wiley and Sons: New York, NY, USA, 1977.

13. Rosell, J.A.; Olson, M.E.; Anfodillo, T.; Martínez-Méndez, N. Exploring the bark thickness-stem diameter relationship: Clues from lianas, successive cambia, monocots and gymnosperms. *New Phytol.* **2017**, *215*, 569–581. [CrossRef] [PubMed]

14. Kotina, E.L.; Oskolski, A.A.; Tilney, P.M.; Van Wyk, B.E. Bark anatomy of *Adansonia digitata* L. (Malvaceae). *Adansonia* **2017**, *39*, 31–40. [CrossRef]

15. Rosell, J.A. Bark thickness across the angiosperms: More than just fire. *New Phytol.* **2016**, *211*, 90–102. [CrossRef] [PubMed]

16. Rosell, J.A. Bark in woody plants: Understanding the diversity of a multifunctional structure. *Integr. Comp. Biol.* **2019**, *59*, 535–547. [CrossRef]

17. Pilarski, J.; Tokarz, K. Chlorophyll distribution in the stems and trunk of beech trees. *Acta Physiol. Plant.* **2006**, *28*, 233–236. [CrossRef]

18. Sun, Q.; Yoda, K.; Suzuki, M.; Suzuki, H. Vascular tissue in the stem and roots of woody plants can conduct light. *J. Exp. Bot.* **2003**, *54*, 1627–1635. [CrossRef]

19. Bloemen, J.; McGuire, M.A.; Aubrey, D.P.; Teskey, R.O.; Steppe, K. Transport of root-respired CO_2 via the transpiration stream affects aboveground carbon assimilation and CO_2 efflux in trees. *New Phytol.* **2013**, *197*, 555–565. [CrossRef]

20. Cernusak, L.A.; Hutley, L.B. Stable isotopes reveal the contribution of corticular photosynthesis to growth in branches of *Eucalyptus miniata*. *Plant Physiol.* **2011**, *155*, 515–523. [CrossRef]

21. Vandegehuchte, M.W.; Bloemen, J.; Vergeynst, L.L.; Steppe, K. Woody tissue photosynthesis in trees: Salve on the wounds of drought? *New Phytol.* **2015**, *208*, 998–1002. [CrossRef]

22. Thornhill, A.H.; Crisp, M.D.; Külheim, C.; Lam, K.E.; Nelson, L.A.; Yeates, D.K.; Miller, J.T. A dated molecular perspective of eucalypt taxonomy, evolution and diversification. *Aust. Syst. Bot.* **2019**, *32*, 29–48. [CrossRef]

23. Gillison, A.N. Latitudinal Variation in Plant Functional Types. In *Geographical Changes in Vegetation and Plant Functional Types*; Greller, A.M., Fujiwara, K., Pedrotti, F., Eds.; Springer Science and Business Media LLC: Cham, Switzerland, 2018; pp. 21–57.

24. Chen, X.; Gao, J.; Zhao, P.; McCarthy, H.R.; Zhu, L.; Ni, G.; Ouyang, L. Tree species with photosynthetic stems have greater nighttime sap flux. *Front. Plant Sci.* **2018**, *9*, 30. [CrossRef] [PubMed]

25. Yu, H.; Shang, H.; Cao, J.; Chen, Z. How important is woody tissue photosynthesis in *EuCahetus dunnii* Maiden and *Osmanthus fragrans* (Thunb.) Lour. under O_3 stress? *Environ. Sci. Pollut. Res.* **2018**, *25*, 2112–2120. [CrossRef] [PubMed]

26. Cerasoli, S.; McGuire, M.A.; Faria, J.; Mourato, M.P.; Schmidt, M.; Pereira, J.S.; Chaves, M.; Teskey, R.O. CO_2 efflux, CO_2 concentration and photosynthetic refixation in stems of *Eucalyptus globulus* (Labill.). *J. Exp. Bot.* **2009**, *60*, 99–105. [CrossRef] [PubMed]

27. Eyles, A.; Pinkard, E.A.; O'Grady, A.P.; Worledge, D.; Warren, C.R. Role of corticular photosynthesis following defoliation in *Eucalyptus globulus*. *Plant Cell Environ.* **2009**, *32*, 1004–1014. [CrossRef]

28. Gao, J.-G.; Zhou, J.; Sun, Z.; Niu, J.; Zhou, C.; Gu, D.; Huang, Y.; Zhao, P. Suppression of nighttime sap flux with lower stem photosynthesis in *Eucalyptus* trees. *Int. J. Biometeorol.* **2016**, *60*, 545–556. [CrossRef]

29. Tausz, M.; Warren, C.R.; Adams, M.A. Is the bark of shining gum (*Eucalyptus nitens*) a sun or a shade leaf? *Trees Struct. Funct.* **2005**, *19*, 415–421. [CrossRef]

30. Johnstone, D.; Tausz, M.; Moore, G.; Nicolas, M. Chlorophyll fluorescence of the trunk rather than leaves indicates visual vitality in *Eucalyptus saligna*. *Trees* **2012**, *26*, 1565–1576. [CrossRef]

31. Johnstone, D.; Tausz, M.; Moore, G.; Nicolas, M. Bark and leaf chlorophyll fluorescence are linked to wood structural changes in *Eucalyptus saligna*. *AoB PLANTS* **2014**, *6*, plt057. [CrossRef]

32. Chattaway, M. The anatomy of bark. I. The genus *Eucalyptus*. *Aust. J. Bot.* **1953**, *1*, 403–433. [CrossRef]

33. Sokolov, S.Y. Chlorophyll in the wood of tree boughs. *Bot. Zhurnal SSSR* **1953**, *38*, 661–668. (In Russian)

34. Dima, E.; Manetas, Y.; Psaras, G.K. Chlorophyll distribution pattern in inner stem tissues: Evidence from epifluorescence microscopy and reflectance measurements in 20 woody species. *Trees* **2006**, *20*, 515–521. [CrossRef]

35. O'Gara, E.; Howard, K.; Colquhoun, I.J.; Dell, B.; McComb, J.; Hardy, G.E.S.J. The development and characteristics of periderm and rhytidome in *Eucalyptus marginata*. *Aust. J. Bot.* **2009**, *57*, 221–228. [CrossRef]

36. Burrows, G.E.; Hornby, S.K.; Waters, D.A.; Bellairs, S.M.; Prior, L.D.; Bowman, D.M.J.S. A wide diversity of epicormic structures is present in Myrtaceae species in the northern Australian savanna biome - implications for adaptation to fire. *Aust. J. Bot.* **2010**, *58*, 493–507. [CrossRef]

37. Mishra, G.; Collings, D.A.; Altaner, C.M. Cell organelles and fluorescence of parenchyma cells in *Eucalyptus bosistoana* sapwood and heartwood investigated by microscopy. *N. Z. J. For. Sci.* **2018**, *48*, 13. [CrossRef]

38. Scott, D.G. On the distribution of chlorophyll in the young shoots of woody plants. *Ann. Bot.* **1907**, *21*, 437–439. [CrossRef]

39. Pearson, L.C.; Lawrence, D.B. Photosynthesis in aspen bark. *Am. J. Bot.* **1958**, *45*, 383–387. [CrossRef]

40. Gómez-Vazquez, B.G.; Engleman, E.M. Bark anatomy of *Bursera longipes* (Rose) Standley and *Bursera copallifera* (Sessé & Moc.) Bullock. *IAWA Bull.* **1984**, *5*, 335–340. [CrossRef]

41. Kauppi, A. Seasonal fluctuations in chlorophyll content in birch stems with special reference to bark thickness and light transmission, a comparison between sprouts and seedlings. *Flora* **1991**, *185*, 107–125. [CrossRef]

42. Van Cleve, B.; Forreiter, C.; Sauter, J.J.; Apel, K. Pith cells of poplar contain photosynthetically active chloroplasts. *Planta* **1993**, *189*, 70–73. [CrossRef]

43. Pilarski, J. Gradient of photosynthetic pigments in the bark and leaves of lilac (*Syringa vulgaris* L.). *Acta Physiol. Plant.* **1999**, *21*, 365–373. [CrossRef]

44. Berveiller, D.; Kierzkowski, D.; Damesin, C. Interspecific variability of stem photosynthesis among tree species. *Tree Physiol.* **2007**, *27*, 53–61. [CrossRef] [PubMed]

45. Kotina, E.L.; Van Wyk, B.-E.; Tilney, P.M.; Oskolski, A.A. The systematic significance of bark structure in southern African genera of tribe Heteromorpheae (Apiaceae). *Bot. J. Linn. Soc.* **2012**, *169*, 677–691. [CrossRef]

46. Schmitz, N.; Egerton, J.J.G.; Lovelock, C.E.; Ball, M.C. Light-dependent maintenance of hydraulic function in mangrove branches: Do xylary chloroplasts play a role in embolism repair? *New Phytol.* **2012**, *195*, 40–46. [CrossRef] [PubMed]

47. Cocoletzi, E.; Angeles, G.; Sosa, V.; Patrón, A. The chloroplasts and unlignified parenchyma of two tropical pioneer forest tree species (Urticaceae). *Bot. Sci.* **2013**, *91*, 251–260. [CrossRef]

48. Kocurek, M.; Kornaś, A.; Pilarski, J.; Tokarz, K.; Lüttge, U.; Miszalski, Z. Photosynthetic activity of stems in two *Clusia* species. *Trees* **2015**, *29*, 1029–1040. [CrossRef]

49. Wittmann, C.; Pfanz, H. The optical, absorptive and chlorophyll fluorescence properties of young stems of five woody species. *Environ. Exp. Bot.* **2016**, *121*, 83–93. [CrossRef]

50. Blagitz, M.; Machado, S.R.; Marcati, C.R. Savanna trees do not have thicker outer bark than forest trees of two tropical species. *Flora* **2019**, *251*, 20–31. [CrossRef]

51. Angyalossy, V.; Pace, M.R.; Evert, R.F.; Marcati, C.R.; Oskolski, A.A.; Terrazas, T.; Kotina, E.; Lens, F.; Mazzoni-Viveiros, S.C.; Ángeles, G.; et al. IAWA list of microscopic bark features. *IAWA J.* **2016**, *37*, 517–615. [CrossRef]

52. Yiotis, C.; Manetas, Y.; Psaras, G.K. Leaf and green stem anatomy of the drought deciduous Mediterranean shrub *Calicotome villosa* (Poiret) Link. (Leguminosae). *Flora* **2006**, *201*, 102–107. [CrossRef]

53. Osmond, C.B.; Smith, S.D.; Gui-Ying, B.; Sharkey, T.D. Stem photosynthesis in a desert ephemeral, *Eriogonum inflatum*: Characterization of leaf and stem CO_2 fixation and H_2O vapor exchange under controlled conditions. *Oecologia* **1987**, *72*, 542–549. [CrossRef]

54. Redondo-Gómez, S.; Wharmby, C.; Moreno, F.J.; De Cires, A.; Castillo, J.M.; Luque, T.; Davy, A.J.; Figueroa, M.E. Presence of internal photosynthetic cylinder surrounding the stele in stems of the tribe Salicornieae (Chenopodiaceae) from SW Iberian Peninsula. *Photosynthetica* **2005**, *43*, 157–159. [CrossRef]

55. Karabourniotis, G.; Papastergiou, N.; Kabanopoulou, E.; Fasseas, C. Foliar sclereids of *Olea europaea* may function as optical fibers. *Can. J. Bot.* **1994**, *72*, 330–336. [CrossRef]

56. Kakuszi, A.; Böddi, B. Light piping activates chlorophyll biosynthesis in the under-soil hypocotyl section of bean seedlings. *J. Photochem. Photobiol. B Biol.* **2014**, *140*, 1–7. [CrossRef] [PubMed]

57.   Kakuszi, A.; Sárvári, É.; Solti, Á.; Czégény, G.; Hideg, É.; Hunyadi-Gulyás, É.; Bóka, K.; Böddi, B. Light piping driven photosynthesis in the soil: Low-light adapted active photosynthetic apparatus in the under-soil hypocotyl segments of bean (*Phaseolus vulgaris*). *J. Photochem. Photobiol. B Biol.* **2016**, *161*, 422–429. [CrossRef]

58.   Schaedle, M. Tree photosynthesis. *Annu. Rev. Plant Physiol.* **1975**, *26*, 101–115. [CrossRef]

59.   Nilsen, E.T.; Bao, Y. The influence of water stress on stem and leaf photosynthesis in *Glycine max* and *Sparteum junceum* (Leguminosae). *Am. J. Bot.* **1990**, *77*, 1007–1015. [CrossRef]

60.   Pfanz, H.; Aschan, G.; Langenfeld-Heyser, R.; Wittmann, C.; Loose, M. Ecology and ecophysiology of tree stems: Corticular and wood photosynthesis. *Naturwissenschaften* **2002**, *89*, 147–162. [CrossRef]

61.   Teskey, R.O.; Saveyn, A.; Steppe, K.; McGuire, M.A. Origin, fate and significance of CO_2 in tree stems. *New Phytol.* **2008**, *177*, 17–32. [CrossRef]

62.   Wang, X.; Anderson, O.; Griffin, K.L. Chloroplast numbers, mitochondrion numbers and carbon assimilation physiology of *Nicotiana sylvestris* as affected by CO_2 concentration. *Environ. Exp. Bot.* **2004**, *51*, 21–31. [CrossRef]

63.   Sharma, N.; Sinha, P.G.; Bhatnagar, A.K. Effect of elevated [CO_2] on cell structure and function in seed plants. *Clim. Chang. Environ. Sustain.* **2014**, *2*, 69. [CrossRef]

64.   Kalachanis, D.; Psaras, G. Structural changes in primary lenticels of *Olea europaea* and *Cercis siliquastrum* during the year. *IAWA J.* **2007**, *28*, 445–456. [CrossRef]

65.   Manetas, Y.; Pfanz, H. Spatial heterogeneity of light penetration through periderm and lenticels and concomitant patchy acclimation of corticular photosynthesis. *Trees* **2004**, *19*, 409–414. [CrossRef]

66.   Roth, I. *Structural Patterns of Tropical Barks*; Gebrüber Borntrager: Berlin, Germany, 1981.

3

Sequencing, Characterization, and Comparative Analyses of the Plastome of *Caragana rosea* var. *rosea*

Mei Jiang, Haimei Chen, Shuaibing He, Liqiang Wang, Amanda Juan Chen and Chang Liu *

Key Laboratory of Bioactive Substances and Resource Utilization of Chinese Herbal Medicine from Ministry of Education, Institute of Medicinal Plant Development, Chinese Academy of Medical Sciences, Peking Union Medical College, Beijing 100193, China; mjiang0502@163.com (M.J.); hmchen@implad.ac.cn (H.C.); wenyuxuan2530@163.com (S.H.); lys832000@163.com (L.W.); amanda_j_chen@163.com (A.J.C.)
* Correspondence: cliu@implad.ac.cn or cliu6688@yahoo.com

Abstract: To exploit the drought-resistant *Caragana* species, we performed a comparative study of the plastomes from four species: *Caragana rosea*, *C. microphylla*, *C. kozlowii*, and *C. Korshinskii*. The complete plastome sequence of the *C. rosea* was obtained using the next generation DNA sequencing technology. The genome is a circular structure of 133,122 bases and it lacks inverted repeat. It contains 111 unique genes, including 76 protein-coding, 30 tRNA, and four rRNA genes. Repeat analyses obtained 239, 244, 258, and 246 simple sequence repeats in *C. rosea*, *C. microphylla*, *C. kozlowii*, and *C. korshinskii*, respectively. Analyses of sequence divergence found two intergenic regions: *trnI-CAU-ycf2* and *trnN-GUU-ycf1*, exhibiting a high degree of variations. Phylogenetic analyses showed that the four *Caragana* species belong to a monophyletic clade. Analyses of Ka/Ks ratios revealed that five genes: *rpl16*, *rpl20*, *rps11*, *rps7*, and *ycf1* and several sites having undergone strong positive selection in the *Caragana* branch. The results lay the foundation for the development of molecular markers and the understanding of the evolutionary process for drought-resistant characteristics.

Keywords: *Caragana*; *Caragana rosea* var. *rosea*; plastome; comparative genomics; molecular markers

1. Introduction

The genus *Caragana* has more than 100 species and it belongs to the family of Leguminosae. The plants mainly grow in arid and semi-arid areas of Asia and Europe. Plants from this genus are well-known in resisting drought, barren, cold, and heat, and have a strong adaptability to the sand environment to prevent wind and fixate sand [1]. Sixty two species of *Caragana* are distributed in China [2], most of them can afforest barren hills and preserve water and soil. The distributions of several *Caragana* species have been well-studied in China. *C. microphylla* and *C. korshinskii* are distributed in Northeast China, North China, and Northwest China [3]. *C. microphylla* is adapted to the typical steppe zone, forest steppe zone, and deciduous broad-leaved forest steppe zone of the Mongolian plateau. *C. korshinskii* is suitable for the fixed and semi fixed sand land in the steppe desert and the typical desert belt zone. *C kozlowii* origins in the Lancang River and Tibet, and was mostly found in riverside with 3600–4000 m altitude [4]. *C rosea* is mainly from Northeast China, North China, East China, Henan, and southern Gansu Province, growing in slopes and valleys [5]. In addition to its drought adaptability, its medicinal value, such as strengthening the spleen and tonifying the kidney, was also well known [1]. Furthermore, it was shown that chemical constituents in *C. rosea* have anti-HIV activities [6]. Until now, the plastome of *C. korshinskii*, *C. microphylla*, and *C. kozlowii* were reported, while the plastome of *C. rosea* has not been studied.

Plants live in constantly changing environments that impose many biotic stress, such as pathogen infection and herbivore attack and abiotic stress, such as drought, heat, cold, nutrient deficiency, and excess of salt or toxic metals. Through the past years, many abiotic stress signaling and response pathways in plants have been discovered, with the core pathways involve protein kinases related to the yeast SNF1 and mammalian AMPK [7]. There is also research from the evolution of chloroplast genes adapted to contrasting habitats. For example, the *Cardamine resedifolia* plastid gene has undergone a more aggressive positive selection than *Cardamine impatiens,* which is located at lower elevations, which is why it is more adapted to the plateau environment [8,9]. For *Caragana* species, dozens of studies have been carried out on the morphological changes, such as stomatal status, leaf water state, cellular carbon metabolism, and etc., in drought responses [3,10]. However, little studies have been reported on the molecular bases for the stress responses in *Caragana* species at this time.

Different species of *Caragana* exhibit significantly varied abilities in drought resistance. For example, *C. rosea, C. microphylla, C. korshinskii,* and *C. kozlowii* have been evaluated for their strength of drought resistance based on leaf microstructure analysis, the drought resistance order from large to small is *C. korshinskii* > *C. microphylla* > *C. rosea* > *C. kozlowii* [11]. Furthermore, *C. microphylla* and *C. korshinskii* are closed related phylogenatically but difficult to differentiate morphologically [1,3]. Therefore, it is important to develop molecular markers to distinguish species accurately and to promote the rational use of species.

The plastome is an ideal choice for the development of molecular markers. It has many biological characteristics when compared with the nuclear genome, such as uni-parental inheritance, simpler structure, and being easier to obtain. Moreover, it provides more genetic information than a single gene/locus, resulting in much higher resolution in distinguishing closely related species. Furthermore, the plastome contains a series of genes that are related to photosynthesis, and the photosystem II (PSII) is a key part of drought stress, high temperature, and many other stresses [12,13]. Leaf physiological characteristics, such as photosynthetic capacity, which is related to the plastome function, and stomatal conductance, are the fixed indicators of water-use efficiency [14]. The water-use efficiency is critical for plants to cope with drought stress [15]. Therefore, a comparative analysis of the plastome of *Caragana* would shed light into the molecular bases for their tolerance drought.

Here, the complete plastome of *C. rosea* was sequenced and analyzed, which complements the plastid genome database of environmental stresses-resistant plants. Comparative analyses of the plastomes from *C. rosea* and other three *Caragana* species e.g., *C. korshinskii, C. microphylla,* and *C. kozlowii* were performed. The results that were obtained here provided valuable resources to illustrate molecular mechanisms that are related to drought resistance of *C. rosea*, to carry out chloroplast genetic engineering experiments and to select for plant individuals with favorable characteristics using molecular breeding. Furthermore, the results also provide useful information for future phylogenetic and taxonomic studies in the *Caragana* species.

2. Results

2.1. General Features of the Plastome

The complete plastome of *C. rosea* var. *rosea* is 133,122 bases in length and it lacks inverted repeat (Figure 1). This genome has been deposited in GenBank (accession number: MF593790). In the legume family, the phenomenon of IR regions loss was commonly found [16,17]. The sequence of protein-coding, tRNA, and rRNA regions accounted for 49.76%, 1.77%, and 3.4% of the whole genome, respectively; and, the rest are intergenic regions (Table 1). Moreover, a total of 111 unique genes were annotated, including 77 protein-coding, 30 tRNA, and four rRNA genes. The functional classification of these genes is shown in Table S1. The *C. rosea* plastome has 16 intron-containing genes, including 10 protein-coding genes and six tRNA genes. A total of 10 genes have only one intron, and only the *ycf3* contains two introns (Table S2). The *trnK-UUU* has the largest intron (2485 bases), which contains the *matK* gene. The *rps12* gene, which has the intron in the plastomes of other legume species [18], does not have the intron in the plastome of *Caragana*. Previous studies have shown that the loss of *rps12* intron may have occurred after the loss of the IR [19].

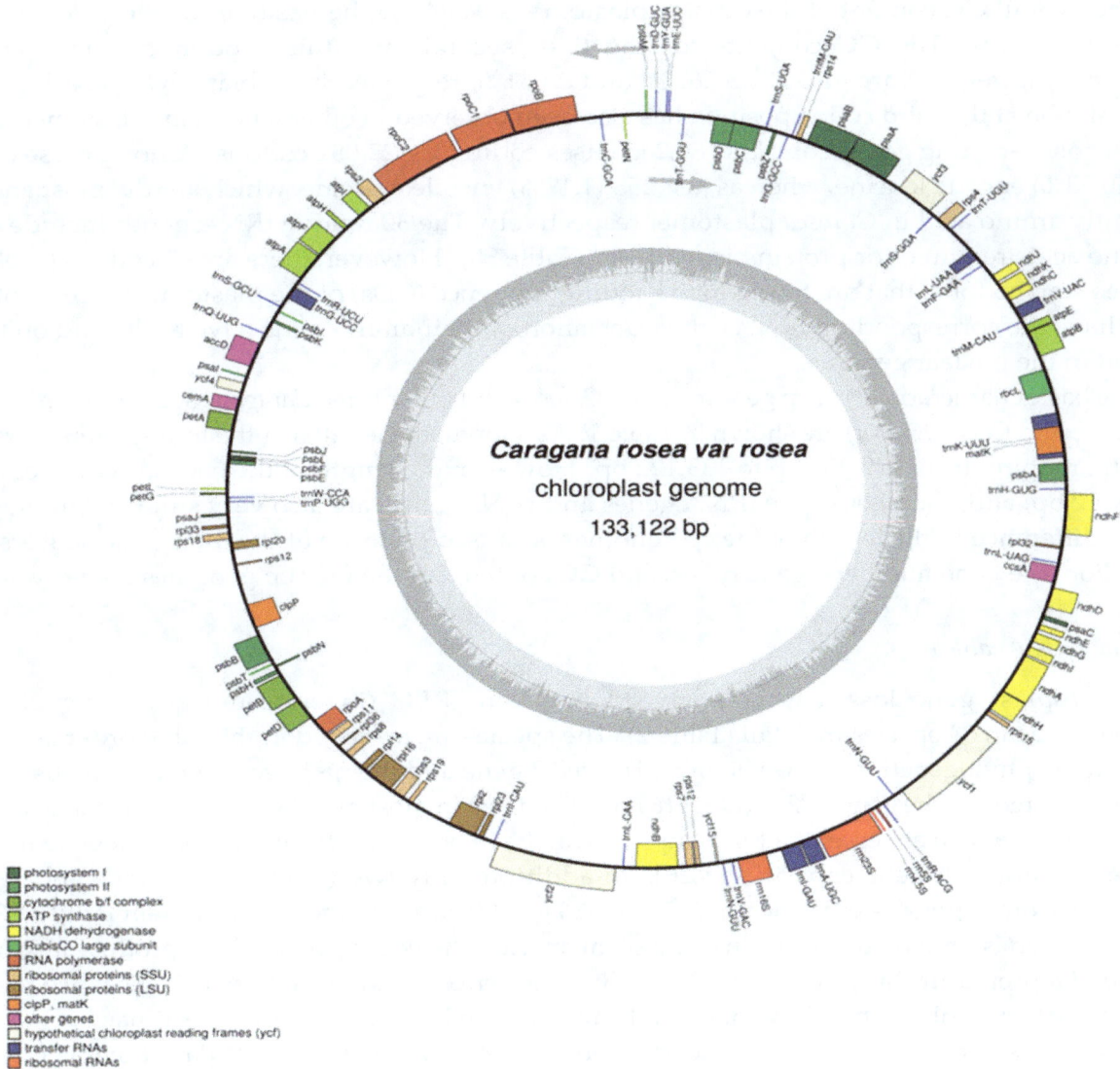

Figure 1. Circular gene map of the *C. rosea* plastome. Genes drawn inside the circle are transcribed clockwise, and those outside the circle are transcribed counterclockwise. Genes belonging to different functional groups are color codes. The inner circle shows the GC content. The two grey arrows represent the direction of transcription.

Table 1. Characteristics of *Caragana* plastome.

Plastome Characteristics	C. rosea	C. microphylla	C. kozlowii	C. korshinskii
complete genome length	133,122 bp	130,029 bp	131,274 bp	129,331 bp
No.of unique genes	110	110	110	110
No.of unique protein-coding genes	76	76	76	76
No.of unique tRNA genes	30	30	30	30
No.of unique rRNA genes	4	4	4	4
Size of protein-coding genes	66,243 bp (49.76%)	66,231 bp (50.94%)	66,234 bp (50.45%)	66,231 bp (51.21%)
Size of tRNA genes	2359 bp (1.77%)	2370 bp (1.82%)	2285 bp (1.74%)	2370 bp (1.83)
Size of rRNA genes	4537 bp (3.4%)	4520 bp (3.48%)	4521 bp (3.44%)	4520 bp (3.49)
Overall GC contents	34.84%	34.26%	34.50%	34.36%
GC contents of protein-coding genes	Coding GC 37.13% #1st position 45.36% #2nd position 37.62% #3rd position 28.41%	Coding GC 36.88% #1st position 44.98% #2nd position 37.67% #3rd position 27.99%	Coding GC 37.03% #1st position 45.30% #2nd position 37.58% #3rd position 28.21%	Coding GC 36.88% #1st position 44.96% #2nd position 37.67% #3rd position 27.99%
GC contents of tRNA genes	52.73%	53.14%	53.15%	53.05%
GC contents of rRNA genes	54.77%	54.82%	54.75%	54.82%

The overall GC content of the *C. rosea* plastome is 34.84%, whereas that for the protein-coding regions is 37.13%. The GC contents for the first, second, and third codon position with the protein-coding regions are 45.36%, 37.62%, and 28.41%, respectively. A bias towards a higher AT representation at the third codon position has also been observed in other land plant plastomes [20–22]. The 77 protein-coding genes comprise 66,243 bases coding for 22,081 codons. Among these codons, 2336 (10.58%) encode leucine, whereas just 258 (1.17%) encode cysteine, which are the most and least frequently amino acid in *C. rosea* plastome, respectively. The 30 unique tRNA genes include all the 20 amino acids required for protein biosynthesis (Table S3). However, there are 61 codons (excluding the three stop codons) that are found in the coding sequence (CDs) of the plastome, Since 31 of them do not have the corresponding tRNAs, the translation of their amino acids have to depend on tRNAs encoded in the nuclear genome.

The basic characteristics of cp genome from *C. rosea* and other three *Caragana* species (*C. microphylla*, *C. kozlowii*, and *C. korshinskii*) are shown in Table 1. As shown, the lengths of the four genomes are quite different, ranging from 129,331 bp to 133,122 bp. However, the length of the coding sequence differs by only 12 bp; and, the sizes of the tRNA genes and rRNA genes are also very similar. This suggests that the difference in the length of the cp genomes is caused by those of the intergenic spacers (IGS). In addition, the gene numbers, gene types, and GC contents in the four cp genomes are very similar.

2.2. Gene Loss Analysis

Chloroplast gene losses were analyzed between IRLC (inverted-repeat-lacking clade) of Papilionoideae in 34 species in detail (Table 2). The species are arranged in the same order as they are shown in the phylogenetic tree (see below). The *rpl22* gene and the *rps16* gene were both absent in all plastome. Moreover, the gene *rpl22* and *rps16* have been lost in most members of the angiosperm [23,24]. The two genes, which are essential for plant survival, have been transferred to the nucleus to maintain the plant's photosynthetic capacity [25,26]. In addition, only two (*Wisteria floribunda* and *Wisteria sinensis*) plastomes possess the gene *ycf15*. The *ycf15* gene belongs to the PFAM protein family PF10705 and its function is unknown. In fact, in some plant species, the *ycf15* gene may not produce any protein because of a premature stop codons in the coding sequences of these species [27]. Because the *ycf15* genes in the GenBank are highly variable in terms of gene length and sequence, it has been difficult to annotate this gene. The absence of the *ycf4* gene occurs in many species of Papilionatae [23,28,29], whereas the plastome of four *Caragana* species contains this gene. *Psal, ycf1, rpl23, rps18,* and *ndhB* genes were absent in 5, 4, 3, 3, 2 species, respectively, and the *rps32* gene was also found to be transferred to nuclear genomes from plastomes in several species [30,31]. The losses of *accD, atpE, ndhA, psbJ, psbL, psbZ,* and *rps2* were only found in *Trifolium boissieri, Astragalus mongholicus* var. *nakaianus, Medicago falcate, Lathyrus littoralis, Medicago falcate, Lens culinaris,* and *Medicago falcate*, respectively; the deletion of these genes rarely occurs in the plastome of angiosperms [23]. Overall, the losses of genes are monophyletic. However, exceptions can be found in the *Medicago falcate* and several species in the genus *Lathyrus*, including *L. sativus, L. odoratus, L. inconspicuus, L. tingitanus, L. davidii,* and *L. pubescens*.

2.3. Repeat Analysis

Repeated units play an important role in genome evolution, such as structural rearrangements and size evolution [32,33]. We analyzed the content and distribution of repeated sequences in the *C. rosea* plastomes. A total of 19 repeated elements that were longer than 30 bases were identified. The similarities between all the repeated elements were greater than 90%. Tandem, forward, and palindromic repeats presented a similar pattern of distribution. Most of them were found in the intergenic spacer region (IGS), around 31.6% were found in the protein-coding region, and only one repeat was found in the tRNA gene. However, the length of repeats among *C. rosea* plastome was obviously longer than those in other legumes [24,34], the largest repeat unit was 291 base long and was located in the spacer between the genes *rps12* and *clpP*. The type, location, and sequence of the repeat units are shown in Table S4.

Table 2. Gene losses in the plastomes from the inverted-repeat-lacking clade (IRLC) of Papilionoideae.

Category	Name of Species	rps16	rpl22	ycf15	ycf4	psaI	ycf1	rpl23	rps18	ndhB
Millettieae	W. floribunda	−	−	+	+	+	−	+	+	+
	W. sinensis	−	−	+	+	+	−	+	+	−
Galegeae	G. glabra	−	−	−	+	+	+	+	+	+
	G. lepidota	−	−	−	+	+	+	+	+	+
	A. mongholicus	−	−	−	+	+	+	+	+	+
	A. mongholicus var. nakaianus	−	−	−	+	+	+	+	+	+
Caraganeae	C. kozlowii	−	−	−	+	+	+	+	+	+
	C. korshinskii	−	−	−	+	+	+	+	+	+
	C. microphylla	−	−	−	+	+	+	+	+	+
	C. rosea var. rosea	−	−	−	+	+	+	+	+	+
Cicereae	C. arietinum	−	−	−	−	+	+	+	+	+
Trifolieae	M. truncatula	−	−	−	−	+	+	+	+	+
	M. papillosa	−	−	−	−	+	+	+	+	+
	M. hybrida	−	−	−	−	+	+	+	+	+
	M. falcata	−	−	−	+	+	−	+	+	−
	T. boissieri	−	−	−	−	+	+	+	+	+
	T. glanduliferum	−	−	−	−	+	+	+	+	+
Fabeae	T. strictum	−	−	−	−	+	+	+	+	+
	L. sativus	−	−	−	+	−	+	−	+	+
	L. odoratus	−	−	−	−	−	+	+	−	+
	L. inconspicuus	−	−	−	−	−	+	+	+	+
	L. ochroleucus	−	−	−	−	+	+	+	+	+
	L. venosus	−	−	−	−	+	+	+	+	+
	L. palustris	−	−	−	−	+	+	+	+	+
	L. tingitanus	−	−	−	−	+	+	+	−	+
	L. davidii	−	−	−	−	−	+	+	+	+
	L. graminifolius	−	−	−	−	+	+	+	+	+
	L. littoralis	−	−	−	−	+	+	+	+	+
	L. japonicus	−	−	−	−	+	+	+	+	+
	L. pubescens	−	−	−	−	−	+	+	+	+
	L. clymenum	−	−	−	−	+	+	+	+	+
	P. sativum	−	−	−	−	+	+	−	+	+
	V. sativa	−	−	−	−	+	+	−	+	+
	L. culinaris	−	−	−	−	+	−	+	−	+
	Number [a]	34	34	32	22	5	4	3	3	2

[a] The number refers to the total number of species that do not have the gene. "+": presence; "−" absence.

More remarkably, we found that 31.6% and 21.1% of these repeats were located in the IGS(*rps12-clpP*) and IGS(*rps19-rpl2*) regions, with the total length of the IGS regions being 2139 bases and 4060 bases, respectively. By contrast, the other three *Caragana* species do not have so many repeat units in the same regions. For example, the length of the corresponding IGS(*rps12-clpP*) regions in *C. microphylla* and *C. korshinskii* are less than 1 kb. Repeats are known to play a major role in plastome size evolution in angiosperm [35], the presence of abundant repeat sequences may be the reason why the genome of *C. rosea* is larger than the other three *Caragana* species.

Simple sequence repeat (SSR) loci are effective molecular markers because of their high variability that are wide distribution throughout the whole plastome [36]. SSR can provide useful information in polymorphism investigations and population genetics [37,38]. We identified SSRs in the plastome of four *Caragana* species. The number of SSRs ranges from 239 to 258 (Figure 2). In *Caragana rosea*, 66.9% are mono-nucleotide repeats, as compared with only 26.4%, 2.5%, and 4.2%, of di-, tri-, and tetra-nucleotide repeats, respectively. Of these SSR loci, 159 contained A or T, whereas only one had G or C; similarly, most dinucleotide repeat sequences were composed of AT/AT repeats. This result is consistent with previous reports that most of the SSR in plastomes are composed of short polyA or polyT repeats, while tandem repeats of G or C are rare [39]. We also analyzed the occurrence of SSRs in the CDs and found that there are fewer SSRs in the protein-coding regions than in the non-coding regions.

The plastomes of the other three *Caragana* species are similar to that of *C. rosea* in terms of the number, distribution, and the GC contents of the SSRs.

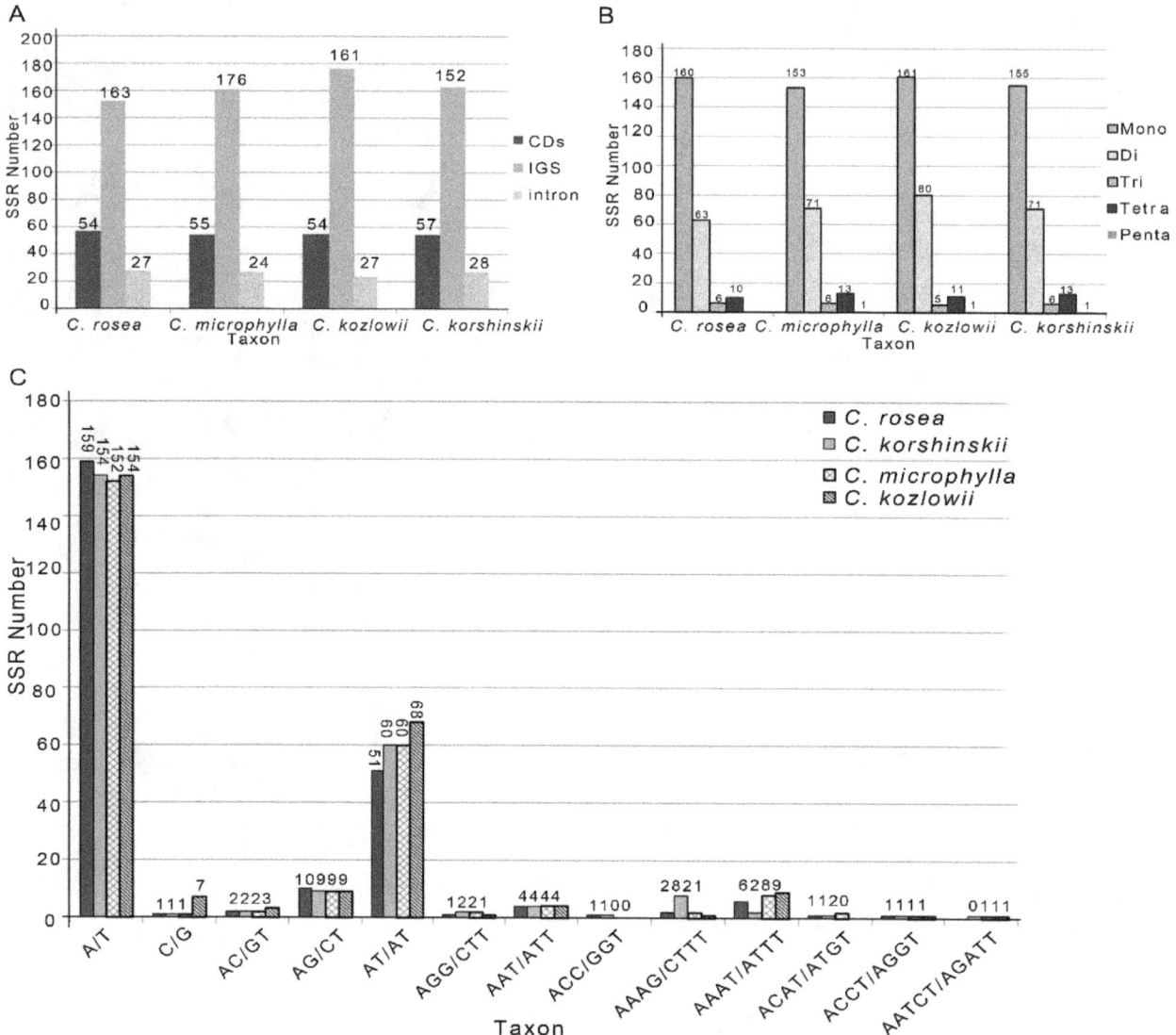

Figure 2. Statistics of simple sequence repeat (SSRs) detected in the plastome of four *Caragana* species. (**A**) Numbers of SSRs found in the coding (CDS), intergenic (IGS), and intronic regions, respectively; (**B**) number of different SSR types identified in the four genomes; and, (**C**) number of identified SSR motifs in different repeat class types.

2.4. Sequence Divergence Analysis among Caragana Species

To elucidate the level of sequence similarities between the *C. rosea* and the other three *Caragana*, the plastome sequences were compared while using the annotated *C. rosea* plastome as the reference (Figure 3). As shown, the four plastome sequences are highly similar. However, there are significant differences between *C. rosea* and other three *Caragana* species in some IGS, such as IGS(*rps12-clpP*) (square A), IGS(*rps19-rpl2*) (square B), and etc., which may be related to the unique large repeat fragments in the *C. rosea* plastome. For the regions IGS(*psaC-ndhD*) (square C), the sequence of *C. rosea* plastome is highly similar to those of *C. microphylla* and *C. korshinskii*, but it is quite different from that of *C. kozlowii*. Overall, the protein-coding regions are highly conserved, while the non-coding regions have different degrees of divergence between the *C. rosea* plastome and those of the other three. This suggests that the IGS of *Caragana* species has evolved rapidly.

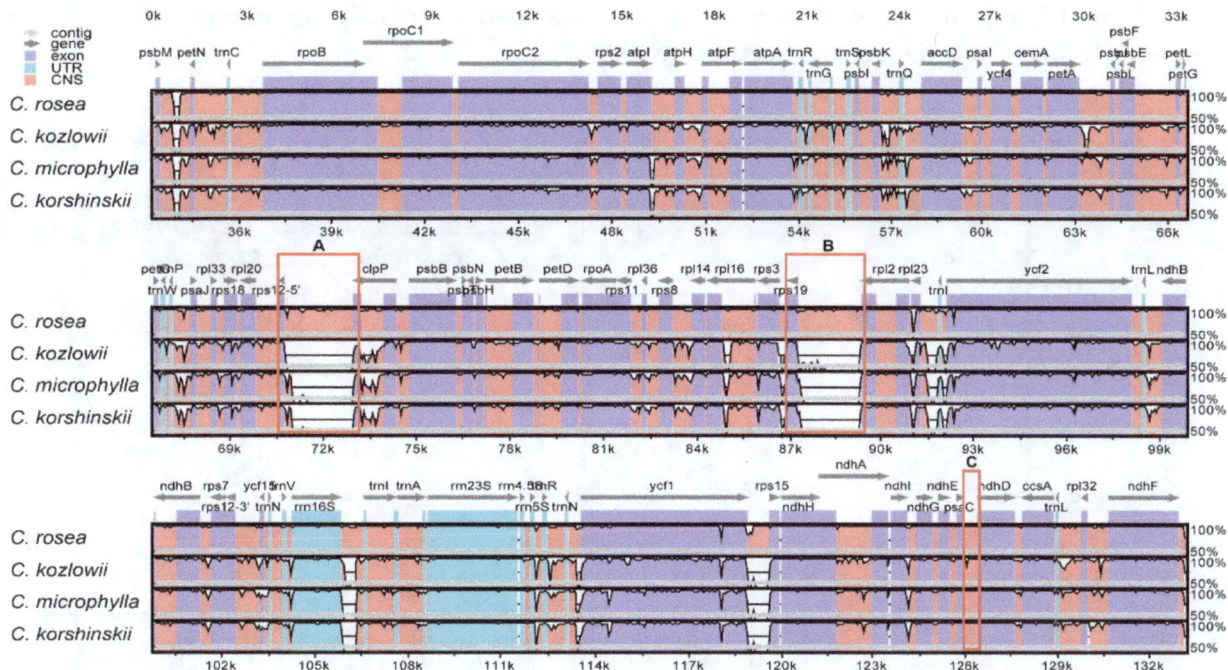

Figure 3. Structure comparison of the four plastomes by using the mVISTA program. Gray arrows and thick black lines above the alignment indicate genes with their orientation and the position of the IRs, respectively. A cut-off value of 70% identity was used for the plots, and the Y-scale represents the percent identity between 50% and 100%. UTR: Untranlated Regions; CNS: Conserved Non-coding Sequences. A: IGS(*rps12-clpP*); B: IGS(*rps19-rp12*); C: IGS(*psaC-ndhD*).

Highly variable sites in the genome can be used to develop molecular markers. We set out to identify the highly variable sites. We conducted pairwise distance comparison analysis for each non-coding region, including the intergenic region and the intron region using Kimura 2-parameter (K2p) model to identify divergence hotspot regions among the *Caragana* species. As expected, the variation of the intron sequence is relatively low, and the K2p distances ranged from 0.000 to 0.0269 (Figure 4, Table S5). The *clpP*, *ndhA* introns and the second intron of *ycf3* show the highest K2p values among the four *Caragana* species.

For IGSs, the K2p values ranged from 0 to 0.6207 (Figure 5). As described in the method section, we calculatecd the mean+2*STD (Standard Deviation) as the threshold for a IGS to be highly variable. It is 0.1649 in this case. The sequence divergences of the IGS regions ranged from 0 to 0.395 between *C. microphylla* and *C. kozlowii*, ranged from 0 to 0.6207 between *C. microphylla* and *C. rosea*, ranged from 0 to 0.5641 between *C. kozlowii* and *C. rosea*, ranged from 0 to 0.068 between *C. korshinskii* and *C. microphylla*, ranged from 0 to 0.3094 between *C. korshinskii* vs C. kozlowii, and ranged from 0 to 0.5176 between *C. korshinskii* vs *C. rosea*. The seven IGS regions (*atpB-atpE*, *ndhC-ndhK*, *ndhH-ndhA*, *psaA-psaB*, *psbC-psbD*, *psbT-psbN*, and *rpoB-rpoC1*) are 100% identical with the K2p values of 0. Meanwhile, we found that the K2p values were particularly high for the five IGS regions: *rpl23-trnI-CAU*, *rps12-clpP*, *rps19-rpl2*, *trnI-CAU-ycf2*, and *ycf1-rps15*.

The high degree of similarity among the plastomes of *C. korshinskii* and *C. microphylla* is consistent with their high degree of similarity in morphological characteristics. To determine whether these species can be distinguished with common molecular markers, such as ITS, *rbcL*, and *matK*, we compared the sequences of these markers from these species. It is found that these regions could not be used to distinguish the two confusable species with 99%, 100%, and 100% identities among these marker sequences, respectively. Interestingly, two regions IGS(*trnI-CAU-ycf2*) and IGS(*trnN-GUU-ycf1*) from *C. korshinskii* and *C. microphylla* have large K2p distances, indicating a high degree of sequence divergences. Moreover, these two regions have relatively high K2p values in the pairwise distance comparison analysis of these four species, which can be used to develop novel molecular markers to distinguish the four *Caragana* species accurately.

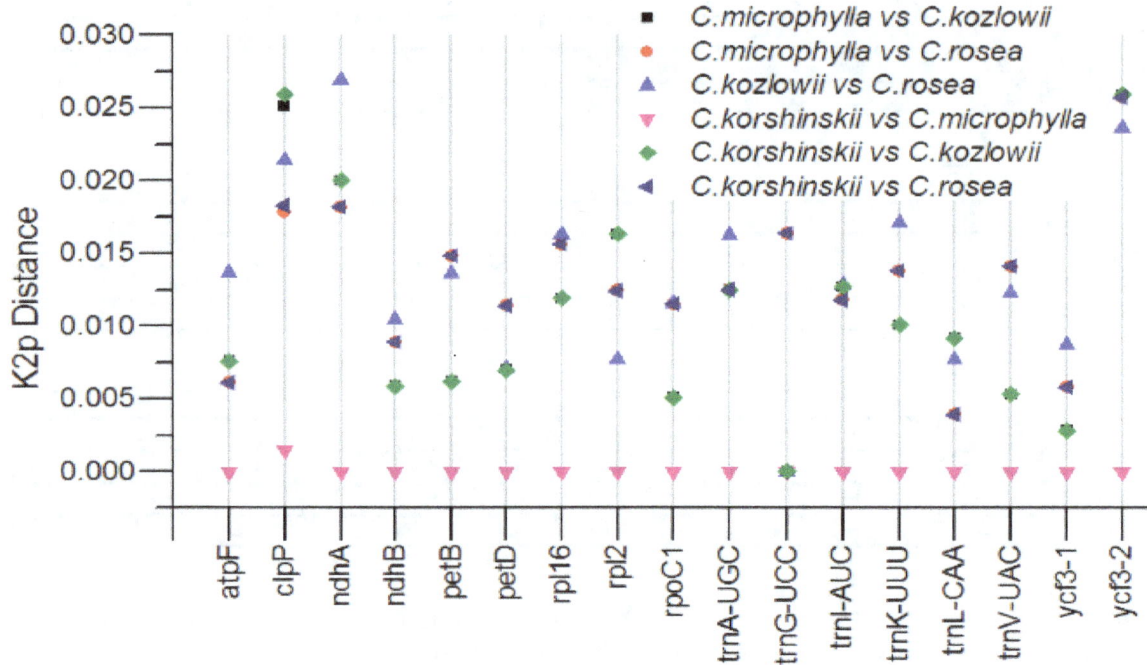

Figure 4. K2p distances for introns among *C. rosea, C. microphylla, C. kozlowii,* and *C. korshinskii.*

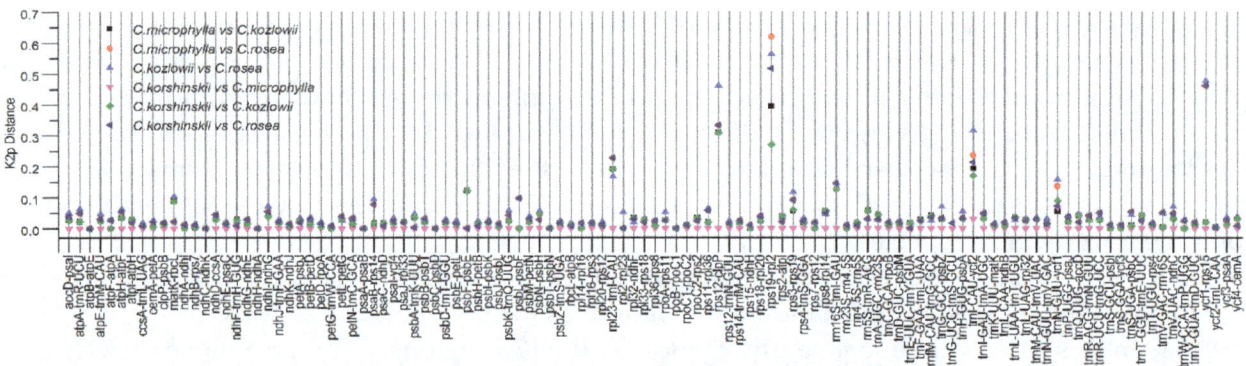

Figure 5. K2p distances for IGS regions among *C. rosea, C. microphylla, C. kozlowii,* and *C. korshinskii.*

2.5. Phylogenetic Analysis

The plastome sequence is an important resource for studying phylogenetic relationships and taxonomic status in the angiosperm [28]. In order to determine the phylogenetic position of *Caragana* in the Papilionoideae, we conducted multiple sequence alignments while using 63 common protein sequences from the plastomes of 36 species. *Arabidopsis thaliana* and *Nicotiana tabacum* were set as outgroup. The other 34 species contained six plant families that belong to the IRLC of Papilionoideae, including *Galegeae* (4), *Caraganeae* (4), *Cicereae* (1), *Fabeae* (16), *Trifolieae* (7), and *Millettieae* (2). The numbers in the parentheses represent the number of species in the corresponding taxa. The final dataset comprised 18745 positions and were subjected to phylogentic analysis using RaxML. Without surprise, *C. rosea* is found to locate in the same branch as the other three *Caragana* species, with 100% bootstrap values (Figure 6).

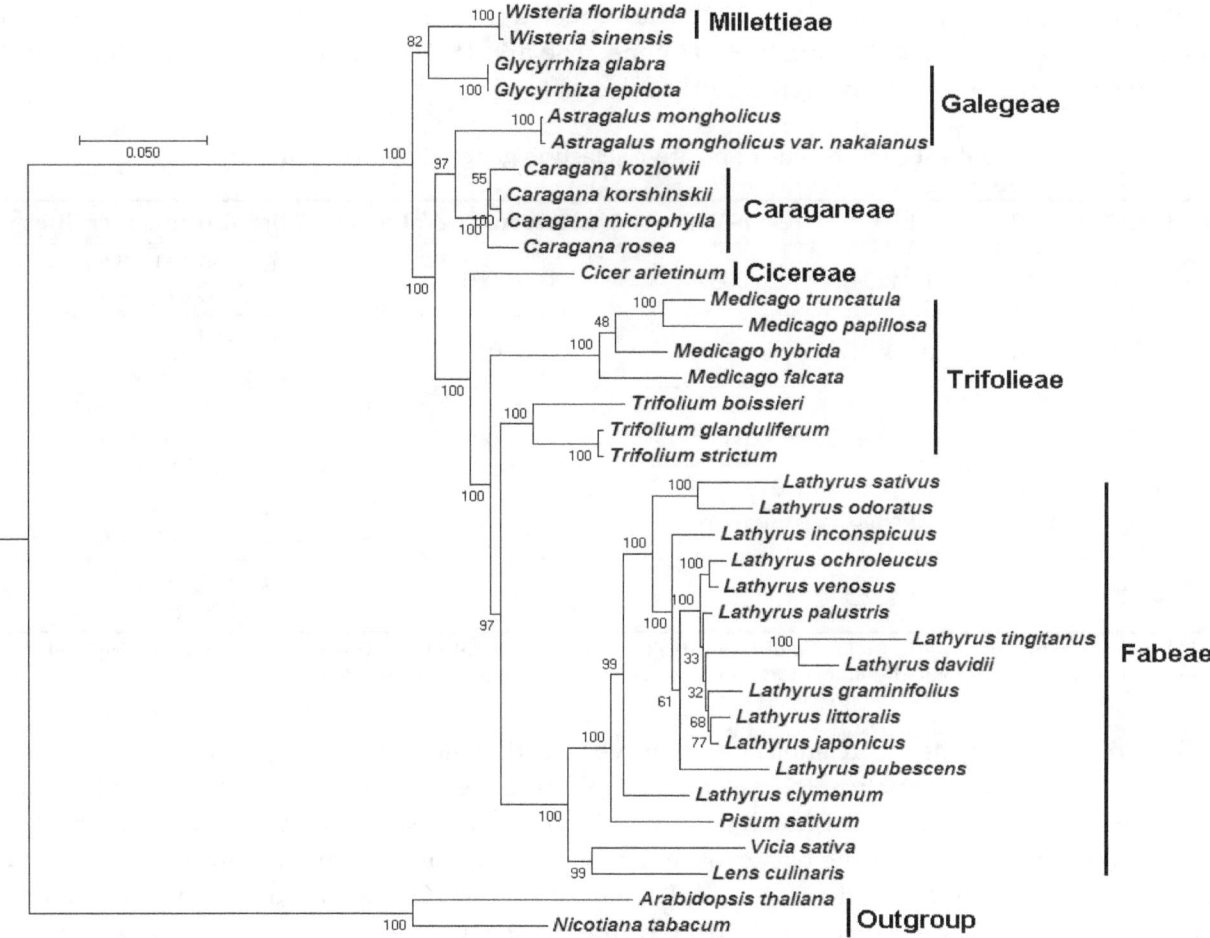

Figure 6. Molecular phylogenetic analyses of plastomes in the inverted-repeat-lacking clade of Papilionoideae. The tree was constructed with the sequences of 63 proteins present in all 36 species by using the maximum likelihood method implemented in RAxML. Bootstrap supports were calculated from 1000 replicates. *Nicotiana tabacum* and *Arabidopsis thaliana* were set as outgroups.

2.6. Selective Pressure Analysis

As synonymous substitutions accumulate nearly neutrally, non-synonyous substitutions are subject to selective pressures of varying degree and direction (positive or negative). In general, the ratio of nonsynonymous to synonymous substitution (ω) measures the levels of selective pressure operating in a protein coding gene. To test which genes were subject to positive selection at the *Caragana* branch, we conducted the selection analysis of the exons of each protein-encoding gene using the adaptive Branch-Site Random Effects Likelihood (aBSREL) model. A total of 69 branches among 36 species (listed in Section 2.5) were tested for diversifying selection. Significance was assessed using the Likelihood Ratio Test at a threshold of $p \leq 0.05$, after correcting for multiple testing.

Five genes (*rpl16, rpl20, rps11, rps7,* and *ycf1*) were found to have evolved under positive selection in *Caragana* branch in the phylogeny, the significance and number of rate categories inferred at the *Caragana* branch are provided in Table 3. The optimized branch length of the *Caragana* branch is 0.0128. The genes can be classified into two groups. The first group contains four ribosomal protein-coding genes. There is very strong evidence in selective pressure for the *rpl20* and *rps11* genes: positive selection was detected in 16 branches for both genes. In contrast, four branches (*Medicago hybrida, Medicago papillosa, Medicago falcata,* and *Caragana*) have experienced positive selection for the *rpl16* gene. Two branches (*Lathyrus clymenum, Caragana*) excluding those for the outgroup have experienced positive selection for the *rps7* gene, with two rate classes per branch. The second group contain the *ycf1* gene. The analysis did not include four species (*Lens culinaris*, *Wisteria floribunda*, *Wisteria sinensis*, and *Medicago falcata*) lacking the *ycf1* gene. The aBSREL model selection procedure identifies 42 branches for *ycf1* to have been significantly selected.

Table 3. The results of positive selection genes at *Caragana* branch.

Gene	B	LRT	Test *p*-Value	Uncorrected *p*-Value	ω Distribution over Sites
rpl16	0.0128	17.4234	0.0037	0.0001	ω1 = 0.121 (98%) ω2 = 781 (1.9%)
rpl20	0.0128	31.0401	0	0	ω1 = 0.147 (99%) ω2 = 290 (0.92%)
rps11	0.0128	12.0303	0.0452	0.0008	ω1 = 0.226 (100%) ω2 = 0.824 (0.20%) ω3 = 10000 (0.14%)
rps7	0.0128	24.3289	0.0001	0	ω1 = 0.00 (100%) ω2 = 286 (0.43%)
ycf1	0.0128	34.4372	0	0	ω1 = 0.384 (99%) ω2 = 91.3 (0.74%)

B: Optimized branch length; LRT: Likelihood ratio test statistic for selection; Test *p*-value: *p*-value corrected for multiple testing; Uncorrected *p*-value: Raw *p*-value without correction for multiple testing.

To find out which sites were subject to positive selection along the *Caragana* branch, codeml from PAML (v4.9) were used to analyze the Ka/Ks for the four genes: *rpl16*, *rpl20*, *rps11*, and *rps7*, using the branch-site model. The *Caragana* branch shown in Figure 6 were set as the foreground branch and all the other branches were set as the background branches. It is found that two sites: 36Y and 67Q was potentially under positive selection for the *rpl16* gene along the *Caragana* branch. In contrast, one site: 97Y were potentially under positive selection for the *rpl20* gene. Seven sites: 32K, 72T, 86N, 92V, 97Q, 103I, 137T were potential positively selected for the *rps11* gene. Four sites: 59E, 60T, 65V, 132V were positively selected for the *rps7* gene. As no three-dimensional (3D) structures are available for *Caragana* proteins, we cannot determine the functional and evolutionary significance of these sites at this time.

3. Discussion

Caragana species are xeromorphic, heat-resistant, and cold-resistant plants. Understanding the underlying molecular mechanism of this genus is of great interest for molecular breeding. As sessile organisms, plants must cope with abiotic stresses, such as soil salinity, drought, and extreme temperatures. Stress signaling pathways in plants have been extensively studied and reviewed [7,40]. All of these signaling pathways involve protein kinases that are related to yeast SNF1 and mammalian AMPK, suggesting that stress signaling in plants evolved from energy sensing.

Signals that are caused by limited water (drought stress) or excessive salt (salt stress) can be divided into primary and secondary types. The primary signal caused by drought is also called hyperosmotic stress. Salt stress exerts both osmotic and ionic effects on cells. The secondary effects are rather complex and they include oxidative stress; such effects include damage to cellular components (such as membrane lipids, proteins, and nucleic acids), chloroplast, mitochondria, ER, and metabolic dysfunction [7].

Chloroplast is an organelle where photosynthetic electron transport and many metabolic reactions occur. Environmental stress can easily perturb the metabolic balance in chloroplasts. The disturbance in chloroplast homeostasis is then passed to the nucleus through retrograde signals; as such, all cellular activities can be adjusted and coordinated. The chloroplast is a major site for the production of reactive oxygen species (ROS), such as superoxide anion, hydrogen peroxide, hydroxyl radical, and singlet oxygen [41]. Various environmental stresses, particularly high light stress, can exacerbate ROS production, thereby disrupting ROS-managing systems and generating various secondary messengers.

The relationship between protein synthesis and stress response has been studied [7]. The accuracy of protein synthesis is critical for life because a high degree of fidelity of the translation of the genetic information is required to accomplish the needs for cellular functions and to preserve variability developed by evolution. Even in simple organisms, this process involves more than 100 macromolecules, such as ribosomal proteins, translation factors, aminoacyl-tRNA synthetases, ribosomal RNAs, and transfer RNAs. Moreover, in vivo or in vitro experiments indicated that several macromolecules that participate in translation are the targets of oxidation; hence, translation is directly targeted by oxidative species. In bacteria, target macromolecules include elongation factors, such as Tu [42,43], Ts (EF-Ts) [44], and G (EF-G) [45–48]; several ribosomal proteins [31,36,37,49–51]; tRNAs [38–42,52,53]; and, aminoacyl-tRNA synthetases (aaRS) [43,49–51,54]. In particular, the disulfide bond formation of ribosomal proteins S7 and L16 are affected by oxidative stress [51]. Furthermore, scholars have identified the covalent binding of ribosomal protein S11 to chaperon/oxido-reductase protein through cysteine bond [43].

Previous studies showed that environmental stress can cause oxidative stress, which in turn affects the translation system, particularly in prokaryote originated systems, such as chloroplast. When considering the axis of environmental stress->oxidative stress->translation system, we can speculate that ribosomal protein-coding genes, such as *rps7*, *rps11*, *rpl16*, and *rpl20*, are strongly selected to maintain the integrity of the protein synthesis machinery under various environmental stresses. This phenomenon might, at least in part, contribute to the strong environmental stress resistance characteristics of the Caragana species. Additional analyses would be needed to confirm this hypothesis.

4. Materials and Methods

4.1. Plant Material, DNA Extraction, and Sequencing

The fresh leaves of the *C. rosea* were collected from the Institute of Medicinal Plant Development, China. After washing, the leaves were kept in the −80 refrigerator until use. DNA from about 100 mg leaves was extracted using the modified CTAB (Cetyltrimethylammonium bromide) method. Subsequently, the extracted DNA integrity and concentration were detected by electrophoresis in 1% (*w/v*) agarose gel and spectrophotometer (Nanodrop 2000, Thermo Fisher Scientific, Waltham, MA, USA). The genomic DNA of *C. rosea* was subjected to high-throughput sequencing using an Illumina Hiseq2000 sequencer (Illumina Inc., San Diego, CA, USA), with insert sizes of 500 bases for the library.

A total of 18,932,846 paired-end reads were obtained with 100 bases long. The other three plastome sequences of Caragana e.g., *C. korshinskii* (Accession number: NC_035229), *C. kozlowii* (Accession number NC_035228), *C. microphylla* (Accession number NC_032691), and ITS (Accession number: FJ537266, FJ537264) were obtained from Genbank.

4.2. Genome Assembly and Gap Filling

In order to extract the reads belonging to the plastome from those for the total DNAs, we downloaded 1688 plastome sequences from GenBank in February 2016, which were used to search against Illumina paired-end reads using BLASTN with an E-value cutoff of 1×10^5 [55]. The genome sequence of *C. kozlowii* was found to have the highest similarity and it was chosen as the reference sequence for the following assembly.

A total of 3076 paired-end reads similar to the *C. kozlowii* plastome sequence were selected and assembled by AbySS (v1.5.2) [56] and CLC Genomics Workbench (v7) Software. Nine and seven contigs were obtained using the two software tools, respectively. Then, the 16 sequences were further assembled by the Seqman module of DNAStar (v6.10.01), resulting in three contigs. The gaps between the contigs were filled with PCR amplification and Sanger sequencing using the sequence-specific primers (Table S6) designed to cross the gaps. Finally, the draft plastome sequence was validated by mapping the raw Illumina paired-end reads against it using Bowtie2 (v2.0.1) with default settings [57].

4.3. Genome Annotation and Characteristics Analysis

The *C. rosea* plastome sequence was annotated by CpGAVAS web service [58], with the default parameter. The tRNA genes were annotated using ARAGORN [59] and tRNAscan-SE [60]; the protein sequences were verified again by BLASP against the GenBank sequences. Subsequently, the intron/exon boundaries and the start/stop codons of predicted genes are manually edited using the Apollo program (v1.11.8) [61]. The circular plastome map of *C. rosea* was drawn using OrganellarGenomeDRAW [62]. Both GC contents and codon usage were calculated using the programs Cusp and Compseq from EMBOSS (v6.3.1) [63].

4.4. Repeat and SSR Analysis

Repeats (palindrome and forward repeats) were identified by REPuter web service [64], with the settings of 3 for the Hamming Distance (sequence identity $\geq$ 90%) and 30 for Minimal Repeat Size, as reported previously [34,65]. The number and location of tandem repeated elements in the *Caragana* genus plastome were determined using the Tandem Repeats Finder [66], with the following parameters: matches, mismatches and indels, minimum alignment score, and the maximum period size were 2, 7, 50, and 500, respectively. We manually verified all of the detected repeats and removed nested and redundant sequences. SSR in the plastome was analyzed by MISA software with the same parameters as reported previously [67]. Briefly, the cutoff for the numbers of units for mono-, di-, tri-, tetra-, penta-, and hexa-nucleotides were 8, 4, 4, 3, 3, and 3, respectively.

4.5. Comparative Genomic Analysis

The complete plastome sequence of *C. rosea* was compared with the those of *C. korshinskii*, *C. kozlowii*, and *C. microphylla*, using the mVISTA program in a Shuffle-LAGAN mode with default parameters [68]. The annotated *C. rosea* plastome was used as the reference. In order to analyze sequence diversity and selective pressure, a total of 112 intergenic regions, 17 introns, and 76 exons were extracted from the four plastomes using custom MatLab scripts. The corresponding nucleotide sequences were aligned using the CLUSTALW2 (v2.0.12) program with options "-type = DNA -gapopen = 10 -gapext = 2" [69]. Pairwise distance were determined with the Distmat program that was implemented in EMBOSS (v6.3.1) [63] using the Kimura 2-parameters (K2p) evolution model [70] for intergenic regions and introns. To determine the threshold for the K2p distance to be highly variable, we calculated the mean and the standard deviation for all the K2p values. The mean + 2*STD were then set as the threshold. The 76 exons sequences were aligned using the RevTrans (v2.0) [71] with the option of CLUSTALW2 program. Subsequently, the selective pressure analysis were conducted using adaptive branch-site random effects likelihood (aBSREL) model [72], implemented in HyPhy (https://veg.github.io/hyphy-site/getting-started/#characterizing-selective-pressures). Finally, we analyzed which sites were subject to positive selection along *Caragana* branch using the Codeml program that was implemented in PAML (v4.9) [73].

4.6. Phylogenetic Analysis

The plastome sequences of 33 species belonging to the IRLC (inverted-repeat-lackingclade) of Papilionoideae and two outgroup species (*Arabidopsis thaliana* and *Nicotiana tabacum*) were downloaded from NCBI RefSeq database, and a total of 63 protein sequences that were present in all of the 35 species and *C. rosea* were obtained by manual detection (ATPA, ATPB, ATPF, ATPH, ATPI, CCSA, CEMA, CLPP, MATK, NDHC, NDHD, NDHE, NDHF, NDHG, NDHH, NDHI, NDHJ, NDHK, PETA, PETB, PETD, PETG, PETL, PETN, PSAA, PSAB, PSAC, PSAJ, PSBA, PSBB, PSBC, PSBD, PSBE, PSBF, PSBH, PSBI, PSBK, PSBM, PSBN, PSBT, RBCL, RPL14, RPL16, RPL2, RPL20, RPL32, RPL33, RPL36, RPOA, RPOB, RPOC1, RPOC2, RPS11, RPS12, RPS14, RPS15, RPS19, RPS3, RPS4, RPS7, RPS8, YCF2, and YCF3) (Table S7). For the phylogenetic analysis, these protein sequences were aligned using the CLUSTALW2 (v2.0.12) program with options "-gapopen = 10 -gapext = 2 -output = phylip".

The Maximum Likelihood method implemented in RaxML (v8.2.4) [69] was used to inferred the evolutionary history, using "raxmlHPC-PTHREADS-SSE3 -f a -N 1000 -m PROTGAMMACPREV -x 551314260 -p 551314260 -o A_thaliana, N_tabacum -T 20". Subsequently, the Bootstrap analysis was also performed with 1000 replicates for the phylogenetic tree.

5. Conclusions

In this study, we sequenced the plastome of *C. rosea* and carried out a comparative study with those from *C. microphylla*, *C. kozlowii*, and *C. korshinskii*. Phylogenetic analyses showed that four *Caragana* species were on a monophyletic clade, with 100% bootstrap values. Analyses of selective pressure revealed that five genes: *rpl16, rpl20, rps11, rps7,* and *ycf1* were evolved undergoing positive selection. Analyses of sequence divergence found two sites: IGS(*trnI-CAU-ycf2*) and IGS(*trnN-GUU-ycf1*) had high degree of variations and might be sources for markers that can be used to distinguish these four species. The results presented in this paper will facilitate the further investigation for these four species in terms the molecular mechanisms for drought-resistance.

Author Contributions: C.L. conceived the study; M.J. collected samples of *C. rosea* var. *rosea*, extracted DNA for next-generation sequencing, assembled the genome, performed data analysis, conducted PCR validation and drafted the manuscript; H.C. annotated the genome; L.W. wrote the matlab scripts. S.H. and A.J.C. reviewed the manuscript critically. All authors have read and agreed the contents of the manuscript.

References

1.   Meng, Q.; Niu, Y.; Niu, X.; Roubin, R.H.; Hanrahan, J.R. Ethnobotany, phytochemistry and pharmacology of the genus *Caragana* used in traditional chinese medicine. *J. Ethnopharmacol.* **2009**, *124*, 350–368. [CrossRef] [PubMed]

2.   Delectis Flora Reipublicae Popularis Sinicae Agendae Academiae Sinicae Edita. *Flora Reipublicae Popularis Sinicae*; Science Press: Beijing, China, 1993; Volume 42, p. 18.

3.   Ma, F.; Na, X.; Xu, T. Drought responses of three closely related *Caragana* species: Implication for their vicarious distribution. *Ecol. Evol.* **2016**, *6*, 2763–2773. [CrossRef] [PubMed]

4.   Delectis Flora Reipublicae Popularis Sinicae Agendae Academiae Sinicae Edita. *Flora Reipublicae Popularis Sinicae*; Science Press: Beijing, China, 1993; Volume 42, p. 31.

5.   Delectis Flora Reipublicae Popularis Sinicae Agendae Academiae Sinicae Edita. *Flora Reipublicae Popularis Sinicae*; Science Press: Beijing, China, 1993; Volume 42, p. 60.

6.   Yang, G.X.; Qi, J.B.; Cheng, K.J.; Hu, C.Q. Anti-HIV chemical constituents of aerial parts of *Caragana rosea*. *Yao Xue Xue Bao (Acta Pharm. Sin.)* **2007**, *42*, 179–182.

7.   Zhu, J.K. Abiotic stress signaling and responses in plants. *Cell* **2016**, *167*, 313–324. [CrossRef] [PubMed]

8.   Hu, S.; Sablok, G.; Wang, B.; Qu, D.; Barbaro, E.; Viola, R.; Li, M.; Varotto, C. Plastome organization and evolution of chloroplast genes in *Cardamine* species adapted to contrasting habitats. *BMC Genom.* **2015**, *16*, 306. [CrossRef] [PubMed]

9.   Ometto, L.; Li, M.; Bresadola, L.; Varotto, C. Rates of evolution in stress-related genes are associated with habitat preference in two *Cardamine* lineages. *BMC Evol. Biol.* **2012**, *12*, 7. [CrossRef] [PubMed]

10. Gong, C.; Bai, J.; Wang, J.; Zhou, Y.; Kang, T.; Wang, J.; Hu, C.; Guo, H.; Chen, P.; Xie, P.; et al. Carbon storage patterns of *Caragana korshinskii* in areas of reduced environmental moisture on the loess plateau, China. *Sci. Rep.* **2016**, *6*, 28883. [CrossRef] [PubMed]

11. Li, M.; Liu, D.; Liu, Y. Evaluation on drought-resistant characteristics of ten *Caragana* species based on leaf micromorphological structure. *J. Desert Res.* **2016**, *3*, 708–717. [CrossRef]

12. Yamamoto, Y.; Aminaka, R.; Yoshioka, M.; Khatoon, M.; Komayama, K.; Takenaka, D.; Yamashita, A.; Nijo, N.; Inagawa, K.; Morita, N.; et al. Quality control of photosystem II: Impact of light and heat stresses. *Photosynth. Res.* **2008**, *98*, 589–608. [CrossRef] [PubMed]

13. Mulo, P.; Sakurai, I.; Aro, E.M. Strategies for psba gene expression in cyanobacteria, green algae and higher plants: From transcription to psii repair. *Biochim. Biophys. Acta* **2012**, *1817*, 247–257. [CrossRef] [PubMed]

14. Wright, I.J.; Reich, P.B.; Westoby, M.; Ackerly, D.D.; Baruch, Z.; Bongers, F.; Cavender-Bares, J.; Chapin, T.; Cornelissen, J.H.; Diemer, M.; et al. The worldwide leaf economics spectrum. *Nature* **2004**, *428*, 821–827. [CrossRef] [PubMed]

15.   Neufeld, H.S. Plant physiological ecology. *Photosynthetica* **1999**, *80*, 1785–1787. [CrossRef]

16.   Sabir, J.; Schwarz, E.; Ellison, N.; Zhang, J.; Baeshen, N.A.; Mutwakil, M.; Jansen, R.; Ruhlman, T. Evolutionary and biotechnology implications of plastid genome variation in the inverted-repeat-lacking clade of legumes. *Plant Biotechnol. J.* **2014**, *12*, 743–754. [CrossRef] [PubMed]

17.   Cardoso, D.; de Queiroz, L.P.; Pennington, R.T.; de Lima, H.C.; Fonty, E.; Wojciechowski, M.F.; Lavin, M. Revisiting the phylogeny of papilionoid legumes: New insights from comprehensively sampled early-branching lineages. *Am. J. Bot.* **2012**, *99*, 1991–2013. [CrossRef] [PubMed]

18.   Dugas, D.V.; Hernandez, D.; Koenen, E.J.; Schwarz, E.; Straub, S.; Hughes, C.E.; Jansen, R.K.; Nageswara-Rao, M.; Staats, M.; Trujillo, J.T.; et al. Mimosoid legume plastome evolution: Ir expansion, tandem repeat expansions, and accelerated rate of evolution in *clpP*. *Sci. Rep.* **2015**, *5*, 16958. [CrossRef] [PubMed]

19.   Jansen, R.K.; Wojciechowski, M.F.; Sanniyasi, E.; Lee, S.B.; Daniell, H. Complete plastid genome sequence of the chickpea (*Cicer arietinum*) and the phylogenetic distribution of *rps12* and *clpP* intron losses among legumes (leguminosae). *Mol. Phylogenet. Evol.* **2008**, *48*, 1204–1217. [CrossRef] [PubMed]

20.   Qian, J.; Song, J.; Gao, H.; Zhu, Y.; Xu, J.; Pang, X.; Yao, H.; Sun, C.; Li, X.; Li, C.; et al. The complete chloroplast genome sequence of the medicinal plant *Salvia miltiorrhiza*. *PLoS ONE* **2013**, *8*, e57607. [CrossRef] [PubMed]

21.   Shen, X.; Wu, M.; Liao, B.; Liu, Z.; Bai, R.; Xiao, S.; Li, X.; Zhang, B.; Xu, J.; Chen, S. Complete chloroplast genome sequence and phylogenetic analysis of the medicinal plant *Artemisia annua*. *Molecules* **2017**, *22*, 1330. [CrossRef] [PubMed]

22.   He, L.; Qian, J.; Li, X.; Sun, Z.; Xu, X.; Chen, S. Complete chloroplast genome of medicinal plant *Lonicera japonica*: Genome rearrangement, intron gain and loss, and implications for phylogenetic studies. *Molecules* **2017**, *22*, 249. [CrossRef] [PubMed]

23.   Daniell, H.; Lin, C.S.; Yu, M.; Chang, W.J. Chloroplast genomes: Diversity, evolution, and applications in genetic engineering. *Genome Biol.* **2016**, *17*, 134. [CrossRef] [PubMed]

24.   Keller, J.; Rousseau-Gueutin, M.; Martin, G.E.; Morice, J.; Boutte, J.; Coissac, E.; Ourari, M.; Ainouche, M.; Salmon, A.; Cabello-Hurtado, F.; et al. The evolutionary fate of the chloroplast and nuclear *rps16* genes as revealed through the sequencing and comparative analyses of four novel legume chloroplast genomes from *Lupinus*. *DNA Res. Int. J. Rapid Publ. Rep. Genes Genomes* **2017**, *24*, 343–358. [CrossRef] [PubMed]

25.   Jansen, R.K.; Saski, C.; Lee, S.B.; Hansen, A.K.; Daniell, H. Complete plastid genome sequences of three rosids (*Castanea, Prunus, Theobroma*): Evidence for at least two independent transfers of *rpl22* to the nucleus. *Mol. Biol. Evol.* **2011**, *28*, 835–847. [CrossRef] [PubMed]

26.   Gantt, J.S.; Baldauf, S.L.; Calie, P.J.; Weeden, N.F.; Palmer, J.D. Transfer of *rpl22* to the nucleus greatly preceded its loss from the chloroplast and involved the gain of an intron. *EMBO J.* **1991**, *10*, 3073–3078. [PubMed]

27.   Steane, D.A. Complete nucleotide sequence of the chloroplast genome from the tasmanian blue gum, *Eucalyptus globulus* (Myrtaceae). *DNA Res. Int. J. Rapid Publ. Rep. Genes Genomes* **2005**, *12*, 215–220. [CrossRef] [PubMed]

28.   Jansen, R.K.; Cai, Z.; Raubeson, L.A.; Daniell, H.; Depamphilis, C.W.; Leebens-Mack, J.; Muller, K.F.; Guisinger-Bellian, M.; Haberle, R.C.; Hansen, A.K.; et al. Analysis of 81 genes from 64 plastid genomes resolves relationships in angiosperms and identifies genome-scale evolutionary patterns. *Proc. Nat. Acad. Sci. USA* **2007**, *104*, 19369–19374. [CrossRef] [PubMed]

29.   Magee, A.M.; Aspinall, S.; Rice, D.W.; Cusack, B.P.; Semon, M.; Perry, A.S.; Stefanovic, S.; Milbourne, D.; Barth, S.; Palmer, J.D.; et al. Localized hypermutation and associated gene losses in legume chloroplast genomes. *Genome Res.* **2010**, *20*, 1700–1710. [CrossRef] [PubMed]

30.   Park, S.; Jansen, R.K.; Park, S. Complete plastome sequence of *Thalictrum coreanum* (ranunculaceae) and transfer of the *rpl32* gene to the nucleus in the ancestor of the subfamily thalictroideae. *BMC Plant Biol.* **2015**, *15*, 40. [CrossRef] [PubMed]

31.   Ueda, M.; Fujimoto, M.; Arimura, S.; Murata, J.; Tsutsumi, N.; Kadowaki, K. Loss of the *rpl32* gene from the chloroplast genome and subsequent acquisition of a preexisting transit peptide within the nuclear gene in *Populus*. *Gene* **2007**, *402*, 51–56. [CrossRef] [PubMed]

32.   Jo, Y.D.; Park, J.; Kim, J.; Song, W.; Hur, C.G.; Lee, Y.H.; Kang, B.C. Complete sequencing and comparative analyses of the pepper (*Capsicum annuum* L.) plastome revealed high frequency of tandem repeats and large insertion/deletions on pepper plastome. *Plant Cell Rep.* **2011**, *30*, 217–229. [CrossRef] [PubMed]

33. Sloan, D.B.; Triant, D.A.; Forrester, N.J.; Bergner, L.M.; Wu, M.; Taylor, D.R. A recurring syndrome of accelerated plastid genome evolution in the angiosperm tribe *Sileneae* (caryophyllaceae). *Mol. Phylogenet. Evol.* **2014**, *72*, 82–89. [CrossRef] [PubMed]

34. Martin, G.E.; Rousseau-Gueutin, M.; Cordonnier, S.; Lima, O.; Michon-Coudouel, S.; Naquin, D.; de Carvalho, J.F.; Ainouche, M.; Salmon, A.; Ainouche, A. The first complete chloroplast genome of the genistoid legume *Lupinus luteus*: Evidence for a novel major lineage-specific rearrangement and new insights regarding plastome evolution in the legume family. *Annu. Bot.* **2014**, *113*, 1197–1210. [CrossRef] [PubMed]

35. Haberle, R.C.; Fourcade, H.M.; Boore, J.L.; Jansen, R.K. Extensive rearrangements in the chloroplast genome of *Trachelium caeruleum* are associated with repeats and tRNA genes. *J. Mol. Evol.* **2008**, *66*, 350–361. [CrossRef] [PubMed]

36. Provan, J.; Corbett, G.; McNicol, J.W.; Powell, W. Chloroplast DNA variability in wild and cultivated rice (*Oryza* spp.) revealed by polymorphic chloroplast simple sequence repeats. *Genome* **1997**, *40*, 104–110. [CrossRef] [PubMed]

37. Xue, J.; Wang, S.; Zhou, S.L. Polymorphic chloroplast microsatellite loci in *Nelumbo* (nelumbonaceae). *Am. J. Bot.* **2012**, *99*, e240–e244. [CrossRef] [PubMed]

38. Pauwels, M.; Vekemans, X.; Gode, C.; Frerot, H.; Castric, V.; Saumitou-Laprade, P. Nuclear and chloroplast DNA phylogeography reveals vicariance among european populations of the model species for the study of metal tolerance, *Arabidopsis halleri* (Brassicaceae). *New Phytol.* **2012**, *193*, 916–928. [CrossRef] [PubMed]

39. Kuang, D.Y.; Wu, H.; Wang, Y.L.; Gao, L.M.; Zhang, S.Z.; Lu, L. Complete chloroplast genome sequence of *Magnolia kwangsiensis* (Magnoliaceae): Implication for DNA barcoding and population genetics. *Genome* **2011**, *54*, 663–673. [CrossRef] [PubMed]

40. Zhu, J.K. Salt and drought stress signal transduction in plants. *Annu. Rev. Plant Biol.* **2002**, *53*, 247–273. [CrossRef] [PubMed]

41. Mignolet-Spruyt, L.; Xu, E.; Idanheimo, N.; Hoeberichts, F.A.; Muhlenbock, P.; Brosche, M.; van Breusegem, F.; Kangasjarvi, J. Spreading the news: Subcellular and organellar reactive oxygen species production and signalling. *J. Exp. Bot.* **2016**, *67*, 3831–3844. [CrossRef] [PubMed]

42. Ichimura, K.; Mizoguchi, T.; Yoshida, R.; Yuasa, T.; Shinozaki, K. Various abiotic stresses rapidly activate *Arabidopsis* MAP kinases ATMPK4 and ATMPK6. *Plant J. Cell Mol. Biol.* **2000**, *24*, 655–665. [CrossRef]

43. Iuchi, S.; Kobayashi, M.; Taji, T.; Naramoto, M.; Seki, M.; Kato, T.; Tabata, S.; Kakubari, Y.; Yamaguchi-Shinozaki, K.; Shinozaki, K. Regulation of drought tolerance by gene manipulation of 9-*cis*-epoxycarotenoid dioxygenase, a key enzyme in abscisic acid biosynthesis in *Arabidopsis*. *Plant J. Cell Mol. Biol.* **2001**, *27*, 325–333. [CrossRef]

44. Ishitani, M.; Liu, J.; Halfter, U.; Kim, C.S.; Shi, W.; Zhu, J.K. Sos3 function in plant salt tolerance requires n-myristoylation and calcium binding. *Plant Cell* **2000**, *12*, 1667–1678. [CrossRef] [PubMed]

45. Chen, H.H.; Li, P.H.; Brenner, M.L. Involvement of abscisic acid in potato cold acclimation. *Plant Physiol.* **1983**, *71*, 362–365. [CrossRef] [PubMed]

46. Choi, H.; Hong, J.; Ha, J.; Kang, J.; Kim, S.Y. ABFs, a family of ABA-responsive element binding factors. *J. Biol. Chem.* **2000**, *275*, 1723–1730. [CrossRef] [PubMed]

47. Gustin, M.C.; Albertyn, J.; Alexander, M.; Davenport, K. MAP kinase pathways in the yeast *Saccharomyces cerevisiae*. *Microbiol. Mol. Biol. Rev. (MMBR)* **1998**, *62*, 1264–1300. [PubMed]

48. Jacob, T.; Ritchie, S.; Assmann, S.M.; Gilroy, S. Abscisic acid signal transduction in guard cells is mediated by phospholipase D activity. *Proc. Nat. Acad. Sci. USA* **1999**, *96*, 12192–12197. [CrossRef] [PubMed]

49. Ingram, J.; Bartels, D. The molecular basis of dehydration tolerance in plants. *Annu. Rev. Plant Physiol. Plant Mol. Biol.* **1996**, *47*, 377–403. [CrossRef] [PubMed]

50. Jaglo-Ottosen, K.R.; Gilmour, S.J.; Zarka, D.G.; Schabenberger, O.; Thomashow, M.F. *Arabidopsis CBF1* overexpression induces *COR* genes and enhances freezing tolerance. *Science* **1998**, *280*, 104–106. [CrossRef] [PubMed]

51. Jonak, C.; Kiegerl, S.; Ligterink, W.; Barker, P.J.; Huskisson, N.S.; Hirt, H. Stress signaling in plants: A mitogen-activated protein kinase pathway is activated by cold and drought. *Proc. Nat. Acad. Sci. USA* **1996**, *93*, 11274–11279. [CrossRef] [PubMed]

52. Knight, H.; Trewavas, A.J.; Knight, M.R. Calcium signalling in *Arabidopsis thaliana* responding to drought and salinity. *Plant J. Cell Mol. Biol.* **1997**, *12*, 1067–1078. [CrossRef]

53. Katagiri, T.; Takahashi, S.; Shinozaki, K. Involvement of a novel *Arabidopsis* phospholipase D, AtPLDδ, in

dehydration-inducible accumulation of phosphatidic acid in stress signalling. *Plant J. Cell Mol. Biol.* **2001**, *26*, 595–605. [CrossRef]

54. Kovtun, Y.; Chiu, W.L.; Tena, G.; Sheen, J. Functional analysis of oxidative stress-activated mitogen-activated protein kinase cascade in plants. *Proc. Nat. Acad. Sci. USA* **2000**, *97*, 2940–2945. [CrossRef] [PubMed]

55. Camacho, C.; Coulouris, G.; Avagyan, V.; Ma, N.; Papadopoulos, J.; Bealer, K.; Madden, T.L. Blast+: Architecture and applications. *BMC Bioinform.* **2009**, *10*, 421. [CrossRef] [PubMed]

56. Simpson, J.T.; Wong, K.; Jackman, S.D.; Schein, J.E.; Jones, S.J.; Birol, I. ABySS: A parallel assembler for short read sequence data. *Genome Res.* **2009**, *19*, 1117–1123. [CrossRef] [PubMed]

57. Langmead, B.; Trapnell, C.; Pop, M.; Salzberg, S.L. Ultrafast and memory-efficient alignment of short DNA sequences to the human genome. *Genome Biol.* **2009**, *10*, R25. [CrossRef] [PubMed]

58. Liu, C.; Shi, L.; Zhu, Y.; Chen, H.; Zhang, J.; Lin, X.; Guan, X. CpGAVAS, an integrated web server for the annotation, visualization, analysis, and genbank submission of completely sequenced chloroplast genome sequences. *BMC Genom.* **2012**, *13*, 715. [CrossRef] [PubMed]

59. Laslett, D.; Canback, B. ARAGORN, a program to detect tRNA genes and tmRNA genes in nucleotide sequences. *Nucleic Acids Res.* **2004**, *32*, 11–16. [CrossRef] [PubMed]

60. Schattner, P.; Brooks, A.N.; Lowe, T.M. The tRNAscan-SE, snoscan and snoGPS web servers for the detection of tRNAs and snoRNAs. *Nucleic Acids Res.* **2005**, *33*, W686–W689. [CrossRef] [PubMed]

61. Misra, S.; Harris, N. Using apollo to browse and edit genome annotations. *Curr. Protocol. Bioinform.* **2006**. Chapter 9, Unit 9 5. [CrossRef]

62. Lohse, M.; Drechsel, O.; Bock, R. Organellargenomedraw (OGDRAW): A tool for the easy generation of high-quality custom graphical maps of plastid and mitochondrial genomes. *Curr. Genet.* **2007**, *52*, 267–274. [CrossRef] [PubMed]

63. Rice, P.; Longden, I.; Bleasby, A. Emboss: The European molecular biology open software suite. *Trends Genet. (TIG)* **2000**, *16*, 276–277. [CrossRef]

64. Kurtz, S.; Choudhuri, J.V.; Ohlebusch, E.; Schleiermacher, C.; Stoye, J.; Giegerich, R. REPuter: The manifold applications of repeat analysis on a genomic scale. *Nucleic Acids Res.* **2001**, *29*, 4633–4642. [CrossRef] [PubMed]

65. Tangphatsornruang, S.; Sangsrakru, D.; Chanprasert, J.; Uthaipaisanwong, P.; Yoocha, T.; Jomchai, N.; Tragoonrung, S. The chloroplast genome sequence of mungbean (*Vigna radiata*) determined by high-throughput pyrosequencing: Structural organization and phylogenetic relationships. *DNA Res. Int. J. Rapid Publ. Rep. Genes Genomes* **2010**, *17*, 11–22. [CrossRef] [PubMed]

66. Benson, G. Tandem repeats finder: A program to analyze DNA sequences. *Nucleic Acids Res.* **1999**, *27*, 573–580. [CrossRef] [PubMed]

67. Lei, W.; Ni, D.; Wang, Y.; Shao, J.; Wang, X.; Yang, D.; Wang, J.; Chen, H.; Liu, C.; Lei, W. Intraspecific and heteroplasmic variations, gene losses and inversions in the chloroplast genome of *Astragalus membranaceus*. *Sci. Rep.* **2016**, *6*, 21669. [CrossRef] [PubMed]

68. Frazer, K.A.; Pachter, L.; Poliakov, A.; Rubin, E.M.; Dubchak, I. Vista: Computational tools for comparative genomics. *Nucleic Acids Res.* **2004**, *32*, W273–W279. [CrossRef] [PubMed]

69. Stamatakis, A. Raxml version 8: A tool for phylogenetic analysis and post-analysis of large phylogenies. *Bioinformatics* **2014**, *30*, 1312–1313. [CrossRef] [PubMed]

70. Kimura, M. A simple method for estimating evolutionary rates of base substitutions through comparative studies of nucleotide sequences. *J. Mol. Evol.* **1980**, *16*, 111–120. [CrossRef] [PubMed]

71. Wernersson, R.; Pedersen, A.G. RevTrans: Multiple alignment of coding DNA from aligned amino acid sequences. *Nucleic Acids Res.* **2003**, *31*, 3537–3539. [CrossRef] [PubMed]

72. Smith, M.D.; Wertheim, J.O.; Weaver, S.; Murrell, B.; Scheffler, K.; Kosakovsky Pond, S.L. Less is more: An adaptive branch-site random effects model for efficient detection of episodic diversifying selection. *Mol. Biol. Evol.* **2015**, *32*, 1342–1353. [CrossRef] [PubMed]

73. Yang, Z.; Nielsen, R. Codon-substitution models for detecting molecular adaptation at individual sites along specific lineages. *Mol. Biol. Evol.* **2002**, *19*, 908–917. [CrossRef] [PubMed]

4

The Complete Plastome Sequence of an Antarctic Bryophyte *Sanionia uncinata* (Hedw.) Loeske

Mira Park [1,2]**, Hyun Park** [1,3]**, Hyoungseok Lee** [1,3,]*****, Byeong-ha Lee** [2,]***** **and Jungeun Lee** [1,3,]*****

[1]   Unit of Polar Genomics, Korea Polar Research Institute, Incheon 21990, Korea; mira0295@kopri.re.kr (M.P.); hpark@kopri.re.kr (H.P.)

[2]   Department of Life Science, Sogang University, Seoul 04107, Korea

[3]   Polar Science, University of Science & Technology, Daejeon 34113, Korea

*   Correspondence: soulaid@kopri.re.kr (H.L.); byeongha@sogang.ac.kr (B.-h.L.); jelee@kopri.re.kr (J.L.)

Abstract: Organellar genomes of bryophytes are poorly represented with chloroplast genomes of only four mosses, four liverworts and two hornworts having been sequenced and annotated. Moreover, while Antarctic vegetation is dominated by the bryophytes, there are few reports on the plastid genomes for the Antarctic bryophytes. *Sanionia uncinata* (Hedw.) Loeske is one of the most dominant moss species in the maritime Antarctic. It has been researched as an important marker for ecological studies and as an extremophile plant for studies on stress tolerance. Here, we report the complete plastome sequence of *S. uncinata*, which can be exploited in comparative studies to identify the lineage-specific divergence across different species. The complete plastome of *S. uncinata* is 124,374 bp in length with a typical quadripartite structure of 114 unique genes including 82 unique protein-coding genes, 37 tRNA genes and four rRNA genes. However, two genes encoding the α subunit of RNA polymerase (*rpoA*) and encoding the cytochrome $b_{6/f}$ complex subunit VIII (*petN*) were absent. We could identify nuclear genes homologous to those genes, which suggests that *rpoA* and *petN* might have been relocated from the chloroplast genome to the nuclear genome.

Keywords: *Sanionia uncinata*; chloroplast genome; Antarctic bryophyte; moss; plastome

---

1. Introduction

Antarctic terrestrial ecosystems are dominated by lichens and bryophytes (including mosses, liverworts and hornworts), encompassing more than 200 lichens and 109 mosses species [1]. Only two vascular plant species have survived and adapted to these extreme environments, with a very limited distribution restricted to the maritime Antarctic, while mosses are common plants on extensive ice-free areas of Antarctica.

Sanionia uncinata is one of the most dominant moss species in Antarctica and is mainly distributed over coastal areas [2,3]. Moreover, *S. uncinata* is distributed across multiple geographic regions, ranging from Northern Hemisphere (Europe, Asia, North America and the Arctic) to Southern Hemisphere (Africa, South America and the Antarctica) and also found at high-altitude mountains in tropical and subtropical areas [4]. A recent phylogeographic study has established the haplotype networks of *S. uncinata* populations by identifying their genetic diversity with massive molecular marker datasets of more than 200 specimens collected from various regions around the world [4].

S. uncinata is a pleurocarpous moss species that form dense and extensive carpets on terrestrial habitats over a wide range of water regimes, from dry rock surfaces to wet areas at the edges of streams or melt pools [5–7]. *S. uncinata* has been extensively used as an experimental model for the study of environmental impacts on plants [8–10]. This species is known to tolerate dehydration by retaining moisture in their tissues for a long period of time by forming a carpet-like community shape that helps to avoid water loss [11]. However, the molecular mechanism and molecular ecology underlying stress tolerance have yet to be elucidated. The dehydration process, although it prevents the moss from freezing, directly affects cell metabolism and as a result, photosynthetic capacity decreases when the water content of moss drops below the optimum level [12]. On the other hand, photosynthesis is essential for the production of energy needed for plant growth and takes place in the chloroplasts. Thus, the integrity and metabolic performance of chloroplasts are very important for photosynthetic activities [13].

Chloroplasts are unique organelles, derived from cyanobacteria through endosymbiosis, that provide essential energy for plants and algae through photosynthesis [14,15]. They contain their own genomes that have a unique mechanism of RNA transcription and are inherited maternally. Chloroplasts are known to play an important role in the synthesis of pigments, starch, fatty acids and amino acids as well as the photosynthesis process [16,17]. In general, chloroplast genomes—namely, plastomes—are highly conserved with regards to gene sequences and gene content in terrestrial plants. Their highly conservative nature is sufficient to perform comparative studies on different species to discuss evolutionary relationships between species in terms of molecular phylogeny and molecular ecology [18]. For instance, plastome sequences provide species-specific information that includes genome size, gene order, genome rearrangements, patterns of base pair composition, codon usage, massive plastid gene losses and various type of nucleotide polymorphism [19,20].

Despite their genetic diversity and evolutionary significance, genetic resources for bryophytes are very limited when compared to angiosperms. Of the 2352 records for chloroplast genome sequences of green plants, only 15 plastomes of bryophytes comprised of 2 from hornworts, 5 from liverworts and 8 from mosses (including *Sanonia uncinata* NC_025668 which was directly submitted by the authors of this study) are available in public repositories [21] (http://www.ncbi.nlm.nih.gov). There are currently only four complete chloroplast genomes fully published for mosses, for example, *Physcomitrella patens* [22], *Tortula ruralis* [13], *Tetraphis pellucida* [23] and *Tetraplodon fuegianus* [24]. The plastome information of more bryophytes and comparative genomic studies are necessary to better understand the molecular evolutionary events or functions of chloroplast genes. In this regard, the plastome information of *S. uncinata* provided in this study will be a very useful resource for future research on the ecology, physiology and molecular evolution of bryophytes.

2. Results and Discussion

2.1. Overall Genome Organization

Illumina MiSeq sequencing produced 4,993,466 raw reads with an average read length of 301 bp and a total number of 1,503,033,266 base pairs. A total of 46,573 chloroplast-related reads were obtained as a result of alignment of quality trimmed reads against other chloroplast genomes publically available in NCBI. Assembly of the nucleotide sequence reads was performed to obtain non-redundant contigs and singletons using CLC Genomics Workbench V7.5 (CLC bio, Aarhus, Denmark). The final *S. uncinata* plastome sequence has been submitted to GenBank (Accession: NC_025668).

The gene map for the *S. uncinata* plastome is shown in Figure 1. The complete plastome of *S. uncinata* is 124,374 base pairs (bp) in length with a typical quadripartite structure including large and small single-copy regions (LSC of 86,570 bp and SSC of 18,430 bp) separated by a pair of identical inverted repeats (IRA and IRB) of 9687 bp each (Figure 1). Most of the chloroplast DNA had a well preserved quadripartite structure in bryophytes [23,25,26] and vascular plants [27,28]. The genome contained 114 unique genes including 82 unique protein-coding genes, 37 tRNA genes and 4 rRNA genes (Table 1). The gene content of the IR regions was conserved among *S. uncinata*, *T. ruralis* and *P. patens*.

The size of the plastome of *S. uncinata* is very similar to those of liverworts (*Marchantia polymorpha* 121,024 bp NC_001319, *Pellia endiviifolia* 120,546 bp NC_019628, *Aneura mirabilis* 108,007 bp NC_010359, *Ptilidium pulcherrimum* 119,007 bp NC_015402) and mosses (*Physcomitrella patens* 122,890 bp NC_005087, *Tortula ruralis* 124,374 bp NC_012052, *Tetraphis pellucida* 127,489 bp NC_024291) and *Tetraplodon fuegianus*

123,670 bp, KU_095851(unverified) but much smaller than hornworts (*Anthoceros formosae* 161,162 bp NC_004543 and *Nothocerosaenigmaticus* 153,208 bp NC_020259), which have an increased length of intragenic spacers in the LSC region or in the identical IR regions [18,29,30].

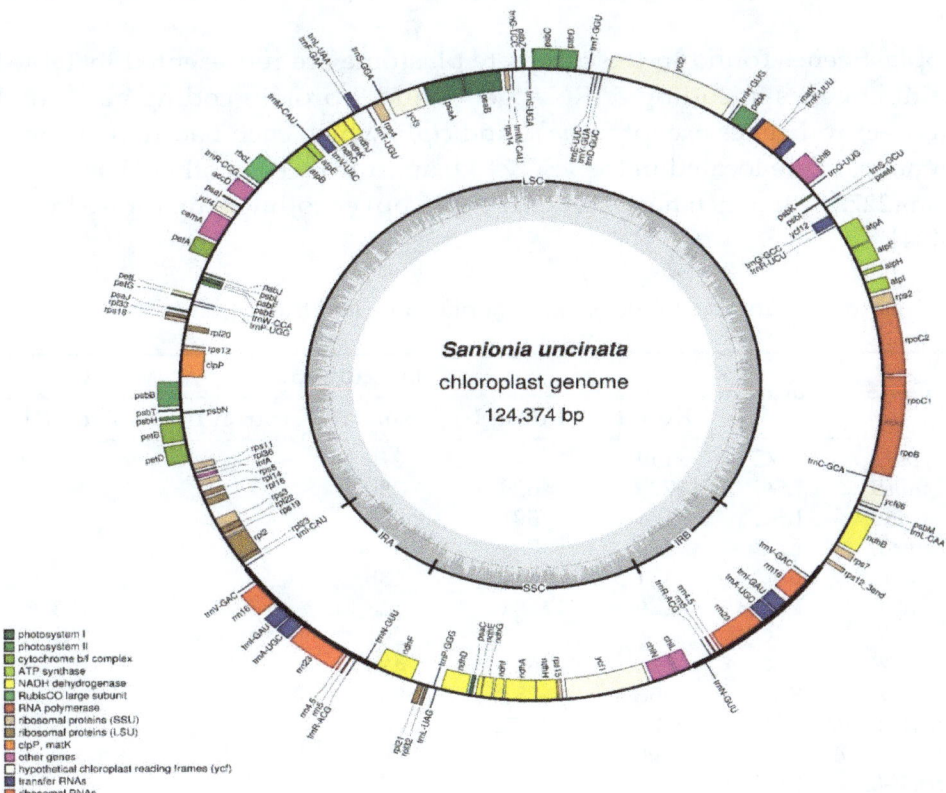

Figure 1. Map of the *Sanionia uncinata* plastome. Complete plastome sequences were obtained from the de novo assembly of Illumina paired-end reads. Genes are color coded by functional group, which are located in the left box. The inner darker gray circle indicates the GC content while the lighter gray corresponds to AT content. IR, inverted repeat; LSC, large single copy region; SSC, small single copy region. Genes shown on the outside of the outer circle are transcribed clockwise and those on the inside counter clockwise. The map was made with OGDraw [31].

Table 1. Genes present in the *S. uncinata* plastome.

Gene Products	Genes
Photosystem I	*psaA, B, C, I, J, M*
Photosystem II	*psbA, B, C, D, E, F, H, I, J, K, L, M, N, T, Z*
Cytochrome b6/f	*petA, B [a], D [a], G, L*
ATP synthase	*atpA, B, E, F [a], H, I*
Translation factor	*infA*
Chlorophyll biosynthesis	*chlB, L, N*
Rubisco	*rbcL*
NADH oxidoreductase	*ndhA [a], B [a], C, D, E, F, G, H, I, J, K*
Large subunit ribosomal proteins	*rpl2 [a], 14, 16 [a], 20, 21, 22, 23, 32, 33, 36*
Small subunit ribosomal proteins	*rps2, 3, 4, 7, 8, 11, 12 [a,b], 14, 15, 18, 19*
RNAP	*rpoB, C1 [a], C2*
Other proteins	*accD, cemA, clpP [c], matK*
Proteins of unknown function	*ycf1, 2, 3 [c], 4, 12, 66 [a]*
Ribosomal	*rrn4.5 [d], 5 [d], 16 [d], 23 [d]*
Transfer RNAs	*trnA(UGC) [a,d], C(GCA), D(GUC), E(UUC), F(GAA), G(UCC) [a], G(UCC), H(GUG), I(CAU), I(GAU) [a,d], K(UUU) [a], L(CAA), L(UAA) [a], L(UAG), fM(CAU), M(CAU), N(GUU) [d], P(UGG), P(GGG), Q(UUG), R(ACG) [d], R(CCG), R(UCU), S(GCU), S(GGA), S(UGA), T(GGU), T(UGU), V(GAC) [d], V(UAC) [a], W(CCA), Y(GUA)*

[a] Gene containing a single intron; [b] Gene divided into two independent transcription units; [c] Gene containing two introns; [d] Two gene copies in the IRs.

The overall G/C content was 29.3% for *S. uncinata*, similar to other known bryophyte plastomes (*P. patens* (28.5%) [22], *T. pellucida* (29.4%) [23], *T. fuegianus* (28.7%) [24]), as well as the liverwort *M. polymorpha* (28.8%) [26], hornwort *A. formosae* (32.9%) [25], the charophyte *Chaetosphaeridium* (29.6%) [32] and algae (30–33%) [22] but significantly less than the 34~40% found in seed plants [33].

The chloroplast genes found in the complete plastome are represented in Table 2. There were 14 intron-containing genes including 5 tRNA genes and 9 protein-coding genes and almost all of which were single-intron genes except for *ycf3* and *clpP*, which each had two introns. Two exons of trans-spliced gene *rps12* are located in the LSC 71 kb apart from each other. The *trnK-UUU* gene has the largest intron (2272 bp), which has *matK* ORF (1548 bp) encoding a maturase involved in splicing type II introns [34].

Table 2. The genes with introns in the *S. uncinata* plastome and the length of the exons and introns.

Gene	Location	Length (bp)				
		Exon I	Intron I	Exon II	Intron II	Exon III
rps12	LSC	114	–	270		
ndhB	LSC	729	629	780		
ycf66	LSC	106	591	320		
rpoC1	LSC	423	789	1614		
atpF	LSC	411	654	135		
ycf3	LSC	126	684	228	739	153
clpP	LSC	69	687	291	483	234
rpl2	LSC	396	637	438		
ndhA	SSC	556	731	551		
trnK-UUU	LSC	37	2272	42		
trnL-UAA	LSC	38	262	50		
trnV-UAC	LSC	37	542	37		
trnI-GAU	IR	42	769	35		
trnA-UGC	IR	38	763	35		

The 82 protein-coding genes in this genome represented nucleotide coding for 40,330 codons. On the basis of the sequences of protein-coding genes and tRNA genes within the plastome, the frequency of codon usage was deduced (Table 3). Among these codons, 4343 (10.85%) encoded for leucine and 455 (1.14%) for Tryptophan, which were the most and the least amino acids, respectively. The codon usage was biased towards a high representation at the third codon position. A biased frequency of codons included the levels of available tRNA, functionally related genes, evolutionary pressures and the rate of gene evolution [35].

Table 3. The codon-anticodon recognition pattern and codon usage for *S. uncinata* plastome.

Amino Acid	Codon	No. *	tRNA	Amino Acid	Codon	No. *	tRNA
Phe	UUU	2862		Tyr	UAU	1492	
Phe	UUC	916	*trnF-GAA*	Tyr	UAC	578	*trnY-GUA*
Leu	UUG	688	*trnL-UAA*	Stop	UAA	1772	
Leu	UUA	1766	*trnL-CAA*	Stop	UAG	577	
Leu	CUG	252		His	CAU	550	
Leu	CUA	591	*trnL-UAG*	His	CAC	224	*trnH-GUG*
Leu	CUU	741		Gln	CAA	685	*trnQ-UUG*
Leu	CUC	305		Gln	CAG	273	
Ile	AUG	1582	*trnI-CAU*	Asn	AAU	1878	
Ile	AUU	1943		Asn	AAC	665	*trnN-GUU*
Ile	AUC	629	*trnI-GAU*	Lys	AAA	2942	*trnK-UUU*
Met	AUG	515	*trnfM-CAU*	Lys	AAG	768	
Val	GUG	244		Asp	GAU	619	

Table 3. *Cont.*

Amino Acid	Codon	No. *	tRNA	Amino Acid	Codon	No. *	tRNA
Val	GUA	588	*trnV-UAC*	Asp	GAC	210	*trnD-GUC*
Val	GUU	652		Glu	GAA	845	*trnE-UUC*
Val	GUC	237	*trnV-GAC*	Glu	GAG	286	
Ser	AGU	579		Cys	UGU	515	
Ser	AGC	470	*trnS-GCU*	Cys	UGC	362	*trnC-GCA*
Ser	UCG	272		Stop	UGA	657	
Ser	UCA	643	*trnS-UGA*	Trp	UGG	455	*trnW-CCA*
Pro	CCG	169		Arg	AGG	387	
Pro	CCA	429	*trnP-UGG*	Arg	AGA	732	*trnR-UCU*
Pro	CCU	423		Arg	CGG	164	*trnR-CCG*
Pro	CCC	230	*trnP-GGG*	Arg	CGA	322	
Thr	ACG	228		Arg	CGU	270	*trnR-ACG*
Thr	ACA	508	*trnT-UGU*	Arg	CGC	143	
Thr	ACU	561		Ser	UCU	735	
Thr	ACC	363	*trnT-GGU*	Ser	UCC	419	*trnS-GGA*
Ala	GCG	155		Gly	GGG	426	
Ala	GCA	367	*trnA-UGC*	Gly	GGA	465	*trnG-UCC*
Ala	GCU	413		Gly	GGU	653	
Ala	GCC	514		Gly	GGC	237	

* Numerals indicate the frequency of usage of each codon in 40,330 in codons in 82 potential protein-coding genes.

2.2. Comparison with Other Bryophyte Plastomes

Multiple complete bryophyte plastomes available provide an opportunity to compare the sequence variation at the genome-level. We therefore compared the whole plastome sequence of *S. uncinata* with those of mosses *T. ruralis*, *P. patens*, *T. pellucida*, the liverwort *M. polymorpha* and hornwort *A. formosae*. The sequence identity between all five bryophyte plastomes was plotted using the mVISTA program with the annotation of *S. uncinata* as a reference (Figure 2).

Sequence similarities of the genes between *S. uncinata* and other bryophytes (mosses, liverwort and hornwort) were compared (Table S1). The *rRNA* genes (*rrn5*, *rrn16*, *rrn4.5* and *rrn23*) in the IRs region showed the highest sequence similarity (average 98.3–96.1%) and PSII-associated genes such as *psbL*, *psbA*, *psbN*, *psbZ*, *psbE*, *psbH*, *psbK*, *psbF*, *psbB*, *psbD* and *psbJ*, which also displayed high levels of sequence similarity (average 93.9–89.8%). Genes for large subunit ribosomal proteins (*rpl32*, *rpl20*, *rpl33* and *rpl23*) and a small subunit ribosomal protein (*rps12*) were relatively more conserved than other coding genes.

Notably, the highest sequence variation occurred in the *matK* gene (average 79.4%), widely known to be evolved rapidly and thus often used as a barcoding marker in phylogenetic and evolutionary studies [34,36], suggesting that this gene has also undergone evolutionary pressure within bryophytes. Following this, genes such as *rpoC2*, *petB atpE*, *atpF* and *rpl22* showed lower similarity (average 80.2–83.5%) than other plastid genes in order (Table S1). Those genes with large sequence variations had evolutionary significance in inferring divergence times and branching patterns among early land plant lineages, while relatively less varied genes such as rRNA and PSII-associated genes were well conserved during the evolution of bryophytes.

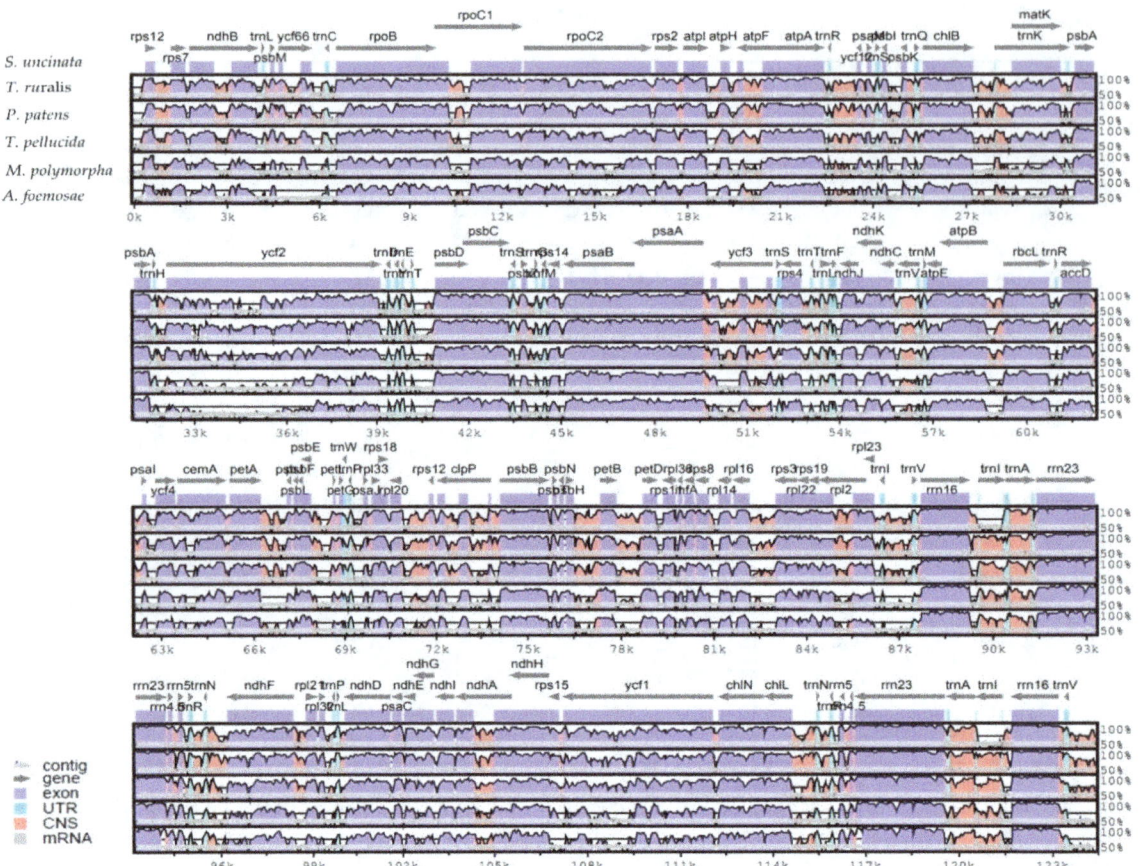

Figure 2. Alignment of complete plastome sequences from six species. Alignment and comparison were performed using mVISTA and the percentage of identity between the plastomes was visualized in the form of an mVISTA plot. The sequence similarity of the aligned regions between *S. uncinata* and other five species is shown as horizontal bars indicating average percent identity between 50–100% (shown on the y-axis of graph). The x-axis represents the coordinate in the plastome. Genome regions are color-coded for protein-coding (exon), rRNA, tRNA and conserved non-coding sequences (CNS) as the guide at the bottom-left.

2.3. Phylogenomic Analysis

Plastome information has provided an important resource for uncovering evolutionary relationships between various plant lineages [20,37]. The whole plastomes and protein-coding genes have been widely used for the reconstruction of phylogenetic relationships among different plant species [38]. The availability of completed *S. uncinata* plastome provided us with the sequence information to study the molecular evolution and phylogeny of *S. uncinata* with closely related species.

Phylogenomic analysis of representatives from the bryophyte subfamily including *S. uncinata*, produced a single, well-supported tree using maximum parsimony (MP) (Figure 3). To do this, a set of 40 protein-coding genes in plastome analyzed in other species were concatenated and these concatenated sequences were used to infer the phylogenetic relationships of 23 taxa including *S. uncinata* using MEGA7. Phylogenetic analysis based on the multigene dataset revealed that mosses, liverwort and hornwort have been resolved as monophyletic in MP tree (Figure 3). Based on molecular phylogeny results, liverworts are placed in a basal position representing the earliest diverging lineage, while hornworts are the closest relatives of extant vascular plants, corroborating a previously reported branch order of "liverworts (mosses(hornworts(vascular plants)))" [37]. *S. uncinata* was most related to *T. ruralis* and formed a sister group with other moss species, which was supported by bootstrap values (100 for both ML and MP). In addition, it provides convincing support for many traditionally recognized genera and identifies higher level phylogenetic structure of mosses [39]. This result suggests that plastome information can effectively resolve phylogenetic positions and evolutionary relationships between different Bryophyte lineages.

The *T. ruralis* and *S. uncinata*, two mosses inhabiting extreme environments of polar and alpine regions, share common features in their plastomes–lack of *petN* and *rpoA* and presence of *trnP-GGG*–which is not the case for other two species of *P. patens* and *T. pellucid* (Table 4). Due to very scarce taxon sampling and the limitation of available plastome data, it is not clear whether the presence or absence of those genes is related to the resistance or adaptation to the extreme environments where they inhabit. To address this, synapomorphic characteristics developed during adaptation should be investigated together with genome evolution.

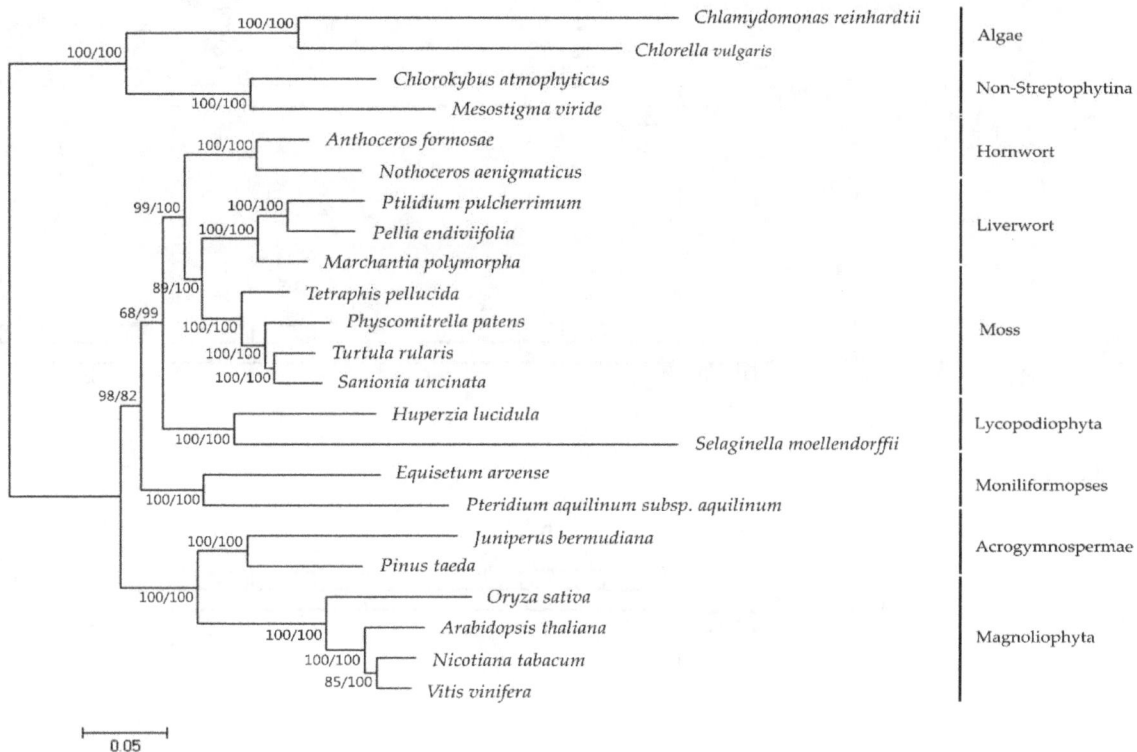

Figure 3. Phylogenetic tree reconstruction of 23 taxa using MEGA7 based on concatenated sequences of 40 protein-coding genes in the plastome. Maximum likelihood (ML) topology is shown with the bootstrap support values (MP/ML) given at nodes. Forty protein-coding sequences were extracted from annotated plastomes found in GenBank [21] (http://www.ncbi.nlm.nih.gov) (Table S2). The nucleotide sequences for each gene were translated into amino acids, aligned in MEGA7 and manually adjusted. Nucleotide sequences were aligned by constraining them to the amino acid sequence alignment. Individual gene alignments were then assembled into a single dataset.

2.4. Loss of rpoA and petN in the S. uncinata Plastome

Previous studies on bryophyte plastomes have shown that the overall structure and gene contents of plastomes are highly conserved among different bryophyte lineages [22,23,25,26]. The plastome of *S. uncinata* also showed a similar level of size and structure when compared to those of other three moss plastomes, *P. patens*, *T. ruralis* and *T. pellucida*. The plastome of *S. uncinata* did not show a large inversion of the ~71 kb fragment found in the LSC region of *P. patens* plastome, which follows the cases of *T. ruralis* and *T. pellucida*, suggesting that this large inversion is limited to the order *Funariales* in Bryophyte [22,40] (Figure 4).

While most mosses and plant groups are conserved in the contents of chloroplast genes, certain lineages undergo apparent and frequent gene loss (e.g., *rpoA* transfer from the chloroplast to the nucleus: [13,22,23]). Two genes encoding α subunit of RNA polymerase (*rpoA*) and encoding cytochrome $b_{6/f}$ complex subunit VIII (*petN*) were absent in the *S. uncinata* plastome (Figure 4). As discussed in previous moss plastome studies, the most prominent and typical feature of the plastomes of moss is the absence of *rpoA*, which is thought to have disappeared together with *ycf5*, *cysA*, *cysT* after the divergence of the mosses from the hepatic bryophytes [22,41]. We could not identify *rpoA* with *cysT* and *cysA* in the *S. uncinata* plastome, which means that the *S. uncinata* plastome follows the typical characteristics of moss plastomes (Table 4, Figure 4). The loss of *petN* reported in other mosses, except in *P. patens* [22], was also observed in the *S. uncinata* plastome (Table 4, Figure 4). The presence of *trnP-GGG* also varied depending on species [13,23], the gene was present in the plastomes of *S. uncinata* and *T. ruralis*, while it was absent in the plastomes of *P. patens* and *T. pellucida* (Table 4).

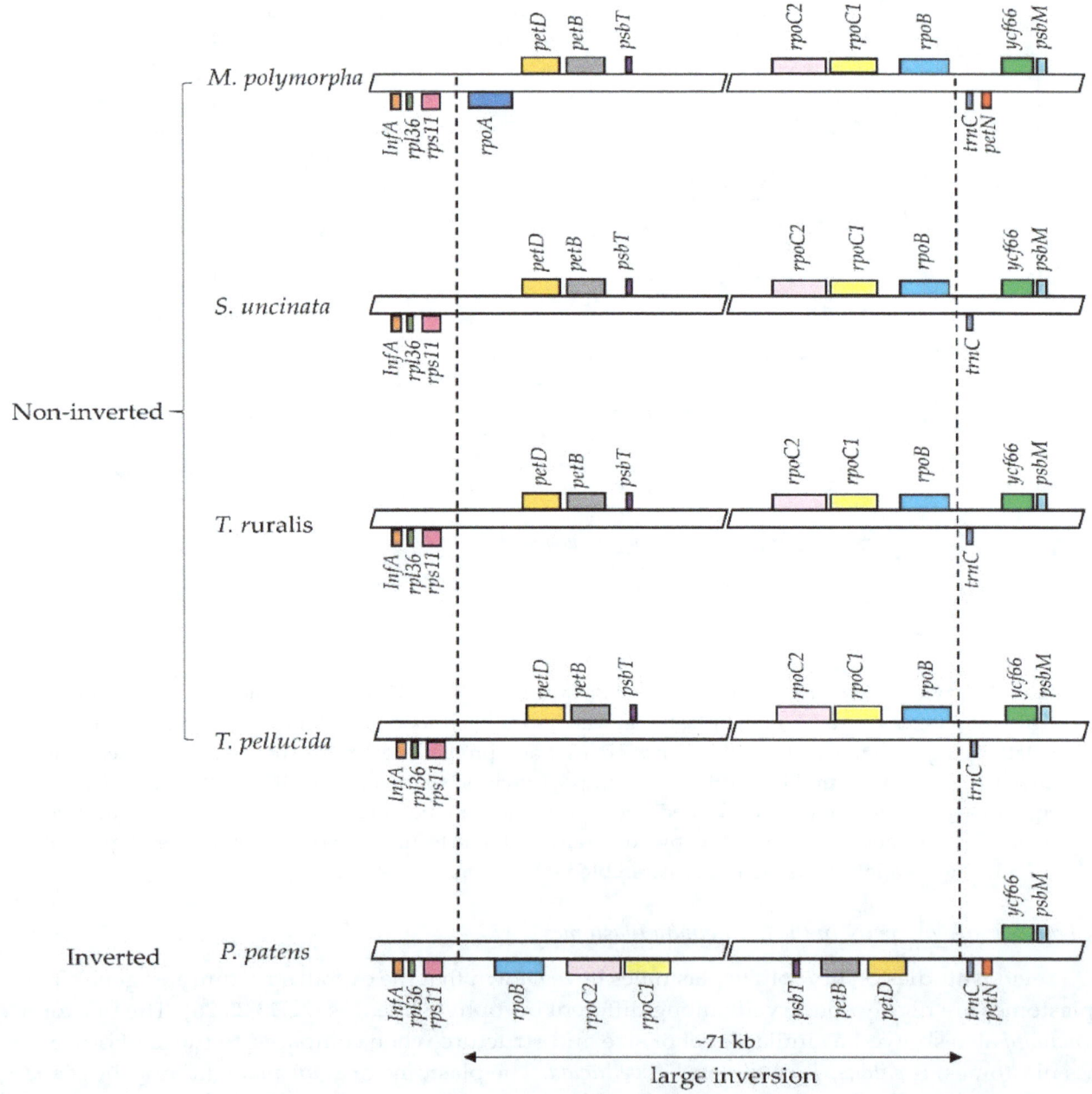

Figure 4. Comparison of the large inversion in the LSC region among six bryophytes plastomes. In comparative LSC region alignment of *rpoA*, *petN* coding regions from *M. polymorpha*, *S. uncinata*, *T. ruralis*, *T. pellucida* and *P. patens*. The inverted-arrangement of 71 kb fragment was only detected for *P. patens*.

Table 4. Gene contents of plastomes from green alga, bryophytes and land plants.

	Plants *	Genome Size (bp)	*petN*	*rpoA*	*ccsA*	*cysA*	*cysT*	*ycf66*	*matK*	*rps15*	*trnP-GGG*
Alga	*Chlorella vulgaris*	150,613	-	+	+	+	+	-	-	-	-
Hornwort	*Anthoceros formosae*	161,162	+	+	+	+	+	-	Ψ	Ψ	+
Liverwort	*Marchantia polymorpha*	121,024	+	+	+	+	+	+	+	+	Ψ
Moss	*Tetraphis pellucida*	127,489	-	-	-	-	-	+	+	+	-
Moss	*Physcomitrella patens*	122,890	+	-	-	-	-	+	+	+	-
Moss	*Tortula rularis*	122,630	-	-	-	-	-	+	+	+	+
Moss	*Sanionia uncinata*	124,374	-	-	-	-	-	+	+	+	+
Lycopodiophyta	*Huperzia lucidula*	154,373	+	+	+	-	-	+	+	+	+
Moniliformopses	*Equisetum arvense*	133,309	+	+	+	-	-	+	+	+	+
Acrogymnospermae	*Pinus thunbergii*	116,635	+	+	+	-	-	-	+	+	+
Magnoliophyta	*Arabidopsis thaliana*	154,515	+	+	+	-	-	-	+	+	-

* The annotated plastomes are listed in Table S2. The presence '+' or absence '-' of each molecular character and pseudogenes are marked 'Ψ'.

2.5. Identification of Nuclear Genes Encoding rpoA and petN

Many cyanobacterial genes have been lost or transferred to the nuclear genomes during the transition from endosymbiont to organelle and the chloroplast genome might have been decreased in size [42,43]. Most gene loss occurred early in the endosymbiotic process; however, some losses have occurred in subsequent evolutionary processes [43]. The gene loss and transfer to the nucleus have been especially frequent, sporadic and temporary during the evolution of embryophytes and bryophytes [44]. Revealing why certain genes lose their functions is of great importance in the genome evolution of chloroplast, enabling the reconstruction of genomic events that occurred after the split of vascular plants and moss.

Therefore, to verify the chance of the nuclear relocation of the lost two chloroplast genes encoding the α subunit of RNA polymerase (*rpoA*) and cytochrome $b_{6/f}$ complex subunit VIII (*petN*), we used blast to search for putative homologues against the draft sequences of the *S. uncinata* nuclear genome (unpublished results), as a result, we found one target sequence for each gene and then verified the sequences of PCR amplified fragments (nucleotide sequences are listed in Table S3), respectively, which might function as the replaced gene for *rpoA* or *petN* in *S. uncinata*. To check whether those nuclear genes would have a conserved function with chloroplastic genes, we performed a comparative alignment of nuc-*rpoA* (nuclear-*rpoA)* and nuc-*petN* (nuclear-*petN*) of *S. uncinata* with the homologous corresponding genes of *P. patens*, *M. polymorpha*, *A. formosae* and *A. thaliana* (Figure 5). Particularly, for *P. patens*, the nuc-*rpoA* sequence (*Pp* nuc-*rpoA*) was used in the alignment [20]. The nuc-*rpoA* sequence showed 41% identity with the nuc-*rpoA* sequence of *P. patens*, 76% of *M. polymorpha*, 66% of *A. formosae* and 67% of *A. thaliana* of the other cp-*rpoA* sequence from other compared species. The nuclear *petN* sequence showed 59–79% identified with the other chloroplastic *petN* sequence from other compared species *P. patens* (79%), *A. formosae* (59%) and *A. thaliana* (76%), respectively. When the sequences were scanned in the signalP 4.1 program [45] (SignalP 4.1 server; http://www.cbs.dtu.dk/services/SignalP), we could identify the signal peptide which targets chloroplast at the 5' terminus of the nuc-*rpoA* gene of *S. uncinata* (Figure 5), suggesting that those genes would be localized in the chloroplast after being translated and functioned as other cp-*rpoA* or cp-*petN* genes in chloroplasts. This is very similar to the case of nuc-*rpoA* of *P. patens*, which has been shown to target the chloroplast organelle by the fluorescence-labelled protein method [22]. It would therefore be very interesting to prove that the proteins encoded by these nuclear genes actually constitute the chloroplastic-RNA polymerase complex or the Cyt $b_{6/f}$ complex in the moss protein expression system.

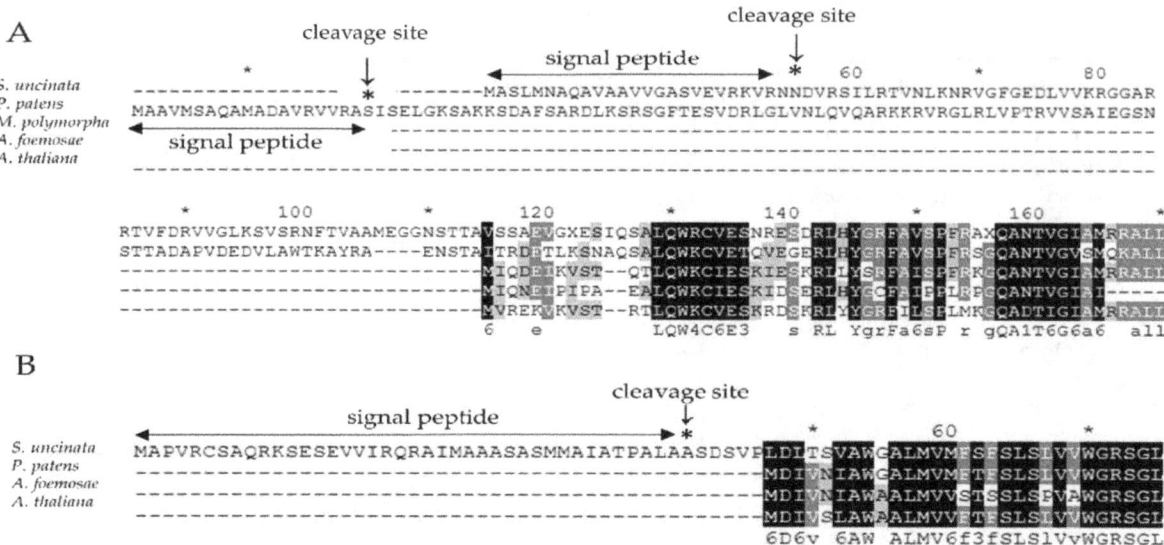

Figure 5. Amino acid alignment of (**A**) nuc-*rpoA* and (**B**) nuc-*petN* genes of *S. uncinata* with the nuc-*rpoA* or cp-*rpoA* and cp-*petN* genes from other green plants. Identical amino acid residues are boxed in black, other residues are printed in grey. Signal peptide sequences were predicted using SignalP [45] and shown as double arrow lines and the asterisk.

2.6. Prediction of RNA Editing Sites of Chloroplast Genes

Genetic information on DNA is sometimes altered at the transcript level by a process known as RNA editing, a sequence-specific post transcriptional modification resulting in the conversion, insertion, or deletion of nucleotides in a precursor RNA [46,47]. Such modifications are observed across organisms and have been reported with the discovery of C to U as well as U to C conversions in mitochondria and chloroplast in plants [48–50]. Since the first evidence of RNA editing in chloroplasts was found 24 years ago [51], RNA editing has been found in chloroplast transcripts from all major lineages of land plants [52]. In general, it is known that extensive RNA-editing occurs in hornworts and ferns when compared to seed plants [53]. The number of shared editing sites increases in closely related taxa, implying that RNA editing sites are evolutionarily conserved.

We predicted 72 editing sites in 22 protein coding genes by analyzing chloroplast coding sequences of *S. uncinata* using the PREP [54] and PREPACT 2.0-chloroplast program (http://www.prepact.de) [55] (Table S4). Most RNA editings are known to occur at the second codon position [56], with a frequency from 58.6% in hornwort to 68% in fern, 80% or 73.1% in *Cycad taitungensis* and *Pinus thunbergii*, respectively, 85.3%, 91.9% and 92% in *Arabidopsis thaliana*, *Nicotiana tabacum* and *Zea mays*, respectively, and up to 95.2% in *Oryza sativa* [57]. The RNA-editing of *S. uncinata* were mostly C-U editing events enriched at the second positions (64/72, 89%) as well, which allowed for the identification of the putative amino acid conversion by RNA editing (A-V, P-L, S-L, S-F, T-M and T-I).

3. Materials and Methods

3.1. Ethics Statement

Sample collection and field activities were carried out for scientific purposes in accordance with the Protocol on Environmental Protection to the Antarctic Treaty and approved by the Ministry of Foreign Affairs and Trade of the Republic of Korea (International Legal Affairs Division document No. 4029, approved on 18 December 2014). Sampling sites were not located within Antarctic Specially Protected Areas and no protected species were sampled in this study.

3.2. Plant Culture Conditions

Sanionia uncinata (Hedw.) Loeske plants growing under natural conditions were collected in the vicinity of the Korean King Sejong Antarctic Station (62°14′29″ S; 58°44′18″ W) on the Barton Peninsula of King George Island during the austral summer (mainly in January 2012) and then transferred to the lab and grown hydroponically in BCDAT solid media under a 16:8 h light:dark cycle at 15 °C.

3.3. Library Preparation and Sequencing

The total genomic DNA was extracted from tissues of plants using the DNeasy Plant Mini Kit (Qiagen). The quality of DNA was checked by Bioanalyzer 2100 (Aligent, Santa Clara, CA, USA). Library quantification was performed using the KAPA Library Quantification Kit (KAPA Biosystems, Boston, MA, USA). Paired-end cluster generation and sequencing were performed on the Illumina MiSeq, with each library being allocated to one lane of a flow cell. For the DNA library, TruSeq DNA sample preparation kits were used and sequenced in one lane of Illumina MiSeq 300 X 2 bp. The files containing the sequences and quality scores of reads were deposited in the NCBI Short Read Archive and the accession numbers are SRR6440975 genomic DNA-Seq (BioProject PRJNA428497).

3.4. Assembly and Annotation

After performing read preprocessing including adapter removal and quality filtering, the high-quality raw reads were aligned to 3 publicly available moss plastomes (*Physcomitrella patens* NC_005087, *Tortula ruralis* NC_012052 and *Tetraphis pellucida* NC_024291) downloaded from the NCBI organelle genome resources. The chloroplast reads were recovered from whole genome sequences by identifying them based on the alignment results and then de novo assembled using CLC Genomics Workbench V7.5 (CLC bio, Aarhus, Denmark). The assembled contigs were ordered based on their position in the reference plastome of *T. ruralis* because *T. ruralis* was identified as the top-hit species when BLAST searches were performed against the nr database. All gaps and junction regions between contigs and highly variable regions were validated by Sanger sequencing. The complete plastome was annotated using the program online software DOGMA [58] with default parameters, tRNAscan-SE [59] and BLAST similarity search tools from NCBI. OGDraw was used for map drawing [31]. The final *S. uncinata* plastome sequence has been submitted to GenBank (Accession: NC_025668).

3.5. Genome Alignment

The complete plastome sequences of 7 species were aligned using the mVISTA online suite [60]. A comparison of *Sanionia uncinata* (NC_025668) plastome structures with *Physcomiterella patens* (NC_005087), *Tortula ruralis* (NC_012052), *Tetraphis geniculate* (NC_024291), *Marchantia polymorpha* (NC_001319) and *Anthoceros formosae* (NC_004543), which are all in the bryophyte, was performed using the mVISTA program in Shuffle-LAGAN mode [60]. Default parameters were applied and the sequence annotation information of *S. uncinata* was used. Percentage identity between each plastome, all relative to that of *S. uncinata*, was subsequently visualized through an mVISTA plot. The mVISTA program only compared the sequence similarity by aligning the entire plastomes of all 6 taxa and the ~71 Kb inversion region of *P. patens* was reversed to match with other plastomes.

3.6. Phylogenetic Analysis

A set of 40 protein-coding genes, which have been analyzed in other species (accession numbers are listed in Table S2 and 40 protein-coding genes sequence of plastomes derived from 23 plants in Table S5), were used to infer the phylogenetic relationships among *S. uncinata*. Sequences were aligned using ClustalW. Phylogenetic analyses using maximum parsimony (MP) and maximum likelihood (ML) were performed, with MEGA7 [61]. For the MP analyses, the Subtree-Pruning-Regrafting algorithm was used with search level 1 in which the initial trees were obtained by the random addition of sequences (10 replicates). Sites with gaps or missing data were excluded from the analysis and statistical support was achieved through bootstrapping using 1000 replicates. For the ML analyses, the "Models" function of MEGA7 was used to find the best model for ML analysis. The Jones-Taylor-Thornton (JTT) + G + I + F model which was estimated to be the Best-Fit substitution model showing the lowest Bayesian Information Criterion, was employed in subsequent analyses. All positions containing gaps and missing data were eliminated. Bootstrap support was estimated with 1000 bootstrap replicates.

3.7. Prediction of RNA-Editing Sites

The RNA editing sites (C-to-U) in protein-coding genes were predicted by the online program Plant RNA Editing Prediction and Analysis Computer Tool (PREPACT 2.0) (http://www.prepact.de) [55] and Predictive RNA Editor for Plants (PREP) suite (http://prep.unl.edu/) [54] with a cutoff value of 0.8.

4. Conclusions

This study provided the whole sequence of the *S. uncinata* plastome, the dominant species of the Antarctic Peninsula, with information on the sequence and regulatory regions of chloroplast genes. We completed the *S. uncinata* plastome using a high-throughput sequencing method. The sequence and structure of *S. uncinata* were well conserved with the moss plastome sequences and was the most similar to the plastome sequence of *T. ruralis*. We confirmed that the two genes coding for *rpoA* and *petN* were lost in the *S. uncinata* plastome through comparative analysis of the plastome contents of the representative species of land plants and the sequence analysis results suggested the possibility of their relocation from plastome to nuclear genome. Our results also suggested that the possibility of post-transcriptional regulation of the chloroplast genes of *S. uncinata* by predicting the RNA-editing site. These results will contribute not only to the functional utilization of chloroplast genes but also to systematic phylogenetic analysis of land plants using whole plastome sequences.

Acknowledgments: This work was supported by "Polar Genome 101 Project: Genome Analysis of Polar Organisms and Establishment of Application Platform (PE18080)" and "Modeling responses of terrestrial organisms to environmental changes on King George Island (PE18090)" funded by the Korea Polar Research Institute.

Author Contributions: Hyoungseok Lee, Hyun Park and Jungeun Lee conceived and designed the experiment; Mira Park performed the experiments; Mira Park, Hyoungseok Lee, Byeong-ha Lee and Jungeun Lee analyzed the data and wrote the paper. All authors read and approved the final version of the manuscript.

Abbreviations

LSC	Large single copy
SSC	Small single copy
IR	Inverted repeat
Cp	Chloroplast
nuc	Nucleus
MP	Maximum parsimony
ML	Maximum likelihood
A	Adenine
T	Thymine
G	Guanine
C	cytosine

References

1.   Bramley-Alves, J.; King, D.H.; Robinson, S.A.; Miller, R.E. Dominating the Antarctic environment: Bryophytes in a time of change. In *Photosynthesis in Bryophytes and Early Land Plants*; Hanson, D., Rice, S., Eds.; Springer Netherlands: Dordrecht, The Netherlands, 2014; pp. 309–324.

2.   Smith, R.L. Introduced plants in Antarctica: Potential impacts and conservation issues. *Biol. Conserv.* **1996**, *76*, 135–146. [CrossRef]

3.   Victoria, F.D.C.; Pereira, A.B.; da Costa, D.P. Composition and distribution of moss formations in the ice-free areas adjoining the Arctowski region, admiralty bay, King George Island, Antarctica. *Iheringia Ser. Bot.* **2009**, *64*, 81–91.

4.   Hedenäs, L. Global phylogeography in *Sanionia uncinata* (amblystegiaceae: Bryophyta). *Bot. J. Linn. Soc.* **2011**, *168*, 19–42. [CrossRef]

5.   Nakatsubo, T. Predicting the impact of climatic warming on the carbon balance of the moss *Sanionia uncinata* on a maritime Antarctic island. *J. Plant Res.* **2002**, *115*, 99–106. [CrossRef] [PubMed]

6.   Hokkanen, P.J. Environmental patterns and gradients in the vascular plants and bryophytes of eastern Fennoscandian herb-rich forests. *For. Ecol. Manag.* **2006**, *229*, 73–87. [CrossRef]

7.   Kushnevskaya, H.; Mirin, D.; Shorohova, E. Patterns of epixylic vegetation on spruce logs in late-successional boreal forests. *For. Ecol. Manag.* **2007**, *250*, 25–33. [CrossRef]

8.   Samecka-Cymerman, A.; Wojtuń, B.; Kolon, K.; Kempers, A. *Sanionia uncinata* (Hedw.) loeske as bioindicator of metal pollution in polar regions. *Polar Biol.* **2011**, *34*, 381–388. [CrossRef]

9.   Lud, D.; Moerdijk, T.; Van de Poll, W.; Buma, A.; Huiskes, A. DNA damage and photosynthesis in Antarctic and Arctic *Sanionia uncinata* (hedw.) Loeske under ambient and enhanced levels of UV-B radiation. *Plant Cell Environ.* **2002**, *25*, 1579–1589. [CrossRef]

10. Lud, D.; Schlensog, M.; Schroeter, B.; Huiskes, A. The influence of UV-B radiation on light-dependent photosynthetic performance in *Sanionia uncinata* (Hedw.) Loeske in Antarctica. *Polar Biol.* **2003**, *26*, 225–232.

11. Zúñiga-González, P.; Zúñiga, G.E.; Pizarro, M.; Casanova-Katny, A. Soluble carbohydrate content variation in *Sanionia uncinata* and polytrichastrum alpinum, two Antarctic mosses with contrasting desiccation capacities. *Biol. Res.* **2016**, *49*, 6. [CrossRef] [PubMed]

12. Van Gaalen, K.E.; Flanagan, L.B.; Peddle, D.R. Photosynthesis, chlorophyll fluorescence and spectral reflectance in sphagnum moss at varying water contents. *Oecologia* **2007**, *153*, 19–28. [CrossRef] [PubMed]

13. Oliver, M.J.; Murdock, A.G.; Mishler, B.D.; Kuehl, J.V.; Boore, J.L.; Mandoli, D.F.; Everett, K.D.; Wolf, P.G.; Duffy, A.M.; Karol, K.G. Chloroplast genome sequence of the moss Tortula ruralis: Gene content, polymorphism and structural arrangement relative to other green plant chloroplast genomes. *BMC Genom.* **2010**, *11*, 143. [CrossRef] [PubMed]

14. Gray, M.W. The evolutionary origins of organelles. *Trends Genet.* **1989**, *5*, 294–299. [CrossRef]

15. Howe, C.J.; Barbrook, A.C.; Koumandou, V.L.; Nisbet, R.E.R.; Symington, H.A.; Wightman, T.F. Evolution of the chloroplast genome. *Philos. Trans. R. Soc. Lond. B Biol. Sci.* **2003**, *358*, 99–107. [CrossRef] [PubMed]

16. Sugiura, M. The chloroplast chromosomes in land plants. *Annu. Rev. Cell Biol.* **1989**, *5*, 51–70. [CrossRef] [PubMed]

17. Neuhaus, H.; Emes, M. Nonphotosynthetic metabolism in plastids. *Annu. Rev. Plant Biol.* **2000**, *51*, 111–140. [CrossRef] [PubMed]

18. Daniell, H.; Lin, C.-S.; Yu, M.; Chang, W.-J. Chloroplast genomes: Diversity, evolution and applications in genetic engineering. *Genome Biol.* **2016**, *17*, 134. [CrossRef] [PubMed]

19. Chumley, T.W.; Palmer, J.D.; Mower, J.P.; Fourcade, H.M.; Calie, P.J.; Boore, J.L.; Jansen, R.K. The complete chloroplast genome sequence of pelargonium× hortorum: Organization and evolution of the largest and most highly rearranged chloroplast genome of land plants. *Mol. Biol. Evol.* **2006**, *23*, 2175–2190. [CrossRef] [PubMed]

20. Karol, K.G.; Arumuganathan, K.; Boore, J.L.; Duffy, A.M.; Everett, K.D.; Hall, J.D.; Hansen, S.K.; Kuehl, J.V.; Mandoli, D.F.; Mishler, B.D. Complete plastome sequences of *Equisetum arvense* and *isoetes flaccida*: Implications for phylogeny and plastid genome evolution of early land plant lineages. *BMC Evol. Biol.* **2010**, *10*, 321. [CrossRef] [PubMed]

21. Coordinators, N.R. Database resources of the National Center for Biotechnology Information. *Nucleic Acids Res.* **2016**, *44*, D7–D19.

22. Sugiura, C.; Kobayashi, Y.; Aoki, S.; Sugita, C.; Sugita, M. Complete chloroplast DNA sequence of the moss Physcomitrella patens: Evidence for the loss and relocation of rpoA from the chloroplast to the nucleus. *Nucleic Acids Res.* **2003**, *31*, 5324–5331. [CrossRef] [PubMed]

23. Bell, N.E.; Boore, J.L.; Mishler, B.D.; Hyvönen, J. Organellar genomes of the four-toothed moss, *Tetraphis pellucida*. *BMC Genom.* **2014**, *15*, 383. [CrossRef] [PubMed]

24.   Lewis, L.R.; Liu, Y.; Rozzi, R.; Goffinet, B. Infraspecific variation within and across complete organellar genomes and nuclear ribosomal repeats in a moss. *Mol. Phylogenet. Evol.* **2016**, *96*, 195–199. [CrossRef] [PubMed]

25.   Kugita, M.; Kaneko, A.; Yamamoto, Y.; Takeya, Y.; Matsumoto, T.; Yoshinaga, K. The complete nucleotide sequence of the hornwort (*Anthoceros formosae*) chloroplast genome: Insight into the earliest land plants. *Nucleic Acids Res.* **2003**, *31*, 716–721. [CrossRef] [PubMed]

26.   Ohyama, K.; Fukuzawa, H.; Kohchi, T.; Shirai, H.; Sano, T.; Sano, S.; Umesono, K.; Shiki, Y.; Takeuchi, M.; Chang, Z. Chloroplast gene organization deduced from complete sequence of liverwort Marchantia polymorpha chloroplast DNA. *Nature* **1986**, *322*, 572–574. [CrossRef]

27.   Wakasugi, T.; Tsudzuki, T.; Sugiura, M. The genomics of land plant chloroplasts: Gene content and alteration of genomic information by RNA editing. *Photosynth Res.* **2001**, *70*, 107–118. [CrossRef] [PubMed]

28.   Lee, J.; Kang, Y.; Shin, S.C.; Park, H.; Lee, H. Combined analysis of the chloroplast genome and transcriptome of the Antarctic vascular plant Deschampsia antarctica Desv. *PLoS ONE* **2014**, *9*, e92501. [CrossRef] [PubMed]

29.   Maul, J.E.; Lilly, J.W.; Cui, L.; Miller, W.; Harris, E.H.; Stern, D.B. The Chlamydomonas reinhardtii plastid chromosome islands of genes in a sea of repeats. *Plant Cell* **2002**, *14*, 2659–2679. [CrossRef] [PubMed]

30.   Turmel, M.; Pombert, J.-F.; Charlebois, P.; Otis, C.; Lemieux, C. The green algal ancestry of land plants as revealed by the chloroplast genome. *Int. J. Plant Sci.* **2007**, *168*, 679–689. [CrossRef]

31.   Lohse, M.; Drechsel, O.; Kahlau, S.; Bock, R. OrganellarGenomeDRAW—A suite of tools for generating physical maps of plastid and mitochondrial genomes and visualizing expression data sets. *Nucleic Acids Res.* **2013**, *41*, W575–W581. [CrossRef] [PubMed]

32.   Turmel, M.; Otis, C.; Lemieux, C. The chloroplast and mitochondrial genome sequences of the charophyte Chaetosphaeridium globosum: Insights into the timing of the events that restructured organelle DNAs within the green algal lineage that led to land plants. *Proc. Natl. Acad. Sci. USA* **2002**, *99*, 11275–11280. [CrossRef] [PubMed]

33.   Jansen, R.K.; Ruhlman, T.A. Plastid genomes of seed plants. In *Genomics of Chloroplasts and Mitochondria*; Bock, R., Knoop, V., Eds.; Springer Netherlands: Dordrecht, The Netherlands, 2012; pp. 103–126.

34.   Barthet, M.M.; Hilu, K.W. Expression of matK: Functional and evolutionary implications. *Am. J. Bot.* **2007**, *94*, 1402–1412. [CrossRef] [PubMed]

35.   Quax, T.E.; Claassens, N.J.; Söll, D.; van der Oost, J. Codon bias as a means to fine-tune gene expression. *Mol. Cell* **2015**, *59*, 149–161. [CrossRef] [PubMed]

36.   Hausner, G.; Olson, R.; Simon, D.; Johnson, I.; Sanders, E.R.; Karol, K.G.; McCourt, R.M.; Zimmerly, S. Origin and evolution of the chloroplast trnK (matK) intron: A model for evolution of group II intron RNA structures. *Mol. Biol. Evol.* **2005**, *23*, 380–391. [CrossRef] [PubMed]

37.   Gao, L.; Su, Y.J.; Wang, T. Plastid genome sequencing, comparative genomics and phylogenomics: Current status and prospects. *J. Syst. Evol.* **2010**, *48*, 77–93. [CrossRef]

38.   Shaw, J.; Lickey, E.B.; Schilling, E.E.; Small, R.L. Comparison of whole chloroplast genome sequences to choose noncoding regions for phylogenetic studies in angiosperms: The tortoise and the hare III. *Am. J. Bot.* **2007**, *94*, 275–288. [CrossRef] [PubMed]

39.   Cox, C.J.; Goffinet, B.; Wickett, N.J.; Boles, S.B.; Shaw, A.J. Moss diversity: A molecular phylogenetic analysis of genera. *Phytotaxa* **2014**, *9*, 175–195. [CrossRef]

40.   Goffinet, B.; Wickett, N.J.; Werner, O.; Ros, R.M.; Shaw, A.J.; Cox, C.J. Distribution and phylogenetic significance of the 71-kb inversion in the plastid genome in Funariidae (Bryophyta). *Am. J. Bot.* **2007**, *99*, 747–753. [CrossRef] [PubMed]

41.   Wicke, S.; Schneeweiss, G.M.; Müller, K.F.; Quandt, D. The evolution of the plastid chromosome in land plants: Gene content, gene order, gene function. *Plant Mol. Biol.* **2011**, *76*, 273–297. [CrossRef] [PubMed]

42.   Martin, W.; Rujan, T.; Richly, E.; Hansen, A.; Cornelsen, S.; Lins, T.; Leister, D.; Stoebe, B.; Hasegawa, M.; Penny, D. Evolutionary analysis of *Arabidopsis*, cyanobacterial and chloroplast genomes reveals plastid phylogeny and thousands of cyanobacterial genes in the nucleus. *Proc. Natl. Acad. Sci. USA* **2002**, *99*, 12246–12251. [CrossRef] [PubMed]

43.   Martin, W.; Stoebe, B.; Goremykin, V.; Hansmann, S.; Hasegawa, M.; Kowallik, K.V. Gene transfer to the nucleus and the evolution of chloroplasts. *Nature* **1998**, *393*, 162–165. [CrossRef] [PubMed]

44.   Rensing, S.A.; Lang, D.; Zimmer, A.D.; Terry, A.; Salamov, A.; Shapiro, H.; Nishiyama, T.; Perroud, P.-F.;

Lindquist, E.A.; Kamisugi, Y. The Physcomitrella genome reveals evolutionary insights into the conquest of land by plants. *Science* **2008**, *319*, 64–69. [CrossRef] [PubMed]

45. Petersen, T.N.; Brunak, S.; von Heijne, G.; Nielsen, H. SignalP 4.0: Discriminating signal peptides from transmembrane regions. *Nat. Methods* **2011**, *8*, 785–786. [CrossRef] [PubMed]

46. Wakasugi, T.; Hirose, T.; Horihata, M.; Tsudzuki, T.; Kössel, H.; Sugiura, M. Creation of a novel protein-coding region at the RNA level in black pine chloroplasts: The pattern of RNA editing in the gymnosperm chloroplast is different from that in angiosperms. *Proc. Natl. Acad. Sci. USA* **1996**, *93*, 8766–8770. [CrossRef] [PubMed]

47. Zandueta-Criado, A.; Bock, R. Surprising features of plastid *ndhD* transcripts: Addition of non-encoded nucleotides and polysome association of mRNAs with an unedited start codon. *Nucleic Acids Res.* **2004**, *32*, 542–550. [CrossRef] [PubMed]

48. Covello, P.S.; Gray, M.W. RNA editing in plant mitochondria. *Nature* **1989**, *341*, 662–666. [CrossRef] [PubMed]

49. Gualberto, J.M.; Lamattina, L.; Bonnard, G.; Weil, J.-H.; Grienenberger, J.-M. RNA editing in wheat mitochondria results in the conservation of protein sequences. *Nature* **1989**, *341*, 660–662. [CrossRef] [PubMed]

50. Hiesel, R.; Wissinger, B.; Schuster, W.; Brennicke, A. RNA editing in plant mitochondria. *Science* **1989**, *246*, 1632–1634. [CrossRef] [PubMed]

51. Hoch, B.; Maier, R.M.; Appel, K.; Igloi, G.L.; Kössel, H. Editing of a chloroplast mRNA by creation of an initiation codon. *Nature* **1991**, *353*, 178–180. [CrossRef] [PubMed]

52. Freyer, R.; Kiefer-Meyer, M.-C.; Kössel, H. Occurrence of plastid RNA editing in all major lineages of land plants. *Proc. Natl. Acad. Sci. USA* **1997**, *94*, 6285–6290. [CrossRef] [PubMed]

53. Stern, D.B.; Goldschmidt-Clermont, M.; Hanson, M.R. Chloroplast RNA metabolism. *Annu. Rev. Plant Biol.* **2010**, *61*, 125–155. [CrossRef] [PubMed]

54. Mower, J.P. The prep suite: Predictive RNA editors for plant mitochondrial genes, chloroplast genes and user-defined alignments. *Nucleic Acids Res.* **2009**, *37*, W253–W259. [CrossRef] [PubMed]

55. Lenz, H.; Knoop, V. PREPACT 2.0: Predicting C-to-U and U-to-C RNA editing in organelle genome sequences with multiple references and curated RNA editing annotation. *Bioinform. Biol. Insights* **2013**, *7*, 1–19. [CrossRef] [PubMed]

56. Bock, R. Sense from nonsense: How the genetic information of chloroplastsis altered by RNA editing. *Biochimie* **2000**, *82*, 549–557. [CrossRef]

57. Chen, H.; Deng, L.; Jiang, Y.; Lu, P.; Yu, J. RNA editing sites exist in protein-coding genes in the chloroplast genome of Cycas taitungensis. *J. Integr. Plant Biol.* **2011**, *53*, 961–970. [CrossRef] [PubMed]

58. Wyman, S.K.; Jansen, R.K.; Boore, J.L. Automatic annotation of organellar genomes with DOGMA. *Bioinformatics* **2004**, *20*, 3252–3255. [CrossRef] [PubMed]

59. Schattner, P.; Brooks, A.N.; Lowe, T.M. The tRNAscan-SE, snoscan and snoGPS web servers for the detection of tRNAs and snoRNAs. *Nucleic Acids Res.* **2005**, *33*, W686–W689. [CrossRef] [PubMed]

60. Frazer, K.A.; Pachter, L.; Poliakov, A.; Rubin, E.M.; Dubchak, I. Vista: Computational tools for comparative genomics. *Nucleic Acids Res.* **2004**, *32*, W273–W279. [CrossRef] [PubMed]

61. Kumar, S.; Stecher, G.; Tamura, K. MEGA7: Molecular evolutionary genetics analysis version 7.0 for bigger datasets. *Mol. Biol. Evol.* **2016**, *33*, 1870–1874. [CrossRef] [PubMed]

5

Candidate Genes for Yellow Leaf Color in Common Wheat (*Triticum aestivum* L.) and Major Related Metabolic Pathways according to Transcriptome Profiling

Huiyu Wu [1], Narong Shi [1], Xuyao An [1], Cong Liu [1], Hongfei Fu [2], Li Cao [1], Yi Feng [1], Daojie Sun [1] and Lingli Zhang [1,*]

[1] College of Agronomy, Northwest A&F University, Yangling 712100, China; huiyuwu@nwafu.edu.cn (H.W.); narongshi@nwafu.edu.cn (N.S.); 16631407278@163.com (X.A.); congliu@nwafu.edu.cn (C.L.); caolinwafu@126.com (L.C.); fengyi92377504@126.com (Y.F.); daojie49124098@126.com (D.S.)
[2] College of Food Science and Engineering, Northwest A&F University, Yangling 712100, China; fuhongfei@nwafu.edu.cn
* Correspondence: zhanglingli@nwafu.edu.cn

Abstract: The photosynthetic capacity and efficiency of a crop depends on the biosynthesis of photosynthetic pigments and chloroplast development. However, little is known about the molecular mechanisms of chloroplast development and chlorophyll (Chl) biosynthesis in common wheat because of its huge and complex genome. *Ygm*, a spontaneous yellow-green leaf color mutant of winter wheat, exhibits reduced Chl contents and abnormal chloroplast development. Thus, we searched for candidate genes associated with this phenotype. Comparative transcriptome profiling was performed using leaves from the yellow leaf color type (Y) and normal green color type (G) of the *Ygm* mutant progeny. We identified 1227 differentially expressed genes (DEGs) in Y compared with G (i.e., 689 upregulated genes and 538 downregulated genes). Gene ontology and pathway enrichment analyses indicated that the DEGs were involved in Chl biosynthesis (i.e., magnesium chelatase subunit H (CHLH) and protochlorophyllide oxidoreductase (POR) genes), carotenoid biosynthesis (i.e., β-carotene hydroxylase (BCH) genes), photosynthesis, and carbon fixation in photosynthetic organisms. We also identified heat shock protein (HSP) genes (*sHSP*, *HSP70*, *HSP90*, and *DnaJ*) and heat shock transcription factor genes that might have vital roles in chloroplast development. Quantitative RT-PCR analysis of the relevant DEGs confirmed the RNA-Seq results. Moreover, measurements of seven intermediate products involved in Chl biosynthesis and five carotenoid compounds involved in carotenoid-xanthophyll biosynthesis confirmed that CHLH and BCH are vital enzymes for the unusual leaf color phenotype in Y type. These results provide insights into leaf color variation in wheat at the transcriptional level.

Keywords: RNA-Seq; transcription factor; chlorophyll biosynthesis precursor; carotenoid composition; wheat; yellow-green leaf color mutant

---

1. Introduction

Photosynthesis is the basis of plant production, and at least 90% of grain yield is determined by photosynthesis [1]. Under irradiation by light, photosynthetic pigments fix light energy and convert it into chemical energy to synthesize carbohydrates. Chlorophyll (Chl) is the primary photosynthetic pigment in higher plants, where it is responsible for light harvesting in the antenna systems and driving electron transport in the reaction centers [2]. The entire Chl biosynthetic pathway from glutamyl-tRNA to Chl *a* and Chl *b* comprises about 20 different enzymatic steps [3]. Mutations in any of these genes may lead to variations in the Chl contents [4], abnormal chloroplast development [5], and decreased photosynthetic efficiency [6], thereby yielding leaf color mutants. Mutants deficient in Chl biosynthesis have been identified in many higher plants, such as rice [7,8], *Brassica napus* [9], *Arabidopsis thaliana* [10], barley [11], and *Camellia sinensis* [12]. Many of the reported chlorotic mutants exhibit reduced Chl biosynthesis due to the lower activity of magnesium chelatase (Mg-chelatase) [11,13–15]. Mg-chelatase (EC 6.6.1.1) is a key regulatory enzyme that catalyzes the insertion of Mg^{2+} into protoporphyrin IX (Proto IX) in an ATP-dependent manner as the first committed step in Chl biosynthesis, where this protein complex comprises magnesium chelatase subunit I (CHLI), D (CHLD), and H (CHLH) in higher plants [16], which are all required for its activity [17]. CHLI and CHLD belong to the large family of AAA^+ (ATPases associated with various cellular activities) proteins, but only the I-subunit has an ATPase activity [18]. The H-subunit binds the porphyrin substrate, and it is regarded as a catalytic subunit without ATPase activity [19]. The *GUN5* gene encodes the CHLH, and it has a specific function in the plastid signaling pathway where its activity is controlled by *GUN4* [20,21]. The *GUN4* gene encodes a protein that regulates Chl biosynthesis in plastids, and it has been implicated in plastid retrograde signaling via the regulation of Mg-protoporphyrin (Mg-Proto) synthesis or transport [21]. The activity of Mg-chelatase has essential regulatory roles in Chl biosynthesis and chloroplast development in higher plants. For example, in peas, the virus-induced gene silencing of *CHLI* and *CHLD* yields plants with yellow leaf phenotypes and reduced Mg-chelatase activities, as well as lower Chl accumulation correlated with undeveloped thylakoid membranes [22]. In addition, *CHLD* and *CHLI* silencing greatly reduces the levels of photosynthetic proteins, as well as being correlated with reactive oxygen species homeostasis [22]. A T-DNA insertion mutant *OsCHLH* in rice also exhibits underdeveloped chloroplasts with a low Chl content [7]. Previous studies have explored the semi-dominant leaf color mutants caused by Mg-chelatase. In barley, the mutants *chlorina-125*, *-157*, and *-161* have the same phenotypic ratio model (i.e., one green wild-type leaf, two light-green chlorina leaves, and one lethal yellow leaf at the seedling stage). The Mg-chelatase activity of the heterozygous chlorina seedlings is 25–50% of that in wild type seedlings [23,24]. In tobacco, the *Sulfur* mutant, a *CHLI* gene mutation due to the formation of inactive Mg-chelatase, is a semi-dominant aurea mutation, where homozygotes of the mutant are yellow seedling lethals, whereas the heterozygotes have reduced Chl contents and a yellow-green phenotype [25]. Moreover, a semi-dominant *CHLI* allele designated as *Oil Yellow1* (*Oy1*) has also been characterized in maize [26]. In rice, *chl1* and *chl9* mutants exhibit a yellowish-green leaf color phenotype where the abnormal leaf color is controlled by a single recessive gene. The *chl1* and *chl9* genes encode the CHLD and CHLI subunits of Mg-chelatase, and their mutation leads to underdeveloped chloroplasts and low Chl contents [13]. However, to the best of our knowledge, only a few *CHLD*, *CHLH*, and *CHLI* genes encoding the Mg-chelatase D, H, and I subunits have been reported in wheat leaf color mutants [27–29].

Common wheat (*Triticum aestivum* L.) is one of the most important food crops in the world. Two main types of Chl-deficient wheat mutant have been identified (i.e., albinism [30,31] and chlorina [32,33]), which have great research value for understanding the mechanisms of Chl biosynthesis and photosynthesis in wheat. However, only a few studies have reported the molecular mechanisms related to the changes in leaf color in wheat because of its large genome and high proportion of repetitive sequences (>80%) [30,34]. Most of these studies have focused on agronomic traits, photosynthetic characteristics, physiological and biochemical characteristics, and genetic mapping [31–33]. For example, controlled by cytogene, the wheat stage albinism mutant *FA85* exhibits albinism in cold early spring and returns to normal green gradually with the increase of temperature [35]. Chlorophyll precursors and key enzyme activities measurement of *FA85* have revealed that the accumulation of Proto IX, Mg-Proto IX, and Pchlide derives from the downregulated transcription level of Pchlide oxidoreductase and Chl synthase [36]. In addition, the five homeologous allelic chlorina mutants Driscoll's chlorina, chlorina-1, CD3, chlorina-214, and CDd-1 exhibit the reduction in Mg-chelatase activity which leads to the accumulation of Proto IX to different extents [28]. The above five homeologous allelic chlorina mutants have been mapped on the long arm of homoeologous group 7 chromosomes [37]. In recent years, due to the development of high-throughput sequencing technology, transcriptome sequencing (RNA-Seq) has emerged as a powerful tool for studying complex biological processes at the molecular level and for identifying candidate genes involved in specific biological functions [38,39]. RNA-Seq has been employed to investigate leaf color mechanisms in various plants, such as tea (*Camellia sinensis* L.) [12], *Anthurium andraeanum* [40], *Lagerstroemia indica* [41], and wheat [34]. These studies showed that the candidate genes related to leaf color mutation were involved in chloroplast development, Chl biosynthesis, photosynthesis, and transcription factors (TFs) such as phytochrome-interacting factor (PIF1 and PIF3), Golden2-like (GLK), and v-myb avian myeloblastosis viral oncogene homolog (MYB), which might participate in the pathways identified. Therefore, RNA-Seq can increase our understanding and provide new insights into wheat leaf color mutation at the genomics level.

Previously, we reported a spontaneous yellow-green leaf color mutant (*Ygm*) derived from a common wheat cultivar Xinong1718, with yellow (Y), yellow-green (Yg), and normal green (G) types under field temperature conditions from the jointing stage to the adult stage (Figure 1). Genetic analysis indicated that the yellow leaf color trait in the *Ygm* mutant was controlled by an incompletely dominant gene *Y1718* [42]. The homozygous dominant genotype (*Y1718Y1718*) of Y is extremely yellow, stunted, and sterile. The homozygous recessive genotype (*y1718y1718*) of G is normal green, whereas the heterozygous genotype (*Y1718y1718*) of Yg is yellow-green. Types Yg and G have similar agronomic traits to the wild type Xinong1718. The mutant is maintained steadily as the heterozygote genotype Yg. The *Y1718* gene was mapped on the short arm of chromosome 2B (2BS) between the simple sequence repeat marker *Xwmc25* and the expressed sequence tag-sequence tagged site marker *BE498358*, with genetic distances of 1.7 cM and 4.0 cM, respectively [42]. Thus, in the present study, the availability of a set of germplasm for the Y, Yg, and G genotypes allowed us to obtain insights into complex metabolic networks and certain biochemical traits, especially leaf color in common wheat.

Figure 1. Phenotypes of the Y, Yg, and G plants among the progeny of the *Ygm* mutant and wild type (WT, Xinong1718). (**A**) Jointing stage (9 April 2016); (**B**) Adult stage (28 April 2016); (**C**) Enlarged views of the leaves in different development states in G type at the jointing stage; (**D**) Enlarged views of the leaves in different development states in Y type at the jointing stage (WT, Xinong1718). G, normal green leaf color plant in the progeny of *Ygm*; Yg, yellow-green leaf color plant in the progeny of *Ygm*; Y, yellow leaf color plant in the progeny of *Ygm*. F_G and F_Y, fully-developed leaves in G and Y plants, respectively. H_G and H_Y, half-developed leaves in G and Y plants, respectively. L_G and L_Y, small leaf buds in G and Y plants, respectively. Bar = 5 cm.

In this study, comparative transcriptome profiles of the Y and G types in the progeny of the *Ygm* mutant were analyzed by RNA-Seq. Based on a combination of biochemical analysis and bioinformatics, we identified the major metabolic pathways related to leaf color and candidate genes for the loss of pigmentation in these plants. The concentrations of seven intermediate products involved in Chl biosynthesis and five carotenoid compounds involved in carotenoid-xanthophyll biosynthesis were measured to further understand the molecular mechanisms related to pigment biosynthesis in the *Ygm* mutant.

2. Results

2.1. Sequence Analysis Using RNA-Seq

To understand the molecular basis of leaf color polymorphism in the progeny of the *Ygm* mutant, cDNA libraries were constructed from the half-developed leaves of the Y and G types based on three biological replicates, which were then sequenced using the Illumina HiSeq^{TM} 2500 platform (Illumina, San Diego, CA, USA). The correlation coefficient values ranged from 0.947 to 0.989 (Table S1), thereby indicating strong correlations between the replicates. After cleaning and checking the read quality, 387,431,412 clean paired-end reads were generated, with 203,880,864 reads from the Y type and 183,550,548 from the G type, where the clean data GC content ranged from 57.29% to 59.25%, and the Q20 percentage exceeded 96.72%. The high-quality clean reads were then aligned with wheat genome sequences in the Unité de Recherche Génomique Info (URGI) database (http://wheat-urgi.versailles.inra.fr/Seq-Repository), where the alignment efficiency ranged from 69.67% to 70.71% (Table 1). Thus, the throughput and sequencing quality were sufficiently high to warrant further analysis.

Table 1. Summary of transcriptome sequencing data obtained using yellow leaves of Y plants and green leaves of G plants.

Groups	Total Reads	Clean Reads	GC (%)	Q20 (%)	Total Mapped Reads	Ratio (%)
G-1	59,480,874	58,804,014	58.37	96.76	41,285,997	70.21
G-2	61,434,418	60,676,246	58.68	96.54	42,662,324	70.31
G-3	64,966,228	64,070,288	58.68	96.65	44935,906	70.14
Y-1	67,083,278	66,278,074	59.25	96.87	46,615,892	70.33
Y-2	58,156,566	57,490,204	57.87	96.95	40,650,136	70.71
Y-3	81,071,740	80,112,586	57.29	96.61	55,817,063	69.67
Total	392,193,104	387,431,412	58.32	96.72	271,967,318	69.35

2.2. Identification and Functional Annotation of Differentially Expressed Genes (DEGs) in G and Y

The FPKM (fragments per kilobase of transcript per million mapped reads) method was used to analyze the gene expression. As a result, 74,937 and 75,211 genes were identified respectively in the cDNA library from G and Y leaves, of which 3879 and 4153 genes were expressed specifically in the leaves of G and Y, respectively (Figure 2). In total, 1227 DEGs (false discovery rate (FDR) <0.05 and |log$_2$Fold Change| (|log$_2$FC|)>1) were detected, where 689 were upregulated and 538 were downregulated in Y. In order to understand the functions of the DEGs and the biological processes involved with leaf color variation, all of the DEGs were searched for in the GenBank non-redundant (Nr) protein database as well as the Gene Ontology (GO), Clusters of Orthologous Groups (COG), and Kyoto Encyclopedia of Genes and Genomes (KEGG) databases. In total, 882 DEGs had BLASTx matches with known proteins and these 882 DEGs were assigned to one or more GO terms in the biological process (685 genes), molecular function (797 genes), and cellular component (247 genes) categories. Among these three categories, the metabolic process sub-category in the biological process category accounted for the majority of the GO annotations, followed by binding and catalytic activity in the molecular function category (Figure 3). Lipid metabolic process (50 genes) and phosphorus metabolic process (20 genes), which are closely related to cell membrane functions, were significantly enriched in the biological process GO term (Table S2). The nucleic acid binding transcription factor (TF) activity and sequence-specific DNA binding TF activity were the most significantly enriched GO terms in the molecular function category (Table S2). In the cellular component category, most of the DEGs were involved with cell, cell part, and membrane components (Figure 3), but the most significantly enriched component was membrane part (Table S2).

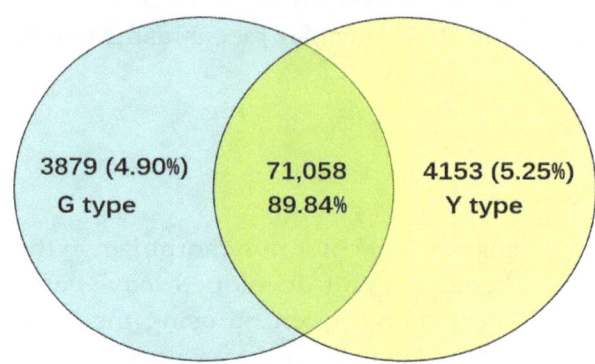

Figure 2. The numbers of specific genes and shared genes between G and Y.

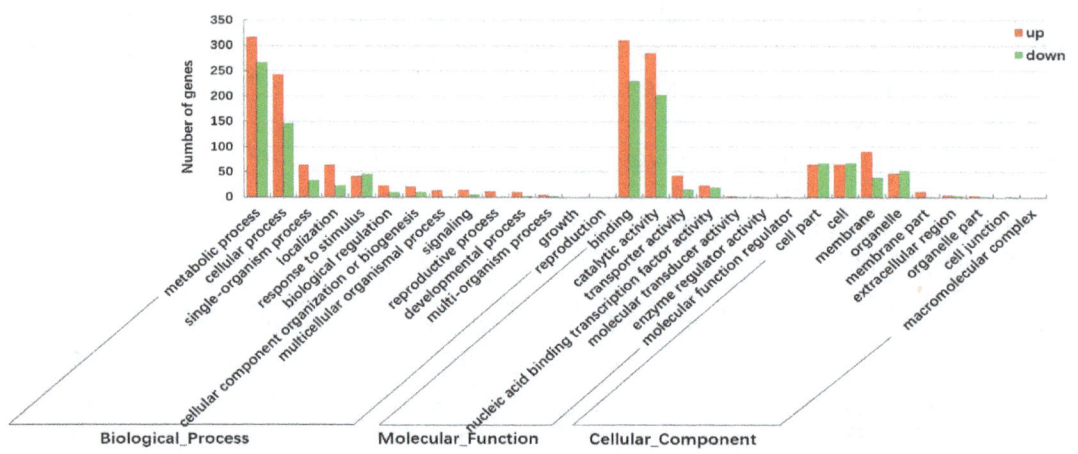

Figure 3. GO classifications of the DEGs in groups G and Y.

All of the 1227 DEGs were further annotated based on the COG database to obtain functional predictions and classifications (Figure 4). In total, 434 (35.37%) of the DEGs were finally mapped onto 22 different COG categories, where the "general function prediction only" cluster represented the largest group (108, 24.88%), followed by "posttranslational modification, protein turnover, chaperones" (97, 22.35%), "signal transduction mechanisms" (66, 15.20%), and "secondary metabolites biosynthesis, transport and catabolism" (59, 13.59%). In addition, "carbohydrate transport and metabolism" (42, 9.67%) and "lipid transport and metabolism" (35, 8.1%) were also annotated, which are closely related to energy metabolism and cell membrane function.

To identify biological pathways, the DEGs were annotated with the corresponding enzyme commission (EC) numbers from BLASTx alignments against the KEGG pathway databases. Among the 1227 DEGs, 323 (26.32%) were assigned to 82 KEGG pathways (Table S3), where the top 11 pathways were considered significant at a cut-off FDR corrected p-value (q-value) < 0.05 (Figure 5A). The 11 enriched pathways comprised protein processing in the endoplasmic reticulum, alpha-linolic acid metabolism, spliceosome, circadian rhythm-plant, linoleic acid metabolism, endocytosis, monoterpenoid biosynthesis, porphyrin and chlorophyll metabolism, brassinosteroid biosynthesis, glutathione metabolism, and thiamine metabolism. Most of the genes mapped in the first eight significantly enriched pathways had downregulated expression trends, except for the circadian rhythm-plant pathway. Chlorophyll and carotenoid biosynthesis are crucial for the leaf color and photosynthesis. We focused our analysis on the major genes related to pigment metabolism and photosynthesis. The results indicated that 33 genes related to either photosynthesis (five genes), Chl metabolism (nine genes), carotenoid biosynthesis (two genes), carbon fixation in photosynthetic organisms (five genes), or carbon metabolism (12 genes) pathways were differentially expressed, and the enrichment of each pathway is shown in Figure 5B. These pathways were investigated in greater detail and they may be important for the unusual leaf color phenotype in Y.

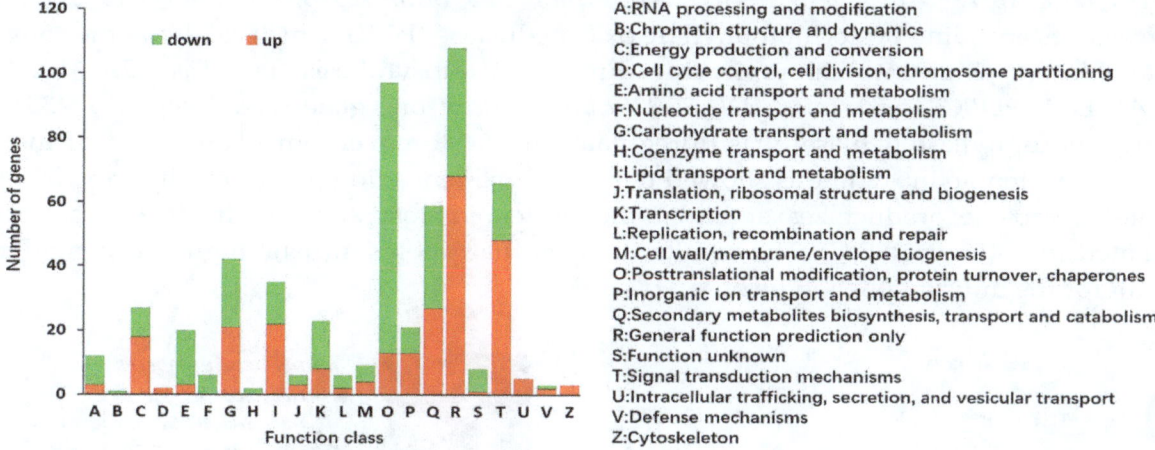

Figure 4. Clusters of Orthologous Groups (COG) classifications of the annotated 434 DEGs. The capital letters on the horizontal axis indicate the COG categories that are listed on the right of the histogram, and those on the vertical axis indicate the number of DEGs.

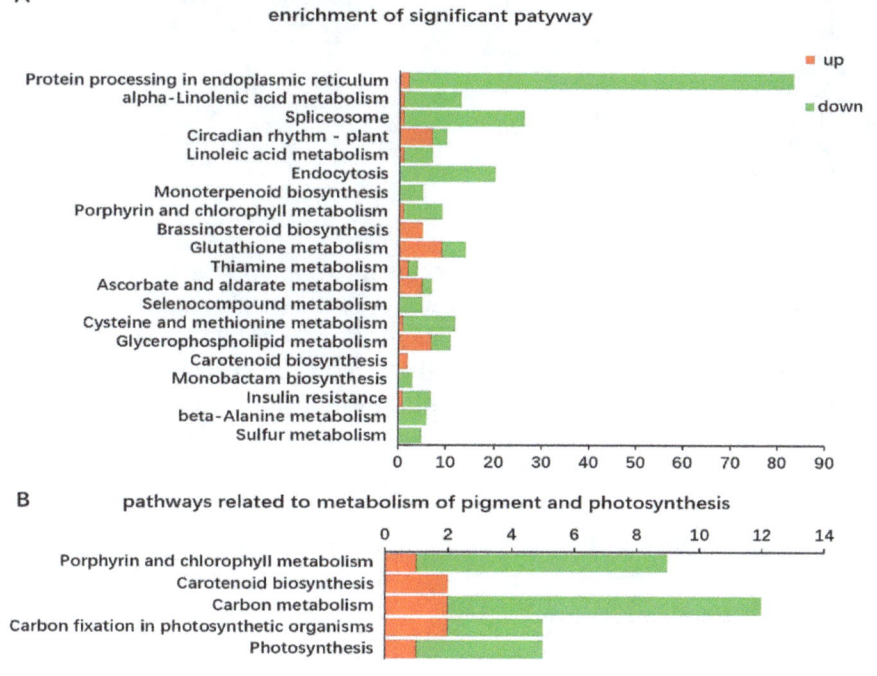

Figure 5. Kyoto Encyclopedia of Genes and Genomes (KEGG) classifications of DEGs. (**A**) Enrichment of the top 20 most significant pathways (*p*-value < 0.05). The vertical axis shows the annotations of the KEGG metabolic pathways. The horizontal axis represents the DEG numbers annotated in each pathway; (**B**) KEGG-based pathway assignments of the 33 DEGs (Y versus G) related to photosynthesis and pigment metabolism: photosynthesis (five genes), porphyrin and chlorophyll metabolism (nine genes), carotenoid biosynthesis (two genes), carbon fixation in photosynthetic organisms (five genes), and carbon metabolism (12 genes).

2.3. Identification of DEGs Related to Chl Metabolism and Carotenoid Biosynthesis in Y and G Plants

To further identify the key transcripts related to yellow leaf color formation in the mutant Y type, the DEGs in two pathways for Chl metabolism and carotenoid biosynthesis were compared in detail between the Y and G transcriptomes. The results demonstrated that eight genes in the porphyrin and Chl metabolism pathway had downregulated expression levels ($q < 0.05$, fold change > 2) (Figure 6A,C), including five genes encoding CHLH (i.e., *Traes_2DS_BB50DEEF8*, *Traes_2DS_DBD06E18F*, *Traes_2AS_BFBD75AB4*, *Traes_2AS_B6BA92570*, and *Traes_2BS_E67494A11*), and three genes encoding protochlorophyllide oxidoreductase (POR), which catalyzes the conversion of proto-chlorophyllide into chlorophyllide during Chl biosynthesis (i.e., *Traes_2BL_F22336B90*, *Traes_2AL_E0AC9DBC7*, and *Traes_2DL_3C229EB92*). Only one gene (i.e., *Traes_3AS_433192E29*) encoding chlorophyllase (Chlase) was upregulated in Y. Moreover, our comparison of the DEGs involved in carotenoid biosynthesis showed that two genes encoding β-carotene hydroxylase (BCH) (substrate: β-carotene; product: zeaxanthin) in the carotenoid biosynthesis pathway were significantly upregulated in Y (Figure 6B,C). The results indicate that photosynthetic pigment biosynthesis is important for the unusual phenotype of Y.

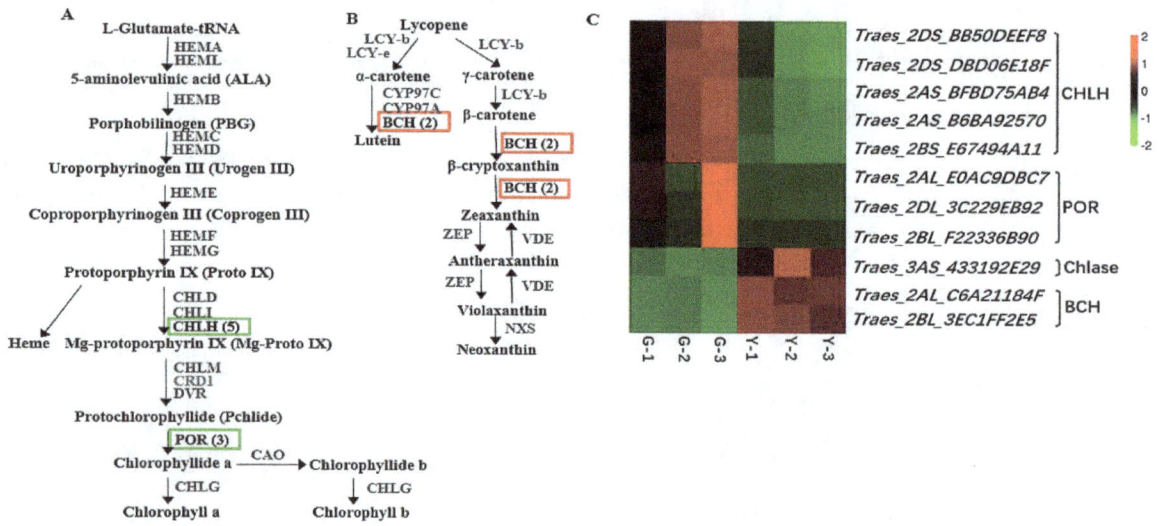

Figure 6. DEGs at the transcript level involved in chlorophyll and carotenoid biosynthesis pathways. (**A**) Chlorophyll biosynthesis pathway; (**B**) Carotenoid–xanthophyll biosynthesis pathway. In (**A**,**B**), upregulated genes are marked by red-line borders and downregulated genes by green-line borders. The numbers following each gene name indicate the number of corresponding DEGs identified in our database; (**C**) Expression profile clustering for chlorophyll and carotenoid biosynthesis pathways. Expression ratios are based on $\log_2$ FPKM values (fragments per kilobase of transcript per million mapped reads), where each vertical column represents a sample (G-1, G-2, and G-3; Y-1, Y-2, and Y-3), and each horizontal row represents a single gene. CHLH, Mg-chelatase H subunit; POR, protochlorophyllide oxidoreductase; Chlase, chlorophyllase; BCH, β-carotene hydroxylase.

2.4. Identification of DEGs Related to Photosynthesis

The Chl content was closely related to photosynthesis. In this study, we identified five DEGs involved in photosynthesis, including two genes that encode photosystem II 47 kDa protein (PsbB), one gene that encodes photosystem II protein D2 (PsbD), one gene encode PsaC in photosystem I, and F-type ATPase β subunit, most of which were significantly downregulated in the Y type (Figure 7A,B). Moreover, most of the genes that mapped to the carbon metabolism and carbon fixation pathways in photosynthetic organisms exhibited decreased expression in Y compared with G, where the genes encoding ribulose-1,5-bisphosphate carboxylase/oxygenase (Rubisco), 6-phosphogluconolactonase (6PGL), glucose-6-phosphate 1-dehydrogenase (G6PDH), and 6-phosphogluconate dehydrogenase

(6PGD) were downregulated. However, two fructose-1,6-bisphosphatase genes were upregulated (Figure 7B). In addition, 18 genes related to early light-inducible proteins (ELIPs) were significantly upregulated in Y (Figure 7B). These results indicate that changes in the expression levels of these genes might have blocked photosynthesis, thereby influencing chlorophyll biosynthesis in the mutant Y type.

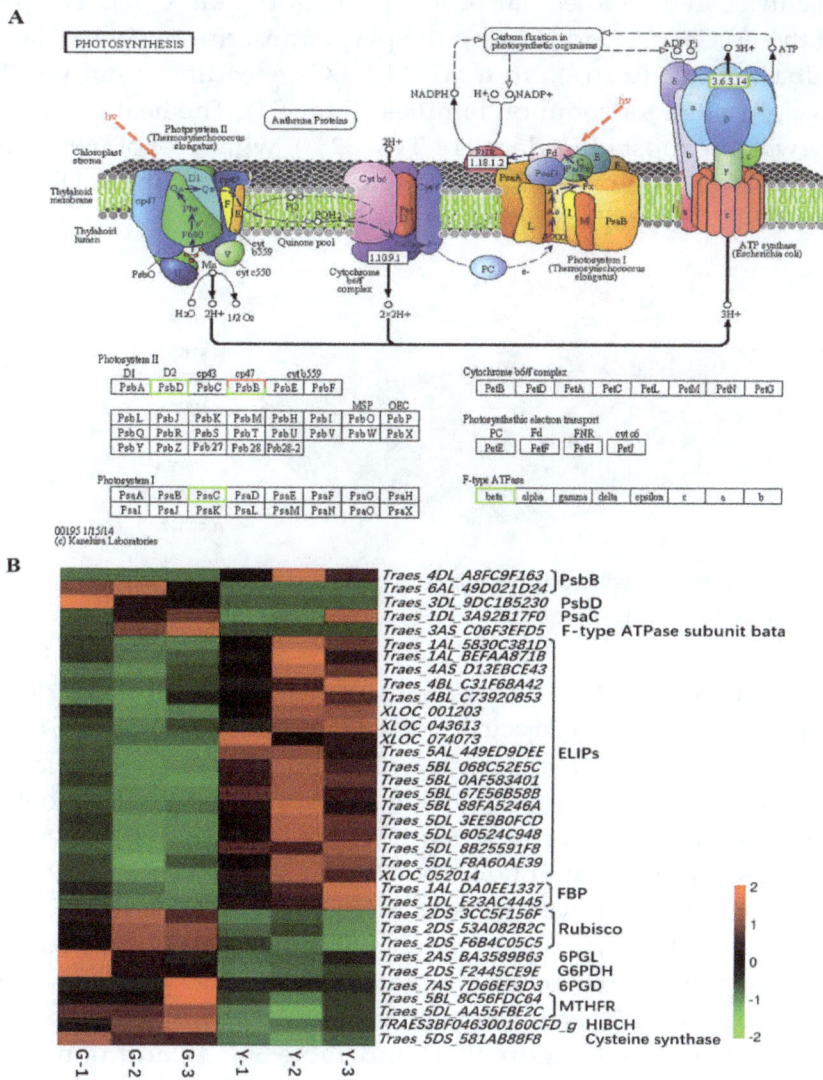

Figure 7. DEGs mapped onto the photosynthesis pathway. (**A**) Photosynthesis pathway. The image of the known photosynthesis pathway was obtained from the freely available KEGG database (http://www.kegg.jp/kegg-bin/show_pathway?ko00195). The green border denotes lower expression in Y compared with G, red color denotes higher expression, and half red/half green donates both up- and downregulated genes in Y compared to G. The blue dashed lines denote photosynthetic electron transport in the thylakoid membrane, red dashed lines denote light irradiation. The black dashed lines denote energy conversion of carbon fixation in photosynthetic organisms and the solid arrows denote molecular interaction or relation; (**B**) Expression profile clustering for genes involved in the photosynthesis and carbon metabolism pathway. Expression ratios are based on $\log_2$ FPKM values, where each vertical column represents a sample (G-1, G-2, and G-3; Y-1, Y-2, and Y-3), and each horizontal row represents a single gene. PsbB, photosystem II 47 kDa protein; PsbD, photosystem II protein D2; PsaC, photosystem I subunit VII; ELIPs, early light-inducible proteins; FBP, fructose-1,6-bisphosphatase; Rubisco, ribulose-1,5-bisphosphate carboxylase/oxygenase; 6PGL, 6-phosphogluconolactonase; G6PDH, glucose-6-phosphate 1-dehydrogenase; 6PGD, 6-phosphogluconate dehydrogenase; MTHFR, methylenetetrahydrofolate reductase; HIBCH, 3-hydroxyisobutyryl-CoA hydrolase-like protein.

2.5. Analysis of TFs and Heat Shock Proteins (HSPs) in Y and G Plants

Transcription factors are important regulators that can activate or repress the expression of genes in a sequence-specific manner, thereby affecting or controlling many biological processes. We found that the GO term "sequence-specific DNA binding transcription factor activity" was most significantly enriched among the entries in the molecular function category (Table S2). Therefore, we identified all of the TFs among the DEGs in Y and G using the plant transcription factor database (PlantTFDB) v4.0 (http://planttfdb.cbi.pku.edu.cn/). In total, 44 DEGs encoding putative TFs were identified and categorized into eight different common families (Table S4). The heat shock transcription factor (HSF) protein family was the most abundant (14 TFs, 32%), where all of exhibited downregulated expression in Y (Table S4), followed by the bZIP (12 TFs, 27%), WRKY (three TFs, 7%), ERF (three TFs, 7%), and GATA (three TFs, 7%) families (Figure 8).

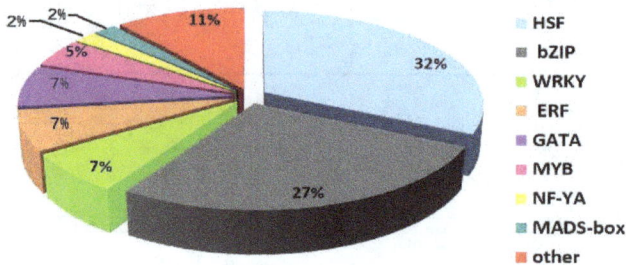

Figure 8. Percentages of different transcription factors involved in the "sequence-specific DNA binding transcription factor activity" GO term.

In order to obtain a more intuitive understanding of the roles of those downregulated HSFs, totally, seven HSF binding motifs were projected in the TFDB database (Figure S1). The binding motifs of these seven HSFs (Figure S1) were then searched to identify the cis-acting elements present in the promoters of the DEGs using FIMO software (version 4.9.0) [43]. The results showed that many genes interacted with these seven HSF genes (Table S5). Thus, we selected 37 genes related to leaf color variations as target nodes and genes annotated as HSFs as source nodes to generate an interaction network diagram using Cytoscape software (version 3.3.0) [44]. We found that genes encoding 30 HSPs (i.e., small HSPs (sHSPs) and HSP70) comprised the main group that interacted with the seven HSFs (Figure 9A). Interestingly, most of the genes annotated as heat shock proteins (HSPs) (e.g., sHSPs, HSP70, and HSP90) in the "protein processing in endoplasmic reticulum" pathway had significantly reduced mRNA levels in the Y type (Figure 9B,C and Table S6). In addition, we found that genes encoding ELIPs, CHLH, and BCH also interacted with the seven HSFs (Figure 9A). Thus, we suggest that downregulation of these seven HSFs may affect the regulation of leaf color formation.

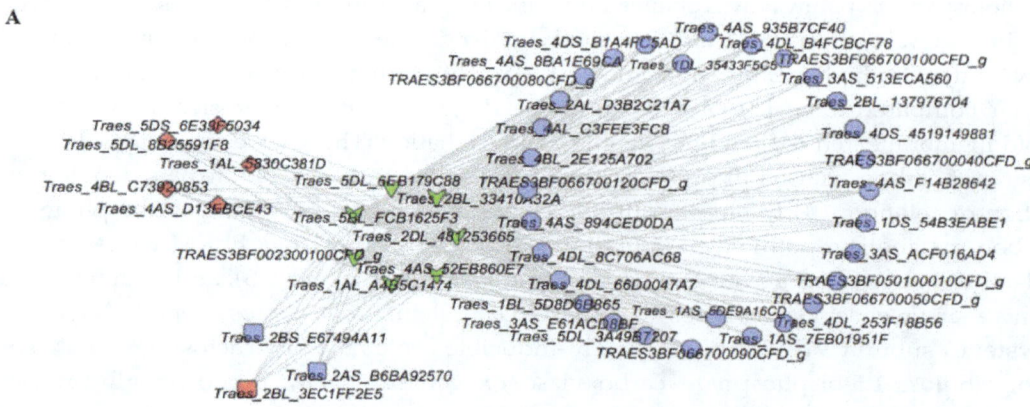

Figure 9. *Cont.*

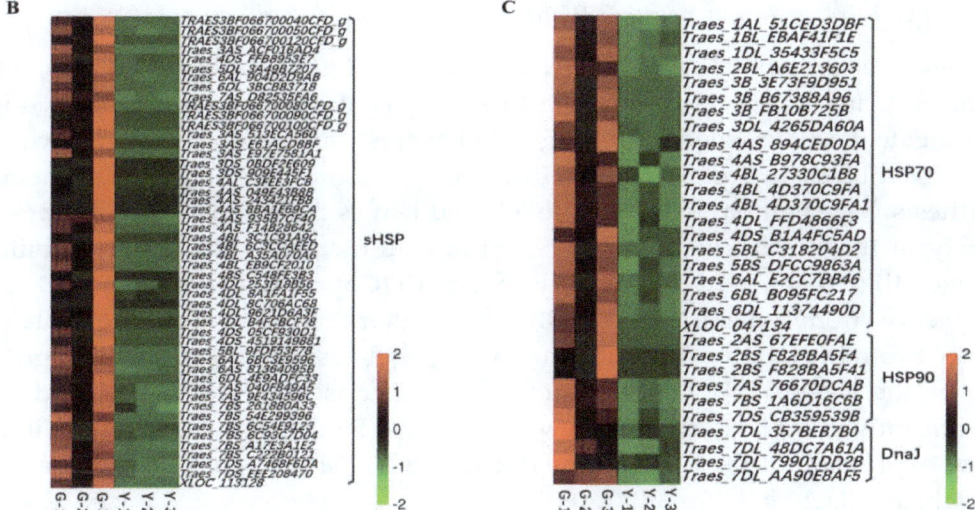

Figure 9. Gene interaction network diagrams and expression profile clustering for genes encoding heat shock proteins (HSPs). (**A**) Interactions between heat shock transcription factors (HSFs) and other genes. Green inverted triangles represent HSFs, HSFA6B, and HSFB2B. Pink rhombuses represent early light-inducible proteins (ELIPs). Blue circles represent HSPs. Blue squares represent CHLH. The pink square represents β-carotene hydroxylase (BCH); (**B,C**) Expression profile clustering for HSP encoding genes. Expression ratios are based on $\log_2$ FPKM values, where each vertical column represents a sample (G-1, G-2, and G-3; Y-1, Y-2, and Y-3), and each horizontal row represents a single gene. sHSP, small heat shock protein; HSP70, heat shock cognate 70 kDa protein; HSP90, heat shock 90 kDa protein; DnaJ, chaperone protein DnaJ.

2.6. Quantitative Real-Time PCR (qRT-PCR) Analysis of Candidate Genes

To validate the reliability of the RNA-Seq data, 15 DEGs that we considered to have strong involvement with leaf color variation in Y were subjected to qRT-PCR analysis. The results showed that the expression patterns of the 14 genes detected by qRT-PCR were consistent with those in the transcriptome data, where only one gene (i.e., *Traes_4DL_A8FC9F163*) encoding PsbB had a different expression pattern (Figure 10). This discrepancy with the PsbB transcripts detected by RNA-Seq and qRT-PCR may have been caused by differences in the sensitivity of the method employed to analyze this gene. qRT-PCR is based on the specific amplification of single gene primers, and it has higher accuracy than RNA-Seq. Moreover, it is possible that this discrepancy is caused by low-expression levels of PsbB. Transcriptome results showed that *Traes_4DL_A8FC9F163* in G was almost no expression, and the expression level in Y was also low. Overall, most of the qRT-PCR data were consistent with the RNA-Seq results, thereby demonstrating the reliability of the RNA-Seq data.

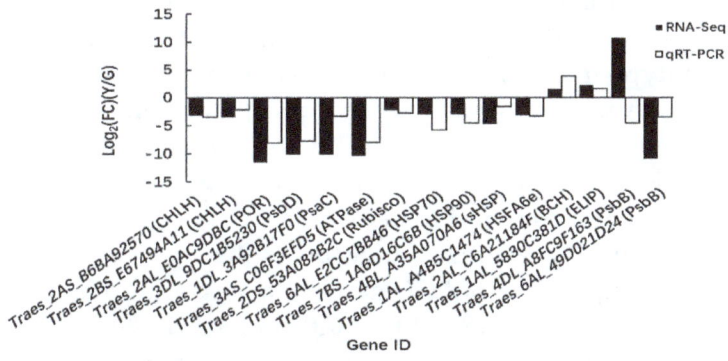

Figure 10. qRT-PCR validation of the RNA-Seq results for the candidate DEGs related to yellow leaf color formation in the Y type. $\log_2$(FC) represents the fold change in Y relative to that in G.

2.7. Comparison of Chl Precursors in Y and G Plants

Mg-chelatase is encoded by the *CHLD*, *CHLI*, and *CHLH* genes in higher plants, which catalyze the insertion of Mg^{2+} into Proto IX to form Mg-Proto IX in chlorophyll biosynthesis [45] (Figure 6A). To further investigate the effects of the downregulated expression of the *CHLH* gene, which is involved in chlorophyll biosynthesis in Y plants, seven intermediate products related to the metabolic process of Chl biosynthesis were compared in half-developed leaves from the Y and G types (i.e., H_Y and H_G, respectively, at the jointing stage). The 5-aminolevulinic acid (ALA), porphobilinogen (PBG), uroporphyrinogen III (Urogen III), coproporphyrinogen III (Coprogen III), and Proto IX contents of the half-developed leaves from Y plants were significantly higher than those from the G plants (Figure 11). In particular, the Proto IX contents (the substrate of Mg-chelatase) were two times higher in Y than G plants. However, the Mg-Proto IX (the product of Mg-chelatase), Pchlide, Chl *a*, and Chl *b* contents, decreased significantly in Y plants compared with G plants (Figure 11). These results indicate that the downregulated expression of the *CHLH* gene decreased the Mg-chelatase activity and reduced the production of Chl *a* and Chl *b* in Y plants.

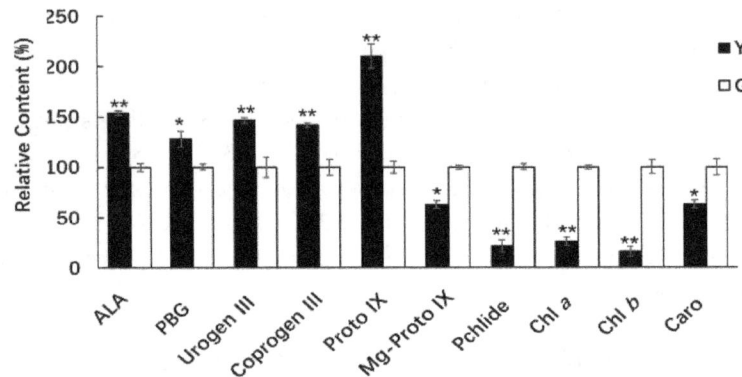

Figure 11. Comparison of the relative contents of chlorophyll precursors, chlorophyll, and carotenoids in H_Y and H_G leaves at the jointing stage. Three individuals were measured for each chlorophyll and chlorophyll precursor. Each plant was extracted once, and the chlorophyll contents were measured three times. Error bars indicate means $\pm$ SD based on three independent experiments. Significant differences were determined using the Student's *t*-test in Y compared with G plants (* $p < 0.05$, ** $p < 0.01$). ALA, 5-aminolevulinic acid; PBG, porphobilinogen; Urogen III, uroporphyrinogen III; Coprogen III, coproporphyrinogen III; Proto IX, protoporphyrin IX; Mg-Proto IX, Mg-protoporphyrin IX; Pchlide, protochlorophyllide; Chl *a*, chlorophyll *a*; Chl *b*, chlorophyll *b*; Caro, carotenoid.

2.8. Analysis of the Carotenoid Composition in Y and G Plants by HPLC

Our results showed that the total carotenoid level was significantly reduced in Y (Figure 11), but upregulated expression of the *BCH* gene related to carotenoid biosynthesis was also identified (Figure 6B,C). To further confirm the involvement of *BCH* in carotenoid-xanthophyll biosynthesis and the yellow leaf color phenotype in Y plants, we quantitatively analyzed five carotenoid compounds (i.e., lutein, zeaxanthin, β-cryptoxanthin, α-carotene, and β-carotene) in the half-developed leaves from the Y and G types using high performance liquid chromatography (HPLC), and the results are shown in Figure 12 and Table 2. The standard curve for carotenoids is shown in Table S7. β-carotene and lutein were the major carotenoids found in G plants, and a very small amount of zeaxanthin was also detected (Figure 12). Yellow plants differed in terms of the composition and levels of carotenoids. Consistent with the increased expression of *BCH* genes, the zeaxanthin level was significantly higher (more than 40 times) in Y, and it was the most abundant carotenoid, whereas the β-carotene and lutein levels were significantly lower than those in G plants (Figure 12 and Table 2). The β-cryptoxanthin level did not differ significantly and α-carotene was below the detection limit in G and Y (Figure 12 and Table 2). In addition, violaxanthin and neoxanthin were identified in Y and G according to individual peaks in the absorption spectra, but their levels were significantly lower in Y (Figure 12). Overall, our results showed that differences in the carotenoid composition could contribute to the yellow leaf phenotype in Y.

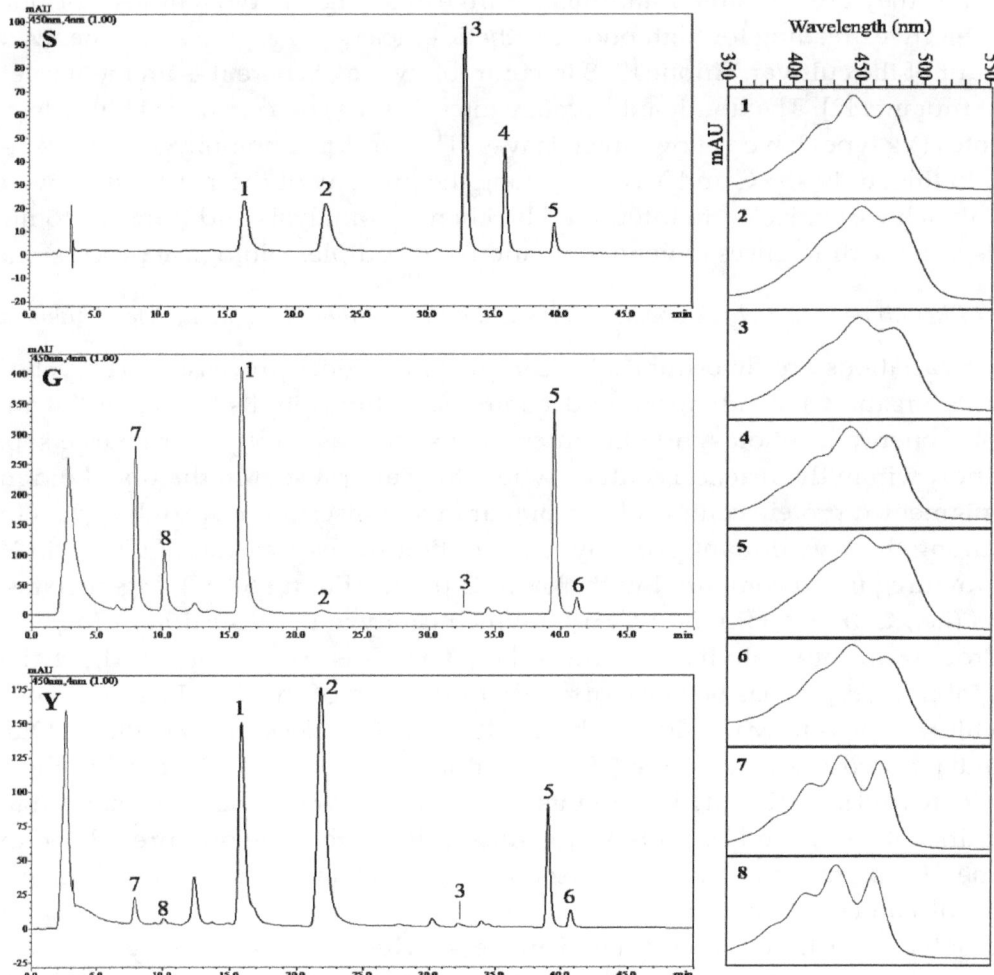

Figure 12. High performance liquid chromatography (HPLC) elution profiles for carotenoids accumulated in G and Y leaves at 450 nm. (**S**) HPLC elution profiles for five carotenoid standards. (**G**) HPLC elution profiles for carotenoids accumulated in G leaves. (**Y**) HPLC elution profiles for carotenoids accumulated in Y leaves. The vertical axis shows the absorbance (mAU) at 450 nm, and the horizontal axis represents the retention time for carotenoids. The right panel is the absorption spectra from peak 1 through 8 at 450 nm in Y and G types. The vertical axis shows the mAU, and the horizontal axis represents absorption wavelength. Peak 1, lutein (absorption peak λmax nm: 444, 472); peak 2, zeaxanthin (450, 476); peak 3, β-cryptoxanthin (451, 476); peak 4, α-carotene (445, 473); peak 5, β-carotene (452, 477); peak 6, 9-cis-β-carotene (446, 472); peak 7, violaxanthin (416, 438, 468); peak 8, neoxanthin (413, 435, 463).

Table 2. Comparison of the leaf carotenoids contents (μg/g Fresh Weight) in G and Y at the jointing stage according to HPLC.

Phenotype	Lutein	Zeaxanthin	β-Cryptoxanthin	α-Carotene	β-Carotene
G	366.60 ± 23.26	3.89 ± 0.66	0.85 ± 0.24	ND	494.94 ± 45.00
Y	140.18 ± 9.43 *	157.62 ± 15.63 **	0.56 ± 0.08	ND	105.73 ± 29.75 *

The carotenoids contents are expressed as the mean ± SD based on three independent replications. Asterisks indicate significant differences in the compound measured in the Y type compared with the G type (t-test, $n = 3$, *: $p < 0.05$; **: $p < 0.01$). ND denotes not detected or below the detection limit for HPLC.

3. Discussion

Leaf color formation is closely related to chloroplast development and photosynthetic pigments, and it is important for photosynthesis. Recently, many leaf color mutants have been identified in higher plants and they are valuable materials for investigating the biosynthesis of photosynthetic pigments and selective breeding for high photosynthetic efficiency. *Ygm* is a spontaneous yellow-green leaf color mutant of the cultivar Xinong1718 in common wheat, where it is an incomplete dominant semidominant mutant [42]. The dominant homozygotes (Y type) of *Ygm* have yellow leaves, whereas the heterozygotes (Yg type) have yellow-green leaves. The recessive homozygotes (G type) of *Ygm* are normal green. In this study, the G and Y types among the progeny of the *Ygm* mutant were selected as the materials, which we subjected to integrated biochemical analysis and transcriptome profiling to obtain insights into the differences in gene regulation and complex biological processes in G and Y.

3.1. Yellow Leaf Phenotype is Closely Associated with Chl and Carotenoid Pigment Metabolism

Leaf color variations are determined by complex biological processes. The yellow leaf color mainly depends on the Chl and carotenoid contents. Chlorophylls are essential molecules for harvesting solar energy in photosynthetic antenna systems, as well as for charge separation and electron transport within the reaction centers, where they are present in the thylakoid membrane in the form of a pigmented protein complex [2]. Comparative transcriptome profiling data for the Y type and G type among the *Ygm* mutant progeny showed that the expression of *CHLH* in Y plants was significantly downregulated compared with that in G plants (Figure 6A,C). Mg-chelatase comprises three subunits (i.e., CHLH, CHLI, and CHLD) which catalyze the insertion of Mg^{2+} into Proto IX to form Mg-Proto IX in chlorophyll biosynthesis [45]. CHLH is crucial for the Mg-chelatase activity as a catalytic subunit [19]. It has been reported that mutation of the CHLH gene leads to defective Chl and the chlorina or yellow phenotype in rice [8] and *Arabidopsis thaliana* [20]. Our analysis of seven intermediate products involved in Chl biosynthesis showed that the level of the substrate for Mg-chelatase increased significantly in Y plants whereas the level of the product from Mg-chelatase decreased significantly (Figure 11). Thus, our results indicate that the downregulated expression of the *CHLH* gene in Y plants may have decreased the Mg-chelatase activity and further reduced the production of Chl *a* and Chl *b* in Y plants.

Chl metabolism is a highly coordinated process, which is catalyzed by numerous enzymes. In addition to CHLH mentioned above, POR catalyzes the photoreduction of protochlorophyllide to chlorophyllide in Chl biosynthesis, which is a light-dependent enzyme that is present in all oxygen-producing photosynthetic organisms [46]. In our study, three genes encoding POR were significantly downregulated in Y compared with G (Figure 6A,C). Previous studies of *Arabidopsis* and rice have suggested that the Chl *a* content is directly related to the expression levels of *PORB and PORC*, and when *PORB* and *PORC* are absent in plants, the Chl *a* content is decreased and the thylakoid is not stacked [47,48]. Moreover, the *pgl10* mutant in rice has phenotypically pale-green leaves with significantly decreased Chl (Chl *a* and Chl *b*) and carotenoid contents, and less grana lamellae. Bioinformatics analysis indicates that *PGL10* encodes PORB [49]. Those results suggest that POR was inhibited in the Y type, which might have reduced the Chl content and led to the yellow leaf phenotype. Moreover, we found that one gene encoding Chlase in the Chl biosynthesis pathway was significantly upregulated in Y plants (Figure 6C). Chlase is considered to be a key enzyme in chlorophyll degradation [50,51]. However, some evidence does not support the involvement of Chlase in chlorophyll breakdown during leaf senescence [52–54]. For example, in *Arabidopsis*, *AtCHL1* and *AtCHL2* encode Chlase, but *CHL1* and *CHL2* single and double knockout mutants are still able to degrade chlorophyll during senescence [52]. Similarly, overexpression of the Chlase-encoding gene *ATHCOR1* in *Arabidopsis* leads to the increased breakdown of Chl *a*, but there are no substantial changes in the total amount of Chl [55]. In fact, Chlase has been shown to participate in chlorophyll breakdown in ethylene-treated citrus fruit, as well as in the tissue damage responses to fungi and bacteria [56–58]. The discovery of the involvement of pheophytin pheophorbide hydrolase (PPH) in early chlorophyll breakdown during leaf senescence [59] has made the biological role of Chlase controversial. Thus, the function and involvement of Chlase in Chl catabolic processes needs to be investigated further.

Carotenoids are essential components of the photosynthetic apparatus and photoprotection system, and their biosynthesis is coordinated with that of Chls in the chloroplasts [60]. In this study, HPLC analysis showed that the β-carotene and lutein levels were significantly lower in Y plants, whereas the zeaxanthin levels were more than 40 times higher than those in G plants (Figure 12 and Table 2). The changes in the abundances of carotenoids were accompanied by the altered expression of carotenoid biosynthesis genes [61]. β-carotene hydroxylase (BCH) is mainly responsible for the β-ring hydroxylation of β-carotene to produce zeaxanthin, and its activity overlaps slightly with the hydroxylation of the β-ring of α-carotene. [62]. In potato tubers, silencing the β-carotene hydroxylase genes CHY1 and CHY2 increases the levels of β-carotene and total carotenoids by up to 38 and 4.5 times, respectively, but reduces that of zeaxanthin [63]. By contrast, downregulating genes that encode BCH increases the β-carotene contents of sweet potato and the maize endosperm [64,65]. In our study, the expression levels of two genes that encode BCH were upregulated in Y plants (by ~1.5 and 1.8 time, respectively) (Figure 6B,C), which presumably enhanced the conversion of β-carotene into zeaxanthin, thereby increasing the zeaxanthin content and decreasing the β-carotene content. However, the increased zeaxanthin content was not converted into vioxanthin and neoxanthin, and their contents actually decreased significantly in Y plants (Figure 12). Zeaxanthin and violaxanthin are involved in the xanthophyll cycle, which is the main mechanism for photoprotection [66]. When plants are exposed to light stress, the photosynthetic organs are damaged and the conversion of violaxanthin into zeaxanthin is increased to protect against further light damage [67]. We suggest that the increase in zeaxanthin and the decrease in violaxanthin were related to damage to the photosynthetic system in Y plants, which enhanced the conversion of violaxanthin into zeaxanthin. We also found that the accumulation of lutein did not accompany the increased expression of BCH, and the lutein content actually decreased greatly in Y. This phenomenon has also been reported in transgenic tobacco, and it may be due to the limited effect of BCH on lutein [68]. In addition, significant reductions in the lutein content were also observed in previous studies in yellow-green tea mutants ZH1 and temperature-sensitive mutant Anji Baicha in the yellow-green stage [69,70].

3.2. Yellow Leaf Phenotype Affected by the Expression of Genes Related to Photosynthesis

Yellow leaf color formation is related to chloroplast development, where chloroplasts comprise the chloroplast membrane, thylakoid, and matrix, which are the main sites for photosynthesis. In higher plants, the multi-subunit pigment–protein complexes (i.e., photosystem (PS)I, PSII, light harvesting complexes, cytochrome b6/f, and ATP synthase) are embedded in the highly folded thylakoid membrane where they are responsible for light absorption and energy transfer [71,72]. The photosynthetic light reaction occurs in the thylakoid in chloroplasts, and thus we suggest that changes in gene expression might have been related to thylakoid development and photosynthesis in the two progeny of Ygm. Similar to the photosynthesis pathway, genes encoding PsbB (the transcriptome data indicated one upregulated gene and one downregulated gene, whereas both were downregulated according to qRT-PCR), PsbD, photosystem I subunit VII, and F-type ATPase β subunit were significantly repressed in Y plants (Figure 7). These results were consistent with our previous chloroplast ultrastructure analysis, which demonstrated that yellow leaf color formation is greatly affected by abnormal chloroplast development [42]. In addition, PSII is distributed mainly over the overlapping regions of the grana lamella [73], and the poorly stacked grana in Y might have been associated with the dramatic downregulation of the PSII protein complex [10,74,75]. In higher plants, ELIPs are light stress-induced, Chl a/b-binding proteins that accumulate in the thylakoid membranes, where their proposed function is in photoprotection [76–78]. The increased accumulation of ELIP transcripts and proteins is correlated with photodamage in the PSII reaction centers [79]. Overexpression of the ELIP2 gene in Arabidopsis downregulates the level and activity of glutamyl tRNA reductase, CHLH, and CHLI, thereby reducing the accumulation of Chl and photosystems assembled in the thylakoid membranes [80]. In yellow leaves, we found that the accumulation of *ELIP* mRNAs was 2–4 times higher compared with that in green leaves (Figure 7B), which suggests that the yellow phenotype may be related to the upregulation of ELIP transcripts. Rubisco is a rate-limiting enzyme that participates in photosynthetic carbon fixation and it is a potential target for genetic manipulation to increase crop yields [81,82]. Albino or pale green phenotypes are observed in *Arabidopsis* transgenic lines due to the co-suppression of *Arabidopsis* Rubisco small subunit gene *RBCS3B*, where among these lines, *RBCS3B-7* exhibits abnormal thylakoid stacking and it is light sensitive under normal light [83]. Furthermore, the abundance of Rubisco is lower in *Brassica napus* and a wheat anther culture Chl-deficient mutant [30,74]. Similarly, we found that genes encoding the Rubisco small chain were significantly downregulated in Y in our study (Figure 7B). The pentose phosphate metabolic (PPP) pathway is part of the carbon metabolism pathway and some of the PPP intermediates such as ribulose-5-phosphate (Ru5P) are required for the synthesis of nucleotides, and they are shared in the carbon fixation pathways of photosynthetic organisms [84,85]. In the carbon metabolism pathway, three genes encoding glucose-6-phosphate 1-dehydrogenase (G6PDH), 6PGL, and 6PGD in the oxidative phase of PPP (product: Ru5P and DADPH) were downregulated in Y (Figure 7B). Thus, the altered expression of genes related to carbon metabolism and photosynthesis in Y may have caused abnormal chloroplast development and decreased the Chl content.

3.3. HSFs, HSPs, and Chloroplasts

In this study, all of the HSFs and HSPs had decreased expression levels in Y (Table S4 and Figure 9B,C). HSPs are among the most abundant protective proteins in plants where there are five conserved protein families: HSP100s, HSP90s, HSP70s, HSP60s, and sHSPs [86]. These HSPs act as molecular chaperones that participate in protein folding, assembly, and the prevention of irreversible protein aggregation to maintain cell homeostasis [87–89]. Therefore, HSPs play key roles in plant development and stress resistance processes. HSP-encoding genes are regulated by HSFs by specifically binding highly conserved characteristic palindromic sequence (5′-AGAAnnTTCT-3′) in the promoters of many HSP genes, thereby causing the accumulation of HSPs [90–92]. Seven HSFs target gene prediction analyses demonstrated that a large number of HSP genes (sHSPs and HSP70) as well as some photosynthesis and Chl biosynthesis related genes were target genes regulated by those seven HSFs (Figure 9A).

Some sHSP family members play roles in chloroplast development and photosynthesis under heat stress in many species. For example, after silencing *sHSP26* in maize, the abundances of four chloroplast proteins comprising ATP synthase subunit β, Chl *a*/*b* binding protein, oxygen-evolving enhancer protein 1, and photosystem I reaction center subunit IV declined greatly under heat stress [93]. The overexpression of *Oshsp26* in tall fescue also enhanced the photochemical efficiency of PSII (*Fv/Fm*) during heat stress [94]. In *Arabidopsis*, the cooperation between HSP21 and pTAC5 is required for chloroplast development under heat stress [95]. These results suggest that the inhibition of sHSPs in Y plants may be caused defective chloroplasts and led to the yellow leaf phenotype. In addition, the involvement of HSP70s with chloroplast development has been observed in *Arabidopsis* and rice. For example, suppressing the HSP70s homologues *cpHsc70-1*/*cpHsc70-2* double genes in *Arabidopsis* resulted in a white and stunted phenotype, and the chloroplasts in these plants had an unusual morphology with few or no thylakoid membranes [96]. The phenotype of the T-DNA inserted heat-sensitive rice mutant *OsHsp70CP1* varies with temperature, where a severe chlorotic phenotype and lower Chl contents are found in the leaves under a constant high temperature (40 °C), whereas plants grown at a constant temperature of 27 °C have a normal phenotype [97]. Furthermore, HSP70 has been implicated in photoprotection and the repair of PSII during and after photoinhibition [98]. DnaJ proteins are chaperones in the HSP40 family, where the J domain is generally used as a DnaJ co-chaperone to activate the HSP70 ATPase domain to allow stable substrate binding and the release of HSP70 [99,100]. In addition, DnaJ proteins are essential for normal chloroplast development. Silencing of DnaJ encoded gene *OsDjA7/8* in rice resulted in an albino lethal phenotype in the seedling stage due to disordered development of chloroplast [101]. The detection of HSPs in our transcriptome data has further proven that the expression levels of *HSPs* are closely related to chloroplast development and Chl biosynthesis in plant species.

Chloroplasts are semi-autonomous organelles and a complex network of regulatory signals exists between the nucleus and plastids. Plastid retrograde signaling is mediated by the tetrapyrrole intermediate Mg-Proto IX, and its methylester (Mg-Proto IX-ME) was identified in an *Arabidopsis* genome-uncoupled mutant (*GUN5*) that encodes the plastid-localized CHLH [20,102,103]. Mg-Proto IX can replace the action of light by inducing two nuclear HSP genes (*HSP70A* and *HSP70B*) [104]. In the present study, the Mg-Proto IX content and the expression of *HSP70* were significantly decreased in Y (Figures 9C and 11), thereby indicating the possible involvement of Mg-Proto IX in the expression of HSP genes. Furthermore, in the photosynthetic system, the responses of HSPs to plastid retrograde signaling have important roles in regulating the expression of nuclear genes involved in photosynthesis. Kindgren et al. [105] found that HSP90 proteins respond to the GUN5-mediated plastid signal to control the expression of photosynthesis-associated nuclear genes (*PhANG*) during the response to oxidative stress. Recently, it was shown that *Arabidopsis* chloroplast HSP21 is activated by the GUN5-dependent retrograde signaling pathway to maintain the stability of the PSII complex and thylakoid membranes under high temperature stress by directly binding to its core subunits, such as D1 and D2 proteins [106]. These results suggest that the responses of HSPs and Mg-Proto IX to plastid signaling might have important roles in photosynthesis.

4. Materials and Methods

4.1. Plant Materials

Ygm was produced from a spontaneous leaf color mutant in the winter wheat cultivar Xinong1718 following 14 generations of self-pollination and direct selection for the yellow-green phenotype in each generation. The progeny of the *Ygm* mutant exhibited three leaf color phenotypes (i.e., yellow leaf plants (Y), yellow-green leaf plants (Yg), and normal green leaf plants (G)) (Figure 1). Yellow leaf plants die after the flowering stage and do not produce seeds, so this genotype was maintained by sowing seeds from the Yg plants each year. Yg plants are similar to the wild type Xinong1718 in terms of their growth period and plant height, but the yield capacity is significantly lower than that of Xinong1718. Green leaf plants are similar to the wild type Xinong1718 in terms of their growth period, plant height, and yield capacity [42]. In total, 66 plant line populations (20 and 46 derived from G and Yg plants, respectively) were sown on 2 October 2015, and 125 seeds from each plant line population were used in a single plot with a spacing of 30 cm between the rows and 8 cm between the plants, where each row measured 2 m. All of the experimental materials were grown in an experimental field at Northwest A&F University, Yangling, China, according to the standard practices employed in the local area.

The colors of the young leaves and leaf developmental states in Y and G were observed visually in the jointing stage in the field (Figure 1A). Three types of young leaves were present in the Y and G seedlings (i.e., fully-developed leaves (F_Y and F_G, respectively), half-developed leaves (H_Y and H_G, respectively), and small leaf buds (L_Y and L_G, respectively)) (Figure 1C,D). We only collected H_Y and H_G in the jointing stage from the Y and G seedlings, respectively, to measure the concentrations of Chl precursors and carotenoids compounds, as well as for transcriptome sequencing analysis with the Illumina HiSeq™ 2500 sequencing system (Illumina, San Diego, CA, USA).

4.2. RNA Extraction, Library Construction, and RNA Sequencing

For transcriptome analysis, H_Y and H_G leaves were collected from Y and G seedlings at the jointing stage (31 March 2017) from 08:00 am to 10:00 am (Figure 1C,D). Six samples from three biological replicates of G and Y were used to construct cDNA libraries designated as G-1, G-2, G-3, Y-1, Y-2, and Y-3, respectively. The samples were frozen immediately in liquid nitrogen and stored at $-80\,°C$ until RNA extraction. Total RNA was extracted using Trizol Reagent (Invitrogen Life Technologies, Shanghai, China) and treated with RNase-free DNase I (TaKaRa, Dalian, China) according to the manufacturer's instructions. The quality of the total RNA was confirmed with a NanoDrop ND1000 spectrophotometer (Thermo Scientific, Wilmington, DE, USA) coupled with 1% agarose gel electrophoresis. The RNA integrity value (>8.0) was also verified using an Agilent 2100 Bioanalyzer (Agilent Technologies, Santa Clara, CA, USA). The cDNA library construction and sequencing of six RNA samples were completed by Guangzhou GENE DENOVO Biotechnology Co., Ltd. (Guangzhou, China).

4.3. Sequence Alignment and Functional Annotation

In order to obtain clean data, it is essential to remove adaptor sequences, more than 10% of the unknown nucleotides, and low quality reads with more than 50% of low quality (q-value ≤ 20) bases. After removing rRNA using the short reads alignment tool Bowtie2 (version 2.2.9) [107], the high-quality clean reads were then mapped to the wheat reference genome sequences in the URGI database (http://wheat-urgi.versailles.inra.fr/Seq-Repository) by TopHat2 (version 2.0.3.12) [108]. In order to identify new genes and new splice variants of known genes, the transcripts were reconstructed using Cufflinks (version 2.2.1) [109] based on reference annotation based transcripts (RABT). Cufflinks constructed faux reads according to references to compensate for the influence of low coverage sequencing. During the last assembly step, all of the reassembled fragments were aligned with reference genes and similar fragments were then removed. Cuffmerge was then employed to combine the assembly results for three samples of each biological duplicate for Y and G for further downstream differential expression analysis. For functional annotation, all of the transcripts including new gene transcripts (≥ 200 bp and exon number >2) were annotated using the BLASTx function with protein databases, including the NCBI Nr protein database (https://ftp.ncbi.nlm.nih.gov/blast/db/FASTA/), GO (http://www.geneontology.org/), and KEGG (http://www.genome.jp/kegg/kegg2.html) databases with a significance threshold of E value $< 10^{-5}$. GO annotation was conducted using Blast2GO (vision 2.5) software [110] and WEGO software (version 2.0) was then applied for gain GO function classification [111]. KEGG pathway analyses were conducted using the KEGG Automatic Annotation Server (KAAS) (http://www.genome.jp/tools/kaas/).

4.4. DEG Analysis

The FPKM values were used as a measure of normalized gene expression, and significance tests of differences in gene expression in Y and G (each with three biological replicates) were performed using the edgeR package (http://www.bioconductor.org/packages/release/bioc/html/edgeR.html). FDR < 0.05 and $|\log_2 (\text{fold change})| \geq 1$ were used to assess the significance of differences in gene expression. Hierarchical clustering of the DEGs was performed using the OmicShare tools, which is a free online platform for data analysis (http://www.omicshare.com/tools).

The DEGs were then annotated with the COG database to predict and classify possible functions using BLASTx (E value $< 10^{-5}$). GO enrichment analysis of DEGs was implemented using the GOseq R package (Bioconductor version: release (3.7)) [112]. The enrichment of the DEGs in KEGG pathways was tested using the KOBAS software (version 2.0) [113]. GO terms and KEGG pathways with corrected q-values < 0.05 were considered significantly enriched for DEGs. The correlations between biological repeats in Y and G were expressed as Pearson's correlation coefficients.

4.5. qRT-PCR Analysis

In order to validate the results of the RNA-Seq and DEG analyses, the color-related DEGs were selected for qRT-PCR with specific primers designed using Primer Premier 5.0 software (Table S8). cDNA was synthesized according to the manufacturer's instructions using a PrimeScript™ RT reagent kit with gDNA Eraser (TaKaRa, Dalian, China). RNA (1 μg) extracted from the half-developed leaves was used as the template. qRT-PCR was performed using a SYBR Premix Ex Taq™ II Kit (TaKaRa,

Dalian, China) according to the manufacturer's instructions with a QuantStudio® 7 Flex Real-Time PCR system (Applied Biosystems, Shanghai, China). Wheat 18S rRNA was used as an internal control for normalization [114]. The amplification efficiencies of the primers were checked based on standard curve analysis. Each of the reactions was performed in triplicate. Dissociation curve analysis was performed after each assay to determine the target specificity. Relative gene expression levels were calculated according to the $2^{-\Delta\Delta Ct}$ comparative C_T method [115].

4.6. Determination of Photosynthetic Pigments and Chl Precursors

The Chl a, Chl b, and total carotenoid contents were determined using an UV-1800 spectrophotometer (Shanghai Mapada Instruments Co. Ltd., Shanghai, China) at 645, 663, and 470 nm according to the method described by Lichtenthaler [116]. The chlorophyll biosynthesis pathway is shown in Figure 6A. Seven precursors of chlorophyll biosynthesis were examined. The ALA contents were extracted and determined as described by Dei [117]. PBG, Urogen III, and Coprogen III were quantified as described by Bogorad [118]. Proto IX, Mg-Proto IX, and Pchlide were extracted according to the methods described by Rebeiz et al. [119]. The Proto IX, Mg-Proto IX, and protochlorophyllide contents were measured with a Hitachi F-4500 fluorescence spectrometer (Hitachi Instrument (Shanghai) Co., Ltd., Shanghai, China). The wavelengths used for detecting each porphyrin were as follows: Proto IX = excitation (Ex) 400 nm and emission (Em) 633 nm; Mg-Proto IX = Ex 440 nm and Em 595 nm; and protochlorophyllide = Ex 440 nm and Em 640 nm. Three individual plants were measured from the Y and G seedlings. The H_Y and H_G leaves from each plant in the jointing stage were extracted once and each sample was measured three times. Statistical analyses were performed using Microsoft Excel 2016 (Microsoft China, Beijing, China) with the one-way ANOVA test. The concentrations of pigments and Chl precursors in the G seedlings were set to 1, and the relative values for pigments and Chl precursors in the Y samples were expressed as fold changes relative to those in the G type samples.

4.7. Isolation and HPLC Analysis of Carotenoid Compounds

Carotenoids were extracted from half-developed wheat leaves from Y and G plants at the jointing stage according to the method described by Norris et al. [120] with appropriate modifications. All of the extraction procedures were conducted on ice with shielding from strong light. Briefly, 2 g of the fresh leaves were ground into a powder with liquid nitrogen and saponification was performed by adding 8 mL 20% w/v KOH and methanol. The homogenates were then transferred into 50 mL centrifuge tubes and heated at 60 °C for 30 min in darkness. After cooling to room temperature, each sample was ultrasonicated with 20 mL acetone:ethyl acetate (v:v = 2:1) for 40 min at 35 °C. The extract was then centrifuged at 8000× g at 4 °C for 5 min. The supernatant was then transferred to a fresh tube and concentrated using a nitrogen blowing instrument. The dried extract was re-suspended in 10 mL of acetone and ethyl acetate (v:v = 2:1) and then filtered through a 0.22 μm organic membrane for HPLC analysis.

Carotenoids were separated by reverse-phase HPLC analysis on a YMC C_{30} carotenoid column (150 mm × 4.6 mm, 3 μm) (YMC Co. Ltd., Shanghai, China) using a Shimadzu LC-20A HPLC system (Shimadzu, Tokyo, Japan). HPLC separation employed (A) methanol:acetonitrile (v:v = 3:1) and (B) methyl tert-butyl ether with a gradient of: 0–5 min, 0% B; 5–30 min, 0–35% B; 30–40 min, 35–45% B; 40–50 min, 45–0% B. The flow rate was 1 mL/min and the column temperature was 25 °C. The detection wavelength was 450 nm. β-carotene and zeaxanthin analytical standards were purchased from Sigma–Aldrich (Shanghai, China). β-cryptoxanthin and lutein analytical standards were purchased from Extrasynthese (Lyon, France). The α-carotene analytical standard was purchased from CaroteNature (Lupsingen, Switzerland). Each standard (1 mg) was dissolved in 1 mL of dimethyl sulfoxide. In order to establish a standard curve, mixed standard solutions of 4 μg/mL, 2 μg/mL, 1 μg/mL, 0.5 μg/mL, and 0.25 μg/mL were then prepared by diluting the stock solutions. Carotenoids were identified based on their retention time relative to known standards and absorption spectra for individual peaks compared with published spectra. Individual carotenoids were quantified based on the individual peak areas by using the standard curves and expressed as μg/g by fresh weight. Three biological replicates were analyzed in Y and G plants, and the carotenoid content was expressed as the mean ± SD based on three independent determinations.

5. Conclusions

In this study, transcriptome sequence analysis and physiological characterization were performed to identify the major molecular mechanisms related to leaf color variation in the mutant progeny of wheat *Ygm*. The transcriptome profiles differed considerably between Y and G, where various genes and pathways associated with yellow leaf formation were identified, including photosynthetic pigment synthesis, photosynthesis, and carbon fixation pathways. In addition, HSPs were shown to have important functions in response to plastid retrograde signaling and chloroplast development. Genes that interact with HSFs were shown to be associated with Chl biosynthesis. The measurements of Chl precursors indicated that the Y phenotype probably exhibited inhibited Chl biosynthesis due to the reduced activity of Mg-chelatase that is caused by the downregulation of *CHLH*. Moreover, the changes in the abundances of carotenoid composition may be associated with the yellow leaf phenotype. Overall, we speculated that the possible formation mechanism of yellow leaf phenotype in Y, which was shown in Figure 13. Our results provide new insights into the molecular mechanisms of yellow leaf formation in common wheat and they may facilitate selective breeding for high photosynthetic efficiency.

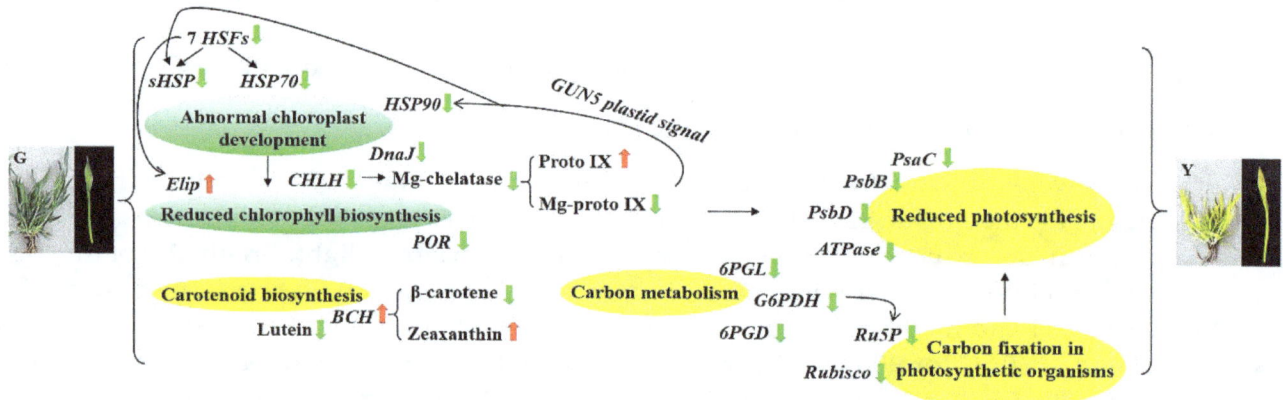

Figure 13. Possible formation pathway of yellow leaf phenotype of Y mutant. The red arrow indicates upregulated expression and the green arrow indicates downregulated expression. The green ovals indicate chlorophyll biosynthesis and chloroplast development. The yellow ovals indicate carotenoid biosynthesis, photosynthesis and energy metabolism.

Author Contributions: L.Z. and H.W. conceived the original screening and research plans; L.Z. supervised the experiments; H.W. performed most of the experiments, analyzed the data, and prepared the figures and tables; L.C., N.S., C.L., and X.A. provided technical assistance to H.W.; H.F. provided help and guidance in carotenoid measurement experiment. Y.F. and D.S. advised on the analysis and interpretation of the results. H.W. wrote the article with contributions from all the authors. All authors approved the final manuscript.

Acknowledgments: This work was sponsored by the National Sci-Tech Support Foundation of China (Grant No. 2013BAD01B02) and the Zhonging Tang Crop Breeding Foundation of Northwest A&F University. We thank Guangzhou GENE DENOVO Biotechnology Co., Ltd. (Guangzhou, China) for their assistance with data processing and bioinformatics analysis.

References

1.   Makino, A. Photosynthesis, grain yield, and nitrogen utilization in rice and wheat. *Plant Physiol.* **2011**, *155*, 125–129. [CrossRef] [PubMed]

2. Tanaka, A.; Tanaka, R. Chlorophyll metabolism. *Curr. Opin. Plant Biol.* **2006**, *9*, 248–255. [CrossRef] [PubMed]

3. Nagata, N.; Tanaka, R.; Satoh, S.; Tanaka, A. Identification of a vinyl reductase gene for chlorophyll synthesis in *Arabidopsis thaliana* and implications for the evolution of Prochlorococcus species. *Plant Cell* **2005**, *17*, 233–240. [CrossRef] [PubMed]

4. Li, W.; Tang, S.; Zhang, S.; Shan, J.; Tang, C.; Chen, Q.; Jia, G.; Han, Y.; Zhi, H.; Diao, X. Gene mapping and functional analysis of the novel leaf color gene SiYGL1 in foxtail millet [*Setaria italica* (L.) P. Beauv]. *Physiol. Plant* **2015**, *157*, 24–37. [CrossRef] [PubMed]

5. Wu, Z.; Zhang, X.; He, B.; Diao, L.; Sheng, S.; Wang, J.; Guo, X.; Su, N.; Wang, L.; Jiang, L.; et al. A chlorophyll-deficient rice mutant with impaired chlorophyllide esterification in chlorophyll biosynthesis. *Plant Physiol.* **2007**, *145*, 29–40. [CrossRef] [PubMed]

6. Zhu, X.; Shuang, G.; Zhongwei, W.; Qing, D.; Yadi, X.; Tianquan, Z.; Wenqiang, S.; Xianchun, S.; Yinghua, L.; Guanghua, H. Map-based cloning and functional analysis of YGL8, which controls leaf colour in rice (*Oryza sativa* L.). *BMC Plant Biol.* **2016**, *16*, 134. [CrossRef] [PubMed]

7. Jung, K.-H.; Hur, J.; Ryu, C.-H.; Choi, Y.; Chung, Y.-Y.; Miyao, A.; Hirochika, H.; An, G. Characterization of a Rice Chlorophyll-Deficient Mutant Using the T-DNA Gene-Trap System. *Plant Cell Physiol.* **2003**, *44*, 463–472. [CrossRef] [PubMed]

8. Zhao, S.; Long, W.; Wang, Y.; Liu, L.; Wang, Y.; Niu, M.; Zheng, M.; Wang, D.; Wan, J. A rice White-stripe leaf3 (wsl3) mutant lacking an HD domain-containing protein affects chlorophyll biosynthesis and chloroplast development. *J. Plant Biol.* **2016**, *59*, 282–292. [CrossRef]

9. Wang, Y.; He, Y.; Yang, M.; He, J.; Xu, P.; Shao, M.; Chu, P.; Guan, R. Fine mapping of a dominant gene conferring chlorophyll-deficiency in *Brassica napus*. *Sci. Rep.* **2016**, *6*, 31419. [CrossRef] [PubMed]

10. Kim, E.H.; Li, X.P.; Razeghifard, R.; Anderson, J.M.; Niyogi, K.K.; Pogson, B.J.; Chow, W.S. The multiple roles of light-harvesting chlorophyll a/b-protein complexes define structure and optimize function of *Arabidopsis* chloroplasts: A study using two chlorophyll b-less mutants. *Biochim. Biophys. Acta* **2009**, *1787*, 973–984. [CrossRef] [PubMed]

11. Braumann, I.; Stein, N.; Hansson, M. Reduced chlorophyll biosynthesis in heterozygous barley magnesium chelatase mutants. *Plant Physiol. Biochem.* **2014**, *78*, 10–14. [CrossRef] [PubMed]

12. Wu, Q.; Chen, Z.; Sun, W.; Deng, T.; Chen, M. De novo sequencing of the Leaf transcriptome reveals complex light-responsive regulatory networks in *Camellia sinensis* cv. *Baijiguan*. *Front. Plant Sci.* **2016**, *7*, 332. [CrossRef] [PubMed]

13. Zhang, H.; Li, J.; Yoo, J.H.; Yoo, S.C.; Cho, S.H.; Koh, H.J.; Seo, H.S.; Paek, N.C. Rice Chlorina-1 and Chlorina-9 encode ChlD and ChlI subunits of Mg-chelatase, a key enzyme for chlorophyll synthesis and chloroplast development. *Plant Mol. Biol.* **2006**, *62*, 325–337. [CrossRef] [PubMed]

14. Campoli, C.; Caffarri, S.; Svensson, J.T.; Bassi, R.; Stanca, A.M.; Cattivelli, L.; Crosatti, C. Parallel pigment and transcriptomic analysis of four barley Albina and Xantha mutants reveals the complex network of the chloroplast-dependent metabolism. *Plant Mol. Biol.* **2009**, *71*, 173–191. [CrossRef] [PubMed]

15. Campbell, B.W.; Mani, D.; Curtin, S.J.; Slattery, R.A.; Michno, J.-M.; Ort, D.R.; Schaus, P.J.; Palmer, R.G.; Orf, J.H.; Stupar, R.M. Identical Substitutions in Magnesium Chelatase Paralogs Result in Chlorophyll-Deficient Soybean Mutants. *G3 Genes Genomes Genet.* **2015**, *5*, 123–131. [CrossRef] [PubMed]

16. Papenbrock, J.; GrMe, S.; Kruse, E.; Hanel, F.; Grimm, B. Mg-chelatase of tobacco: Identification of a Chl D cDNA sequence encoding a third subunit, analysis of the interaction of the three subunits with the yeast two-hybrid system, and reconstitution of the enzyme activity by co-expression of recombinant CHL D, CHL H and CHL I. *Plant J.* **1997**, *12*, 981–990. [CrossRef] [PubMed]

17. Gibson, L.C.; Willows, R.D.; Kannangara, C.G.; Wettstein, D.V.; Hunter, C.N. Magnesium-protoporphyrin chelatase of Rhodobacter sphaeroides: Reconstitution of activity by combining the products of the bchH, -I, and -D genes expressed in *Escherichia coli*. *Proc. Natl. Acad. Sci. USA* **1995**, *92*, 1941–1944. [CrossRef] [PubMed]

18. Fodje, M.N.; Hansson, A.; Hansson, M.; Olsen, J.G.; Gough, S.; Willows, R.D.; Al-Karadaghi, S. Interplay Between an AAA Module and an Integrin I Domain May Regulate the Function of Magnesium Chelatase. *J. Mol. Biol.* **2001**, *311*, 111–122. [CrossRef] [PubMed]

19. Sirijovski, N.; Olsson, U.; Lundqvist, J.; Al-Karadaghi, S.; Willows, R.D.; Hansson, M. ATPase activity associated with the magnesium chelatase H-subunit of the chlorophyll biosynthetic pathway is an artefact. *Biochem. J.* **2006**, *400*, 477–484. [CrossRef] [PubMed]

20. Mochizuki, N.; Brusslan, J.A.; Larkin, R.; Nagatani, A.; Chory, J. *Arabidopsis* genomes uncoupled 5 (GUN5) mutant reveals the involvement of Mg-chelatase H subunit in plastid-to-nucleus signal transduction. *Proc. Natl. Acad. Sci. USA* **2000**, *98*, 2053–2058. [CrossRef] [PubMed]

21. Larkin, R.M.; Alonso, J.M.; Ecker, J.R.; Chory, J. GUN4, a Regulator of Chlorophyll Synthesis and Intracellular Signaling. *Science* **2003**, *299*, 902–906. [CrossRef] [PubMed]

22. Luo, T.; Luo, S.; Araujo, W.L.; Schlicke, H.; Rothbart, M.; Yu, J.; Fan, T.; Fernie, A.R.; Grimm, B.; Luo, M. Virus-induced gene silencing of pea *CHLI* and *CHLD* affects tetrapyrrole biosynthesis, chloroplast development and the primary metabolic network. *Plant Physiol. Biochem.* **2013**, *65*, 17–26. [CrossRef] [PubMed]

23. Hansson, A.; Willows, R.D.; Roberts, T.H.; Hansson, M. Three semidominant barley mutants with single amino acid substitutions in the smallest magnesium chelatase subunit form defective AAA$^+$ hexamers. *Proc. Natl. Acad. Sci. USA* **2002**, *99*, 13944–13949. [CrossRef] [PubMed]

24. Hansson, A.; Kannangara, C.G.; Wettstein, D.V.; Hansson, M. Molecular basis for semidominance of missense mutations in the XANTHA-H (42-kDa) subunit of magnesium chelatase. *Proc. Natl. Acad. Sci. USA* **1999**, *96*, 1744–1749. [CrossRef] [PubMed]

25. Fitzmaurice, W.P.; Nguyen, L.V.; Wernsman, E.A.; Thompson, W.F.; Conkling, M.A. Transposon tagging of the sulfur gene of tobacco using engineered maize *Ac/Ds* elements. *Genetics* **1999**, *153*, 1919–1928. [PubMed]

26. Sawers, R.J.; Viney, J.; Farmer, P.R.; Bussey, R.R.; Olsefski, G.; Anufrikova, K.; Hunter, C.N.; Brutnell, T.P. The maize *Oil yellow1 (Oy1)* gene encodes the I subunit of magnesium chelatase. *Plant Mol. Biol.* **2006**, *60*, 95–106. [CrossRef] [PubMed]

27. Williams, N.D.; Joppa, L.R.; Duysen, M.E.; Freeman, T.P. Inheritance of Three Chlorophyll-Deficient Mutants of Common Wheat. *Crop Sci.* **1985**, *25*, 1023–1025. [CrossRef]

28. Watanabe, N.; Koval, S.F. Mapping of chlorina mutant genes on the long arm of homoeologous group 7 chromosomes in common wheat with partial deletion lines. *Euphytica* **2003**, *129*, 259–265. [CrossRef]

29. Ansari, M.J.; Al-Ghamdi, A.; Kumar, R.; Usmani, S.; Al-Attal, Y.; Nuru, A.; Mohamed, A.A.; Singh, K.; Dhaliwal, H.S. Characterization and gene mapping of a chlorophyll-deficient mutant clm1 of *Triticum monococcum* L. *Biol. Plant.* **2013**, *57*, 442–448. [CrossRef]

30. Zhao, P.; Wang, K.; Zhang, W.; Liu, H.Y.; Du, L.P.; Hu, H.R.; Ye, X.G. Comprehensive analysis of differently expressed genes and proteins in albino and green plantlets from a wheat anther culture. *Biol. Plant.* **2017**, *61*, 255–265. [CrossRef]

31. Zhao, H.B.; Guo, H.J.; Zhao, L.S.; Gu, J.Y.; Zhao, S.R.; Li, J.H.; Liu, L.X. Agronomic Traits and Photosynthetic Characteristics of Chlorophyll-Deficient Wheat Mutant Induced by Spaceflight Environment. *Acta Agron. Sin.* **2011**, *37*, 119–126. [CrossRef]

32. Kosuge, K.; Watanabe, N.; Kuboyama, T. Comparative genetic mapping of homoeologous genes for the chlorina phenotype in the genus Triticum. *Euphytica* **2011**, *179*, 257–263. [CrossRef]

33. Brestic, M.; Zivcak, M.; Kunderlikova, K.; Sytar, O.; Shao, H.; Kalaji, H.M.; Allakhverdiev, S.I. Low PSI content limits the photoprotection of PSI and PSII in early growth stages of chlorophyll b-deficient wheat mutant lines. *Photosynth. Res.* **2015**, *125*, 151–166. [CrossRef] [PubMed]

34. Shi, K.; Gu, J.; Guo, H.; Zhao, L.; Xie, Y.; Xiong, H.; Li, J.; Zhao, S.; Song, X.; Liu, L. Transcriptome and proteomic analyses reveal multiple differences associated with chloroplast development in the spaceflight-induced wheat albino mutant mta. *PLoS ONE* **2017**, *12*, e0177992. [CrossRef] [PubMed]

35. Hou, D.Y.; Xu, H.; Du, G.Y.; Lin, J.T.; Duan, M.; Guo, A.G. Proteome analysis of chloroplast proteins in stage albinism line of winter wheat (*Triticum aestivum*) FA85. *BMB Rep.* **2009**, *42*, 450–455. [CrossRef] [PubMed]

36. Liu, X.G.; Xu, H.; Zhang, J.Y.; Liang, G.W.; Liu, Y.T.; Guo, A.G. Effect of low temperature on chlorophyll biosynthesis in albinism line of wheat (*Triticum aestivum*) FA85. *Physiol. Plant.* **2012**, *145*, 384–394. [CrossRef] [PubMed]

37. Falbel, T.G.; Staehelin, L.A. Characterization of a family of chlorophyll-deficient wheat (*Triticum*) and barley (*Hordeum vulgare*) mutants with defects in the magnesium-Insertion step of chlorophyll biosynthesis. *Plant Physiol.* **1994**, *104*, 639–648. [CrossRef] [PubMed]

38. Zhang, N.; Zhang, H.J.; Zhao, B.; Sun, Q.Q.; Cao, Y.Y.; Li, R.; Wu, X.X.; Weeda, S.; Li, L.; Ren, S.; et al. The RNA-seq approach to discriminate gene expression profiles in response to melatonin on cucumber lateral root formation. *J. Pineal Res.* **2014**, *56*, 39–50. [CrossRef] [PubMed]

39. Bellieny-Rabelo, D.; De Oliveira, E.A.; Ribeiro, E.S.; Costa, E.P.; Oliveira, A.E.; Venancio, T.M. Transcriptome analysis uncovers key regulatory and metabolic aspects of soybean embryonic axes during germination. *Sci. Rep.* **2016**, *6*, 36009. [CrossRef] [PubMed]

40. Yang, Y.; Chen, X.; Xu, B.; Li, Y.; Ma, Y.; Wang, G. Phenotype and transcriptome analysis reveals chloroplast development and pigment biosynthesis together influenced the leaf color formation in mutants of *Anthurium andraeanum* 'Sonate'. *Front. Plant Sci.* **2015**, *6*, 139. [CrossRef] [PubMed]

41. Li, Y.; Zhang, Z.; Wang, P.; Wang, S.A.; Ma, L.; Li, L.; Yang, R.; Ma, Y.; Wang, Q. Comprehensive transcriptome analysis discovers novel candidate genes related to leaf color in a *Lagerstroemia indica* yellow leaf mutant. *Genes Genom.* **2017**, *37*, 851–863. [CrossRef]

42. Zhang, L.L.; Liu, C.; An, X.Y.; Wu, H.Y.; Feng, Y.; Wang, H.; Sun, D.J. Identification and genetic mapping of a novel incompletely dominant yellow leaf color gene, *Y1718*, on chromosome 2BS in wheat. *Euphytica* **2017**, *213*, 141. [CrossRef]

43. Grant, C.E.; Bailey, T.L.; Noble, W.S. Fimo: Scanning for occurrences of a given motif. *Bioinformatics* **2011**, *27*, 1017–1018. [CrossRef] [PubMed]

44. Shannon, P.; Markiel, A.; Ozier, O.; Baliga, N.S.; Wang, J.T.; Ramage, D.; Amin, N.; Schwikowski, B.; Ideker, T. Cytoscape: A software environment for integrated models of biomolecular interaction networks. *Genome Res.* **2003**, *13*, 2498–2504. [CrossRef] [PubMed]

45. Willows, R.D.; Gibson, L.C.; Kanangara, C.G.; Hunter, C.N.; Wettstein, D. Three separate proteins constitute the magnesium chelatase of *Rhodobacter Sphaeroides*. *Eur. J. Biochem.* **1996**, *235*, 438–443. [CrossRef] [PubMed]

46. Masuda, T.; Takamiya, K. Novel insights into the enzymology, regulation and physiological functions of light-dependent protochlorophyllide oxidoreductase in angiosperms. *Photosynth. Res.* **2004**, *81*, 1–29. [CrossRef] [PubMed]

47. Su, Q.; Frick, G.; Armstrong, G.; Apel, K. POR C of *Arabidopsis thaliana*: A third light- and NADPH-dependent protochlorophyllide oxidoreductase that is differentially regulated by light. *Plant Mol. Biol.* **2001**, *47*, 805–813. [CrossRef] [PubMed]

48. Sakuraba, Y.; Rahman, M.L.; Cho, S.H.; Kim, Y.S.; Koh, H.J.; Yoo, S.C.; Paek, N.C. The rice faded green leaf locus encodes protochlorophyllide oxidoreductase B and is essential for chlorophyll synthesis under high light conditions. *Plant J.* **2013**, *74*, 122–133. [CrossRef] [PubMed]

49. Yang, Y.L.; Xu, J.; Rao, Y.C.; Zeng, Y.J.; Liu, H.J.; Zheng, T.T.; Zhang, G.-H.; Hu, J.; Guo, L.B.; Qian, Q.; et al. Cloning and functional analysis of pale-green leaf (PGL10) in rice (*Oryza sativa* L.). *Plant Growth Regul.* **2016**, *78*, 69–77. [CrossRef]

50. Takamiya, K.-I.; Tsuchiya, T.; Ohta, H. Degradation pathway(s) of chlorophyll: What has gene cloning revealed? *Trends Plant Sci.* **2000**, *5*, 426–431. [CrossRef]

51. Harpaz-Saad, S.; Azoulay, T.; Arazi, T.; Ben-Yaakov, E.; Mett, A.; Shiboleth, Y.M.; Hortensteiner, S.; Gidoni, D.; Gal-On, A.; Goldschmidt, E.E.; et al. Chlorophyllase is a rate-limiting enzyme in chlorophyll catabolism and is posttranslationally regulated. *Plant Cell* **2007**, *19*, 1007–1022. [CrossRef] [PubMed]

52. Schenk, N.; Schelbert, S.; Kanwischer, M.; Goldschmidt, E.E.; Dormann, P.; Hortensteiner, S. The chlorophyllases ATCLH1 and ATCLH2 are not essential for senescence-related chlorophyll breakdown in *Arabidopsis thaliana*. *FEBS Lett.* **2007**, *581*, 5517–5525. [CrossRef] [PubMed]

53. Hu, X.; Makita, S.; Schelbert, S.; Sano, S.; Ochiai, M.; Tsuchiya, T.; Hasegawa, S.F.; Hortensteiner, S.; Tanaka, A.; Tanaka, R. Reexamination of chlorophyllase function implies its involvement in defense against chewing herbivores. *Plant Physiol.* **2015**, *167*, 660–670. [CrossRef] [PubMed]

54. Hortensteiner, S. Chlorophyll degradation during senescence. *Annu. Rev. Plant Biol.* **2006**, *57*, 55–77. [CrossRef] [PubMed]

55. Benedetti, C.E.; Arruda, P. Altering the expression of the chlorophyllase gene ATHCOR1 in transgenic *Arabidopsis* caused changes in the chlorophyll-to-chlorophyllide ratio. *Plant Physiol.* **2002**, *128*, 1255–1263. [CrossRef] [PubMed]

56. Kariola, T.; Brader, G.; Li, J.; Palva, E.T. Chlorophyllase 1, a damage control enzyme, affects the balance between defense pathways in plants. *Plant Cell* **2005**, *17*, 282–294. [CrossRef] [PubMed]

57. Jacob-Wilk, D.; Goldschmidt, D.H.E.E.; Riov, J.; Eyal, Y. Chlorophyll breakdown by chlorophyllase: Isolation and functional expression of the Chlase1 gene from ethylene-treated Citrus fruit and its regulation during development. *Plant J.* **1999**, *20*, 653–661. [CrossRef] [PubMed]

58. Azoulay Shemer, T.; Harpaz-Saad, S.; Belausov, E.; Lovat, N.; Krokhin, O.; Spicer, V.; Standing, K.G.; Goldschmidt, E.E.; Eyal, Y. Citrus chlorophyllase dynamics at ethylene-induced fruit color-break: A study of chlorophyllase expression, posttranslational processing kinetics, and in situ intracellular localization. *Plant Physiol.* **2008**, *148*, 108–118. [CrossRef] [PubMed]

59. Schelbert, S.; Aubry, S.; Burla, B.; Agne, B.; Kessler, F.; Krupinska, K.; Hortensteiner, S. Pheophytin pheophorbide hydrolase (pheophytinase) is involved in chlorophyll breakdown during leaf senescence in *Arabidopsis*. *Plant Cell* **2009**, *21*, 767–785. [CrossRef] [PubMed]

60. Demmig-Adams, B.; Adams, W.W., III. The role of xanthophyll cycle carotenoids in the protection of photosynthesis. *Trends Plant Sci.* **1996**, *1*, 21–26. [CrossRef]

61. Zhang, L.; Ma, G.; Kato, M.; Yamawaki, K.; Takagi, T.; Kiriiwa, Y.; Ikoma, Y.; Matsumoto, H.; Yoshioka, T.; Nesumi, H. Regulation of carotenoid accumulation and the expression of carotenoid metabolic genes in citrus juice sacs in vitro. *J. Exp. Bot.* **2012**, *63*, 871–886. [CrossRef] [PubMed]

62. Kim, J.; Smith, J.J.; Tian, L.; Dellapenna, D. The evolution and function of carotenoid hydroxylases in *Arabidopsis*. *Plant Cell Physiol.* **2009**, *50*, 463–479. [CrossRef] [PubMed]

63. Diretto, G.; Welsch, R.; Tavazza, R.; Mourgues, F.; Pizzichini, D.; Beyer, P.; Giuliano, G. Silencing of beta-carotene hydroxylase increases total carotenoid and beta-carotene levels in potato tubers. *BMC Plant Biol.* **2007**, *7*, 11. [CrossRef] [PubMed]

64. Kim, S.H.; Ahn, Y.O.; Ahn, M.J.; Lee, H.S.; Kwak, S.S. Down-regulation of beta-carotene hydroxylase increases beta-carotene and total carotenoids enhancing salt stress tolerance in transgenic cultured cells of sweetpotato. *Phytochemistry* **2012**, *74*, 69–78. [CrossRef] [PubMed]

65. Berman, J.; Zorrilla-Lopez, U.; Sandmann, G.; Capell, T.; Christou, P.; Zhu, C. The silencing of carotenoid beta-hydroxylases by RNA interference in different maize genetic backgrounds increases the beta-carotene content of the endosperm. *Int. J. Mol. Sci.* **2017**, *18*. [CrossRef] [PubMed]

66. Latowski, D.; Grzyb, J.; Strzałka, K. The xanthophyll cycle-molecular mechanism and physiological significance. *Acta Physiol. Plant.* **2004**, *26*, 197. [CrossRef]

67. Demmig-Adams, B.; Adams, W.W. Photoprotection and other responses of plants to high light stress. *Annu. Rev. Plant Physiol. Plant Mol. Biol.* **1992**, *43*, 599–626. [CrossRef]

68. Zhao, Q.; Wang, G.; Ji, J.; Jin, C.; Wu, W.; Zhao, J. Over-expression of *Arabidopsis thaliana* β-carotene hydroxylase (*chyB*) gene enhances drought tolerance in transgenic tobacco. *J. Plant Biochem. Biotechnol.* **2014**, *23*, 190–198. [CrossRef]

69. Wang, L.; Cao, H.; Chen, C.; Yue, C.; Hao, X.; Yang, Y.; Wang, X. Complementary transcriptomic and proteomic analyses of a chlorophyll-deficient tea plant cultivar reveal multiple metabolic pathway changes. *J. Proteom.* **2016**, *130*, 160–169. [CrossRef] [PubMed]

70. Li, C.F.; Xu, Y.X.; Ma, J.Q.; Jin, J.Q.; Huang, D.J.; Yao, M.Z.; Ma, C.L.; Chen, L. Biochemical and transcriptomic analyses reveal different metabolite biosynthesis profiles among three color and developmental stages in 'Anji Baicha' (*Camellia sinensis*). *BMC Plant Biol.* **2016**, *16*, 195. [CrossRef] [PubMed]

71. Bashir, H.; Qureshi, M.I.; Ibrahim, M.M.; Iqbal, M. Chloroplast and photosystems: Impact of cadmium and iron deficiency. *Photosynthetica* **2015**, *53*, 321–335. [CrossRef]

72. Allen, J.F.; de Paula, W.B.; Puthiyaveetil, S.; Nield, J. A structural phylogenetic map for chloroplast photosynthesis. *Trends Plant Sci.* **2011**, *16*, 645–655. [CrossRef] [PubMed]

73. Albertsson, P. A quantitative model of the domain structure of the photosynthetic membrane. *Trends Plant Sci.* **2001**, *6*, 349–354. [CrossRef]

74. Chu, P.; Yan, G.X.; Yang, Q.; Zhai, L.N.; Zhang, C.; Zhang, F.Q.; Guan, R.Z. iTRAQ-based quantitative proteomics analysis of *Brassica napus* leaves reveals pathways associated with chlorophyll deficiency. *J. Proteom.* **2015**, *113*, 244–259. [CrossRef] [PubMed]

75. Ma, C.; Cao, J.; Li, J.; Zhou, B.; Tang, J.; Miao, A. Phenotypic, histological and proteomic analyses reveal multiple differences associated with chloroplast development in yellow and variegated variants from *Camellia sinensis*. *Sci. Rep.* **2016**, *6*, 33369. [CrossRef] [PubMed]

76. Hutin, C.; Nussaume, L.; Moise, N.; Moya, I.; Kloppstech, K.; Havaux, M. Early light-induced proteins protect *Arabidopsis* from photooxidative stress. *Proc. Natl. Acad. Sci. USA* **2003**, *100*, 4921–4926. [CrossRef] [PubMed]

77. Beck, J.; Lohscheider, J.N.; Albert, S.; Andersson, U.; Mendgen, K.W.; Rojas-Stutz, M.C.; Adamska, I.; Funck, D. Small One-Helix Proteins Are Essential for Photosynthesis in *Arabidopsis*. *Front. Plant Sci.* **2017**, *8*, 7. [CrossRef] [PubMed]

78. Adamska, I.; Kruse, E.; Kloppstech, K. Stable insertion of the early light-induced proteins into etioplast membranes requires chlorophyll A. *J. Biol. Chem.* **2001**, *276*, 8582–8587. [CrossRef] [PubMed]

79. Heddad, M.; Noren, H.; Reiser, V.; Dunaeva, M.; Andersson, B.; Adamska, I. Differential expression and localization of early light-induced proteins in *Arabidopsis*. *Plant Physiol.* **2006**, *142*, 75–87. [CrossRef] [PubMed]

80. Tzvetkova-Chevolleau, T.; Franck, F.; Alawady, A.E.; Dall'Osto, L.; Carrière, F.; Bassi, R.; Grimm, B.; Nussaume, L.; Havaux, M. The light stress-induced protein ELIP2 is a regulator of chlorophyll synthesis in *Arabidopsis thaliana*. *Plant J.* **2007**, *50*, 795–809. [CrossRef] [PubMed]

81. Parry, M.; Andralojc, P.J.; Scales, J.C.; Salvucci, M.E.; Carmo-Silva, A.E.; Alonso, H.; Whitney, S.M. Rubisco activity and regulation as targets for crop improvement. *J. Exp. Bot.* **2013**, *64*, 717–730. [CrossRef] [PubMed]

82. Andersson, I.; Backlund, A. Structure and function of Rubisco. *Plant Physiol. Biochem.* **2008**, *46*, 275–291. [CrossRef] [PubMed]

83. Zhan, G.M.; Li, R.J.; Hu, Z.Y.; Liu, J.; Deng, L.B.; Lu, S.Y.; Hua, W. Cosuppression of RBCS3B in *Arabidopsis* leads to severe photoinhibition caused by ROS accumulation. *Plant Cell Rep.* **2014**, *33*, 1091–1108. [CrossRef] [PubMed]

84. Esposito, S. Nitrogen assimilation, Abiotic stress and glucose 6-phosphate dehydrogenase: The full circle of reductants. *Plants* **2016**, *5*, 24. [CrossRef] [PubMed]

85. Berg, J.; Tymoczko, J.; Stryer, L. 20.3 the pentose phosphate pathway generates NADPH and synthesizes five-carbon sugars. In *Biochemistry*, 5th ed.; W H Freeman: New York, NY, USA, 2002; ISBN-10 0-7167-3051-0.

86. Waters, E.R.; Lee, G.J.; Vierling, E. Evolution, structure and function of the small heat shock proteins in plants. *J. Exp. Bot.* **1996**, *47*, 325–338. [CrossRef]

87. Waters, E.R. The evolution, function, structure, and expression of the plant sHSPs. *J. Exp. Bot.* **2013**, *64*, 391–403. [CrossRef] [PubMed]

88. Wang, W.; Vinocur, B.; Shoseyov, O.; Altman, A. Role of plant heat-shock proteins and molecular chaperones in the abiotic stress response. *Trends Plant Sci.* **2004**, *9*, 244–252. [CrossRef] [PubMed]

89. Timperio, A.M.; Egidi, M.G.; Zolla, L. Proteomics applied on plant abiotic stresses: Role of heat shock proteins (HSP). *J. Proteom.* **2008**, *71*, 391–411. [CrossRef] [PubMed]

90. Xue, G.P.; Drenth, J.; McIntyre, C.L. TaHsfA6f is a transcriptional activator that regulates a suite of heat stress protection genes in wheat (*Triticum aestivum* L.) including previously unknown Hsf targets. *J. Exp. Bot.* **2015**, *66*, 1025–1039. [CrossRef] [PubMed]

91. Westerheidea, S.D.; Raynesa, R.; Powella, C.; Xueb, B.; Uversky, V.N. HSF transcription factor family, heat shock response, and protein intrinsic disorder. *Curr. Protein Pept. Sci.* **2012**, *13*, 86–103. [CrossRef]

92. Nover, L.; Scharf, K.D. Heat stress proteins and transcription factors. *Cell. Mol. Life Sci.* **1997**, *53*, 80–103. [CrossRef] [PubMed]

93. Hu, X.; Yang, Y.; Gong, F.; Zhang, D.; Zhang, L.; Wu, L.; Li, C.; Wang, W. Protein sHSP26 improves chloroplast performance under heat stress by interacting with specific chloroplast proteins in maize (*Zea mays*). *J. Proteom.* **2015**, *115*, 81–92. [CrossRef] [PubMed]

94. Kim, K.H.; Alam, I.; Kim, Y.G.; Sharmin, S.A.; Lee, K.W.; Lee, S.H.; Lee, B.H. Overexpression of a chloroplast-localized small heat shock protein OsHSP26 confers enhanced tolerance against oxidative and heat stresses in tall fescue. *Biotechnol. Lett.* **2012**, *34*, 371–377. [CrossRef] [PubMed]

95. Zhong, L.; Zhou, W.; Wang, H.; Ding, S.; Lu, Q.; Wen, X.; Peng, L.; Zhang, L.; Lu, C. Chloroplast small heat shock protein HSP21 interacts with plastid nucleoid protein pTAC5 and is essential for chloroplast development in *Arabidopsis* under heat stress. *Plant Cell* **2013**, *25*, 2925–2943. [CrossRef] [PubMed]

96. Latijnhouwers, M.; Xu, X.M.; Møller, S.G. *Arabidopsis* stromal 70-kDa heat shock proteins are essential for chloroplast development. *Planta* **2010**, *232*, 567–578. [CrossRef] [PubMed]

97. Kim, S.R.; An, G. Rice chloroplast-localized heat shock protein 70, OsHsp70CP1, is essential for chloroplast development under high-temperature conditions. *J. Plant Physiol.* **2013**, *170*, 854–863. [CrossRef] [PubMed]

98. Schroda, M.; Vallon, O.; Wollman, F.A.; Beck, C.F. A chloroplast-targeted heat shock protein 70 (HSP70) contributes to the photoprotection and repair of photosystem II during and after photoinhibition. *Plant Cell* **1999**, *11*, 1165–1178. [CrossRef] [PubMed]

99. Greene, M.K.; Maskos, K.; Landry, S.J. Role of the J-domain in the cooperation of HSP40 with HSP70. *Proc. Natl. Acad. Sci. USA.* **1998**, *95*, 6108–6113. [CrossRef] [PubMed]

100. Cheetham, M.E.; Caplan, A.J. Structure, function and evolution of DnaJ: Conservation and adaptation of chaperone function. *Cell Stress Chaperones* **1998**, *3*, 28–36. [CrossRef]

101. Zhu, X.; Liang, S.; Yin, J.; Yuan, C.; Wang, J.; Li, W.; He, M.; Wang, J.; Chen, W.; Ma, B.; et al. The DnaJ OsDjA7/8 is essential for chloroplast development in rice (*Oryza sativa*). *Gene* **2015**, *574*, 11–19. [CrossRef] [PubMed]

102. Strand, A.; Asami, T.; Alonso, J.; Chory, J. Chloroplast to nucleus communication triggered by accumulation of Mg-ProtoporphyrinIX. *Nature* **2003**, *421*, 79–83. [CrossRef] [PubMed]

103. Chi, W.; Sun, X.; Zhang, L. Intracellular signaling from plastid to nucleus. *Annu. Rev. Plant Biol.* **2013**, *64*, 559–582. [CrossRef] [PubMed]

104. Kropat, J.; Oster, U.; Rüdiger, W.; Beck, C.F. Chlorophyll precursors are signals of chloroplast origin involved in light induction of nuclear heat-shock genes. *Proc. Natl. Acad. Sci. USA* **1997**, *94*, 14168–14172. [CrossRef] [PubMed]

105. Kindgren, P.; Noren, L.; Lopez Jde, D.; Shaikhali, J.; Strand, A. Interplay between heat shock protein 90 and HY5 controls *phANG* expression in response to the GUN5 plastid signal. *Mol. Plant* **2012**, *5*, 901–913. [CrossRef] [PubMed]

106. Chen, S.T.; He, N.Y.; Chen, J.H.; Guo, F.Q. Identification of core subunits of photosystem II as action sites of HSP21, which is activated by the GUN5-mediated retrograde pathway in *Arabidopsis*. *Plant J.* **2017**, *89*, 1106–1118. [CrossRef] [PubMed]

107. Langmead, B.; Salzberg, S.L. Fast gapped-read alignment with Bowtie 2. *Nat. Methods* **2012**, *9*, 357–359. [CrossRef] [PubMed]

108. Kim, D.; Pertea, G.; Trapnell, C.; Pimentel, H.; Kelley, R.; Salzberg, S.L. TopHat2: Accurate alignment of transcriptomes in the presence of insertions, deletions and gene fusions. *Genome Biol.* **2013**, *14*, R36. [CrossRef] [PubMed]

109. Trapnell, C.; Williams, B.A.; Pertea, G.; Mortazavi, A.; Kwan, G.; van Baren, M.J.; Salzberg, S.L.; Wold, B.J.; Pachter, L. Transcript assembly and quantification by RNA-Seq reveals unannotated transcripts and isoform switching during cell differentiation. *Nat. Biotechnol.* **2010**, *28*, 511–515. [CrossRef] [PubMed]

110. Conesa, A.; Gotz, S. Blast2GO: A comprehensive suite for functional analysis in plant genomics. *Int. J. Plant Genom.* **2008**, *2008*. [CrossRef] [PubMed]

111. Ye, J.; Fang, L.; Zheng, H.; Zhang, Y.; Chen, J.; Zhang, Z.; Wang, J.; Li, S.; Li, R.; Bolund, L.; et al. WEGO: A web tool for plotting GO annotations. *Nucleic Acids Res.* **2006**, *34*, W293–W297. [CrossRef] [PubMed]

112. Young, M.D.; Wakefield, M.J.; Smyth, G.K.; Oshlack, A. Gene ontology analysis for RNA-seq: Accounting for selection bias. *Genome Biol.* **2010**, *11*, R14. [CrossRef] [PubMed]

113. Wu, J.; Mao, X.; Cai, T.; Luo, J.; Wei, L. Kobas server: A web-based platform for automated annotation and pathway identification. *Nucleic Acids Res.* **2006**, *34*, W720–W724. [CrossRef] [PubMed]

114. Wang, G.P.; Hou, W.Q.; Zhang, L.; Wu, H.Y.; Zhao, L.F.; Du, X.Y.; Ma, X.; Li, A.F.; Wang, H.W.; Kong, L.R. Functional analysis of a wheat pleiotropic drug resistance gene involved in Fusarium head blight resistance. *J. Integr. Agric.* **2016**, *15*, 2215–2227. [CrossRef]

115. Schmittgen, T.D.; Livak, K.J. Analyzing real-time PCR data by the comparative C_T method. *Nat. Protoc.* **2008**, *3*, 1101–1108. [CrossRef] [PubMed]

116. Lichtenthaler, H.K. Chlorophylls and carotenoids: Pigments of photosynthetic biomembranes. *Methods Enzymol.* **1987**, *148*, 350–382. [CrossRef]

117. Dei, M. Benzyladenine-induced stimulation of 5-aminolevulinic acid accumulation under various light intensities in levulinic acid-treated cotyledons of etiolated cucumber. *Physiol. Plant.* **1985**, *64*, 153–160. [CrossRef]

118. Bogorad, L. Porphyrin synthesis. In *Methods in Enzymology*; Colowick, S.P., Kaplan, N.O., Eds.; Academic Press: New York, NY, USA, 1962; Volume 5, pp. 885–895, ISBN 0076-6879.

119. Rebeiz, C.A.; Mattheis, J.R.; Smith, B.B.; Rebeiz, C.C.; Dayton, D.F. Chloroplast biogenesis: Biosynthesis and accumulation of protochlorophyll by isolated etioplasts and developing chloroplasts. *Arch. Biochem. Biophys.* **1975**, *171*, 549–567. [CrossRef]

120. Norris, S.R.; Barrette, T.R.; DellaPenna, D. Genetic dissection of camtenoid synthesis in *Arabidopsis* defines plastoquinone as an essential component of phytoene desaturation. *Plant Cell* **1995**, *7*, 2139–2149. [CrossRef] [PubMed]

6

Bacterial Heterologous Expression System for Reconstitution of Chloroplast Inner Division Ring and Evaluation of its Contributors

Hiroki Irieda [1],* **and Daisuke Shiomi** [2],*

[1]   Academic Assembly, Institute of Agriculture, Shinshu University, Nagano 399-4598, Japan
[2]   Department of Life Science, College of Science, Rikkyo University, Tokyo 171-8501, Japan
*   Correspondence: irieda@shinshu-u.ac.jp (H.I.); dshiomi@rikkyo.ac.jp (D.S.)

Abstract: Plant chloroplasts originate from the symbiotic relationship between ancient free-living cyanobacteria and ancestral eukaryotic cells. Since the discovery of the bacterial derivative *FtsZ* gene—which encodes a tubulin homolog responsible for the formation of the chloroplast inner division ring (Z ring)—in the *Arabidopsis* genome in 1995, many components of the chloroplast division machinery were successively identified. The knowledge of these components continues to expand; however, the mode of action of the chloroplast dividing system remains unknown (compared to bacterial cell division), owing to the complexities faced in in planta analyses. To date, yeast and bacterial heterologous expression systems have been developed for the reconstitution of Z ring-like structures formed by chloroplast FtsZ. In this review, we especially focus on recent progress of our bacterial system using the model bacterium *Escherichia coli* to dissect and understand the chloroplast division machinery—an evolutionary hybrid structure composed of both bacterial (inner) and host-derived (outer) components.

Keywords: chloroplast division; Z ring; membrane-tethering; heterologous expression; *E. coli*; AtFtsZ1; AtFtsZ2; ARC6; ARC3

---

1. Introduction

Plant chloroplasts evolved from free-living cyanobacteria through primary endosymbiosis, which started with an engulfment of ancient cyanobacteria by ancestral eukaryotic cells approximately a billion years ago [1,2]. As in bacteria, the proliferation of chloroplasts is achieved by binary fission via a hybrid division machinery comprised of inner (stromal) bacterial and outer (cytosolic) host-derived elements [3–5]. This machinery mainly consists of four rings: two (the Z ring and the inner plastid-dividing (PD) ring) are inside, while the other two (the dynamin-related protein5B (DRP5B) ring and the outer PD ring) are outside the chloroplast [3,6,7]. Of those, the Z ring is broadly conserved from bacteria to chloroplasts [3,4,8–11]. Each Z ring is composed of FtsZ homolog tubulin-like GTPase, and is believed to generate a constrictive force for division both in bacteria and chloroplasts of *Arabidopsis thaliana* (although it has also been reported that the motive force was provided by the outer DRP5B ring, but not inner Z ring, in chloroplast of the red alga *Cyanidioschyzon merolae*) [8–10,12–15]. While the four rings and other division-related components might coordinately function to constrict the mid-chloroplast, the initial and critical event is the assembly of FtsZ into the Z ring just beneath the inner envelope membrane (IEM) at the future site of division. The Z ring then works as a scaffold, to which other components are recruited in a specific order to drive a division complex [4].

In contrast to the bacterial Z ring that is composed of a single FtsZ protein, the components of the chloroplast Z ring are two phylogenetically distinct FtsZ proteins, FtsZ1 and FtsZ2, which heteropolymerize into FtsZ filaments in vivo and in vitro [15–19]. FtsZ1 probably emerged from FtsZ2 through a gene duplication event, because the C-terminal amino acid sequence (which is critical for the membrane-tethering in bacterial FtsZ—see below) is conserved only in FtsZ2 [20,21]. FtsZ1 and FtsZ2 show high amino acid sequence identity and similarity regarding their GTPase core domain with bacterial FtsZ, but play distinct roles in the formation of the FtsZ polymers; FtsZ2 dominantly forms the backbone of the filament, while FtsZ1 assists in its remodeling [16,19,22]. In *A. thaliana*, FtsZ2 was additionally duplicated into two functionally redundant paralogs: AtFtsZ2-1 and AtFtsZ2-2. These two paralogs are functionally interchangeable with respect to in vivo chloroplast division activity, although the distinct contributions by these two AtFtsZ2 conforming to the shape of the chloroplast have also been reported [22,23].

The assembly and dynamics of the chloroplast Z ring is elaborately regulated by many components that negatively or positively affect the Z ring formation [3–5]. The stromal protein Accumulation and Replication of Chloroplasts 3 (ARC3) directly interacts with both AtFtsZ1 and AtFtsZ2 and inhibits the assembly of the Z ring at non-division sites [19,24–26], resembling the function of bacterial division inhibitor protein MinC for the positioning of the Z ring [27], whereas the IEM-spanning protein Accumulation and Replication of Chloroplasts 6 (ARC6) directly interacts only with AtFtsZ2 and promotes Z ring assembly in the stroma [22,26,28,29]. ARC3 and ARC6 were identified in the native AtFtsZ1-AtFtsZ2 complex isolated from *Arabidopsis* chloroplast [30]. Another IEM-spanning protein Paralog of ARC6 (PARC6) also directly interacts with AtFtsZ2, and in addition to the stromal proteins MinD and MinE, indirectly affects the Z ring formation through a direct interaction with the inhibitor protein ARC3 [25,26,31–34]. Briefly, in the working model of Z ring regulation in chloroplast division, ARC6 promotes the formation of a Z ring composed of AtFtsZ1-AtFtsZ2 heteropolymer, possibly by the tethering of AtFtsZ2 to the IEM. ARC3, MinD, and MinE act together as a Z ring positioning system and accurately confine the Z ring to the mid-chloroplast. During the remodeling and constriction of the Z ring, ARC3 may also function as an inhibitor of Z ring assembly after being recruited by PARC6 to the division site [3–5]. For an in-depth review of the many contributors to the chloroplast Z ring dynamics that include bacterial- and host-derivatives, we refer readers to previously published reviews [3–5].

To understand the chloroplast division comprehensively, in planta molecular analysis of Z ring assembly and dynamics is important. However, it can be challenging owing to the complexity of the plant cell, wherein many division-related components act together. Furthermore, plant breeding and genetic manipulation require more time compared to model microorganisms, even in the model plant *A. thaliana*. This situation has led to the development of some heterologous expression systems for *Arabidopsis* FtsZ proteins and other related components using single-celled model microorganisms, such as the yeasts *Schizosaccharomyces pombe* and *Pichia pastoris*, as well as the bacterium *Escherichia coli* [15,19,26,35]. The fission yeast *S. pombe* system was established as a cellular model for the functional analysis of bacterial actin-related protein MreB and FtsZ ahead of chloroplast FtsZ [36,37]. At present, together with recent methylotrophic yeast *P. pastoris* system, the yeast systems have shown the value of using heterologous expression systems for chloroplast division-related proteins—particularly filament and ring formation by FtsZs—to analyze their inherent functions [15,19,26,35]. On the other hand, based on the evolutionary background of the chloroplast and the fact that the Z ring-driven division system indeed functions in bacteria, as well as other practical advantages of a model bacterium, the *E. coli* system could be a good tool for the research of chloroplast FtsZ. However, the previous report showed that the chloroplast FtsZ produced in *E. coli* cells did not successfully form the Z ring or Z ring-like structure, but only formed long filaments and aberrant clusters; therefore, this system is lagging behind yeast expression systems [15,19,26,35].

Recently, we progressively developed the *E. coli* system to reconstitute Z ring or Z ring-like structures composed of the *A. thaliana* FtsZ protein AtFtsZ2-1 (hereafter called AtFtsZ2) [38]. Our system plausibly reflects the dynamic properties of AtFtsZ2, where the AtFtsZ2 assembles into long filaments or Z ring-like structures depending on the conditions. In Figure 1, we summarize the proposed molecular mechanism of action of AtFtsZ2 and its positive contributor ARC6, which has been demonstrated to anchor the chloroplast Z ring to the membrane in our system. In the following sections, we describe the development of this system and the important factors contributing to the filament morphology of AtFtsZ2, which includes N-terminal extended region of AtFtsZ2, membrane-tethering of the AtFtsZ2 filament, the negative regulator ARC3, and the positive regulators ARC6 and AtFtsZ1.

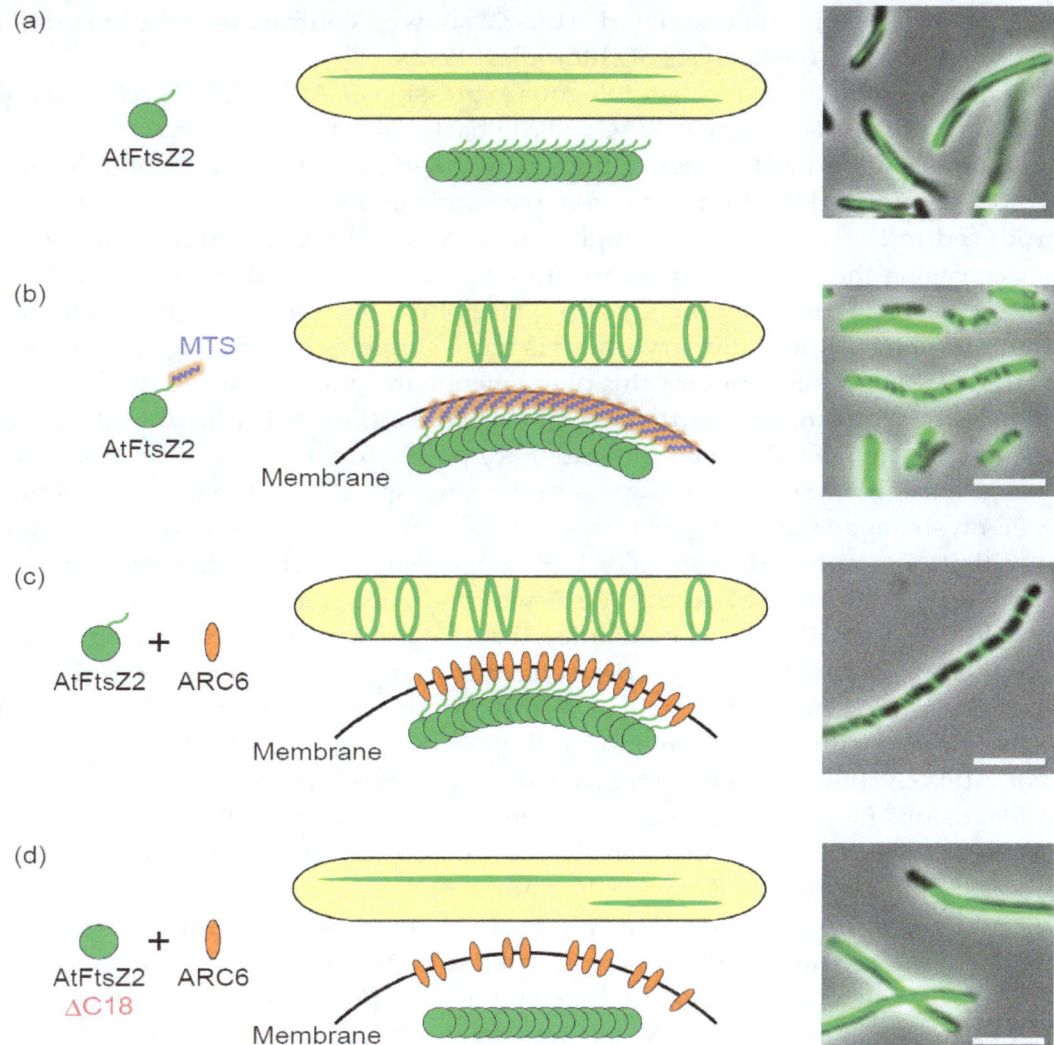

Figure 1. The formation of Z ring-like structures of *Arabidopsis* chloroplast FtsZ2 in the bacterial heterologous expression system. Schematic illustration of the proposed molecular behavior of chloroplast division-related components in *E. coli* cells and its merged microscopic image of phase-contrast and GFP are shown when expressing (**a**) super folder GFP (sfGFP)-AtFtsZ2, (**b**) sfGFP-AtFtsZ2-2MTS, (**c**) sfGFP-AtFtsZ2, and Accumulation and Replication of Chloroplasts 6 (ARC6) and (**d**) sfGFP-AtFtsZ2ΔC18 (C-terminal 18-residue truncated form of AtFtsZ2) and ARC6. *E. coli* cells were grown in L broth (1% bactotryptone, 0.5% yeast extract, 0.5% NaCl) to the stationary phase at 22 °C. Scale bars: 5 μm. MTS: membrane-targeting sequence. To reduce the complexity in the diagrams, bundling of the FtsZ2 filaments was omitted.

2. Optimization of *E. coli* System for Heterologous Expression of Chloroplast FtsZ

2.1. Fluorescent Tagging of AtFtsZ2 and Culture Condition

FtsZ proteins can polymerize, assemble into filament bundles, and eventually form the Z ring in the division system [3–5,8,9,39]. In each organism (bacterium or plant), visualizing the FtsZ protein using immunoelectron microscopy, immunofluorescence, or fluorescent protein (FP) labeling techniques clearly showed the Z ring formation at the division site [40–47]. In particular, FP labeling enables us to monitor the FtsZ protein dynamics in live cells. As for heterologous expression systems for chloroplast FtsZs tagged with FPs, in *S. pombe* yeast cells, AtFtsZ1 and/or AtFtsZ2 formed linear and ring-shaped filaments that were free-floating in the cytosol [19,35]. Furthermore, AtFtsZ1 and/or forcibly membrane-targeted AtFtsZ2 expressed in *P. pastoris* yeast cells assembled into ring structures, and the rings including membrane-tethered AtFtsZ2 showed contractible ability [15]. Thus, yeast systems can be available for the analysis of chloroplast FtsZs.

By contrast, as mentioned above, heterologous expression of AtFtsZ2 in *E. coli* cells only showed long filaments with aberrant clusters at 42 °C [19]. In bacterial cells, recombinant proteins frequently aggregate into inclusion-bodies because of high growth temperature and/or high-level expression of the heterologous protein [48]. In this context, we confirmed that the expression level of AtFtsZ2 protein introduced in *E. coli* was not too high, and successfully removed aberrant aggregation of AtFtsZ2 by decreasing the growth temperature to 22 °C—an optimal temperature for *A. thaliana* (Figure 1a) [38]. We also found that the growth phase of the bacteria strongly affected the AtFtsZ2 filamentation; sampling in the stationary phase is more suitable to form long filaments than in the logarithmic phase [38]. Some theories for this phenomenon include (i) reduced dynamics and a lower turnover rate of AtFtsZ2 compared with those of *E. coli* FtsZ (EcFtsZ) [19], and (ii) less competition with actively assembling EcFtsZ during the stationary phase. In the paper, we critically investigated the fusion terminus of FP [38]. In most papers to-date, irrespective of the derivatives from bacteria and plants, FPs were tagged at the C-terminus of FtsZ. On the other hand, as previously reported, we confirmed that N-terminal FP fusions of EcFtsZ were also precisely localized at the middle of the *E. coli* cell [38,49,50]. As for chloroplast FtsZ, we conducted heterologous expression of both N- and C-terminal FP-fused AtFtsZ2 and concluded that—at least in our *E. coli* system—filamentation of C-terminally FP-tagged AtFtsZ2 was repressed compared with that of an N-terminally fused one (Figure 1a) [38]. Since the C-terminal domain of FtsZ is important for its function, C-terminal FP-tagging might partially interfere with the filamentation ability of AtFtsZ2 [19,51]. Furthermore, the C-terminal FP fusions of AtFtsZ2 showed aberrant aggregations at higher temperature (37 °C) compared to N-terminal FP fusions [38]. As a consequence, we selected N-terminal FP-fused AtFtsZ2 for our expression system. Importantly, C-terminal FP fusions of FtsZ are generally assembly-competent, and the AtFtsZ2-FP fusion forms a Z ring at the middle of the chloroplast in *A. thaliana* [19,52]. Thus, this issue might need to be discussed. Many chloroplast proteins—including FtsZs—are encoded in the nuclear genome and are eventually transported into the chloroplast via its N-terminal transit peptide [12,45,46,53]. The transit peptide is cleaved upon its import into the chloroplast, and this is one reason why the C-terminal FP-fusion of AtFtsZ was preferred for in planta analysis [12,47]. Therefore, if the N-terminal FP-fused FtsZ is expressed in planta, it is necessary that the FP does not disturb the function of the transit peptide. In heterologous expression systems, the transit peptide-lacking chloroplast FtsZs are generally used, and this issue does not need to be considered.

2.2. N-Terminal Region of AtFtsZ2

Besides transit peptide, the chloroplast AtFtsZ2 protein harbors an extended N-terminal region compared with eubacterial FtsZ proteins [35,38]. AtFtsZ1 lacks this "extended" region, but many cyanobacterial and chloroplast FtsZ proteins exhibit N-terminal extension regardless of their amino acid sequence conservation, possibly implying the additional trait(s) in these FtsZs [38]. In the *S. pombe* yeast system, it has been reported that the N-terminal-extended region of AtFtsZ2 promotes its polymer bundling and turnover, suggesting that the N-terminus of AtFtsZ2—as with its C-terminus—is also important for its function [35]. Consistent with this, we confirmed the dependency of AtFtsZ2 filamentation on its N-terminus in *E. coli* cells, where the N-terminally-truncated AtFtsZ2 (here we deleted amino acid residues from 49 to 112 in addition to the transit peptide, AtFtsZ2ΔN) with

N-terminal FP showed considerably shorter filaments than did full-length FP-AtFtsZ2 fusion [38]. This suggests that the behavior of AtFtsZ2 produced in *E. coli* cytosol reflects its inherent properties.

2.3. Membrane-Tethering of AtFtsZ2

To establish the *E. coli* reconstitution system completely for the analysis of chloroplast division-related components, the formation of the Z ring composed of AtFtsZ2 in *E. coli* cells is critical. The FtsZ protein itself has no membrane spanning or anchoring domains and requires the cognate membrane associating and/or transmembrane proteins FtsA and ZipA in *E. coli* and Ftn2/ZipN in cyanobacteria, which interact with the C-terminus of FtsZ and target it to the membrane [42,54–62]. In green lineage chloroplasts, IEM protein ARC6—an Ftn2/ZipN ortholog—was believed to tether the Z ring, though no direct evidence has been reported for this so far [22,28,29,63,64]. *E. coli* has no ARC6 homolog, and FtsA and ZipA might not target AtFtsZ2 to the membrane, despite partial conservation of the FtsA-interacting sequence in the AtFtsZ2 C-terminus, which was supported by the fact that AtFtsZ2 did not form ring-like structures (Figure 1a) [38].

Previous reconstitution systems using liposomes revealed that the membrane-tethering of FtsZ is required for Z ring formation, in which reconstitutions of contractile EcFtsZ ring were achieved by artificial membrane-tethering with the C-terminal membrane-targeting sequence (MTS) or natural tethering through a co-introduced FtsA. This membrane anchored EcFtsZ could actually constrict a liposome, indicating that membrane-tethering is critical to form the *E. coli* Z ring and generate a constriction force [14,62]. Similarly, in our *E. coli* system, an artificial membrane-tethering of AtFtsZ2 by MTS gave AtFtsZ2 the ability to form multiple Z ring-like structures in both wild-type and *ftsZ*-depleted *E. coli*, indicating the intrinsic property of AtFtsZ2 to form Z ring-like structures in *E. coli* cells (Figure 1a,b). However, these Z ring-like structures did not constrict a cell (probably because AtFtsZ2 could not interact with FtsA and ZipA, which stabilize Z ring composed of EcFtsZ) [38]. Around the same time, it was independently shown that the MTS-tagged chloroplast FtsZ derived from *Galdieria sulphuraria* thermophilic red alga (GsFtsZ) formed multiple Z ring-like structures in *E. coli* cells [65]. Together with the recent success in the *P. pastoris* yeast system that showed reconstitution of the ring-like structure of AtFtsZ2 by MTS-tagging [15], these reports strongly demonstrated that membrane-tethering is a necessary and sufficient factor for bacterial and chloroplast FtsZ proteins to form Z ring or Z ring-like structures.

Interestingly, the diameter of the ring-like structure formed by membrane-tethered AtFstZ2 was much larger than that of EcFtsZ when reconstituted in *P. pastoris* cells, resembling the size of their corresponding rings in vivo [15]. This has suggested that the structure of each FtsZ protein determines the curvature, and consequently the size, of each ring [15]. However, we successfully reconstituted the Z ring-like structures of AtFtsZ2 in *E. coli* cells [38]. Furthermore, in *A. thaliana*, the chloroplast size increases during leaf development, and leaf epidermal cells contained small chloroplasts with smaller AtFtsZ1 and AtFtsZ2 rings compared to leaf mesophyll cells [52,66]. Thus, we presume that FtsZ proteins have the potential to form Z rings with various diameters according to the cell or chloroplast diameter.

3. The Function of Negative and Positive Contributors in Bacterial Reconstitution Systems

3.1. The Negative Regulator ARC3

In rod-shaped bacteria such as *E. coli* and *Bacillus subtilis*, mid-cell positioning of the Z ring is tightly regulated by the Min system, in which the negative regulator MinC inhibits FtsZ polymerization at cell poles [67,68]. Spatial regulation of the Z ring by MinC is also conserved in cyanobacteria [69]. In contrast, except for certain algal lineages and the moss *Physcomitrella patens*, the chloroplast has no MinC homolog, but instead acquired the plant-specific stromal protein ARC3 as a functional analog of bacterial MinC [19,24–26,52,70,71]. Indeed, *Arabidopsis arc3* mutants showed multiple Z rings and nonuniform chloroplast size and number, whereas ARC3-overexpressing mutants exhibited a small number of enlarged chloroplasts with fragmented AtFtsZ filaments [25,26,72,73]. ARC3 directly interacts with both AtFtsZ1 and AtFtsZ2, and these interactions were inhibited by the C-terminal membrane-occupation-and-recognition nexus (MORN) domain of ARC3 [25,26]. The MORN domain is a binding site of PARC6, which is believed to recruit and activate ARC3 at the chloroplast division site [32,34]. In a yeast heterologous expression system which does not contain the PARC6 homolog, recombinant ARC3 lacking the MORN domain was used to analyze the ARC3 inhibitory effects on AtFtsZ filaments [19,26]. We also co-expressed this mutant ARC3 and AtFtsZ2 with an FP in *E. coli* cells and evaluated its function in our bacterial system. Consistent with previous reports, we confirmed the inhibition of the AtFtsZ2 assembly by ARC3 regardless of the presence or absence of the MTS tag (Figure 2a,b) [38]. Since it has already been reported in the yeast system that ARC3 inhibited the assembly of cytosolic free-floating AtFtsZ filaments (linear and ring-shaped structures), our bacterial system presented the first example of ARC3 in inhibiting AtFtsZ filaments in membrane-tethered Z ring-like structures in a heterologous expression system (Figure 2b) [19,26,38]. It is worth noting that in our *E. coli* system— like the yeast systems—there might be no factors that affect ARC3 behavior. The consistency of the inhibitory effects of ARC3 on FtsZ filament assembly among yeast, *E. coli*, and in planta analyses strongly demonstrates the clear function of ARC3 in chloroplast Z ring regulation. This is further supported by a recent study in which in vitro assays showed that ARC3 promoted AtFtsZ2 debundling and disassembly by enhancing its GTPase activity and 3D reconstruction using single-particle analysis, suggesting that PARC6 mediated ARC3–AtFtsZ2 interaction [74]. Bacterial MinC also promotes the debundling and disassembly of FtsZ, but does not affect its GTPase activity [27]. In addition, chloroplast ARC3 binds to both MinD and MinE, but bacterial MinC only binds to MinD [25,75]. The analogous function of ARC3 to MinC is indisputable, but there might be differences between ARC3 and MinC in their mode of action in each division system.

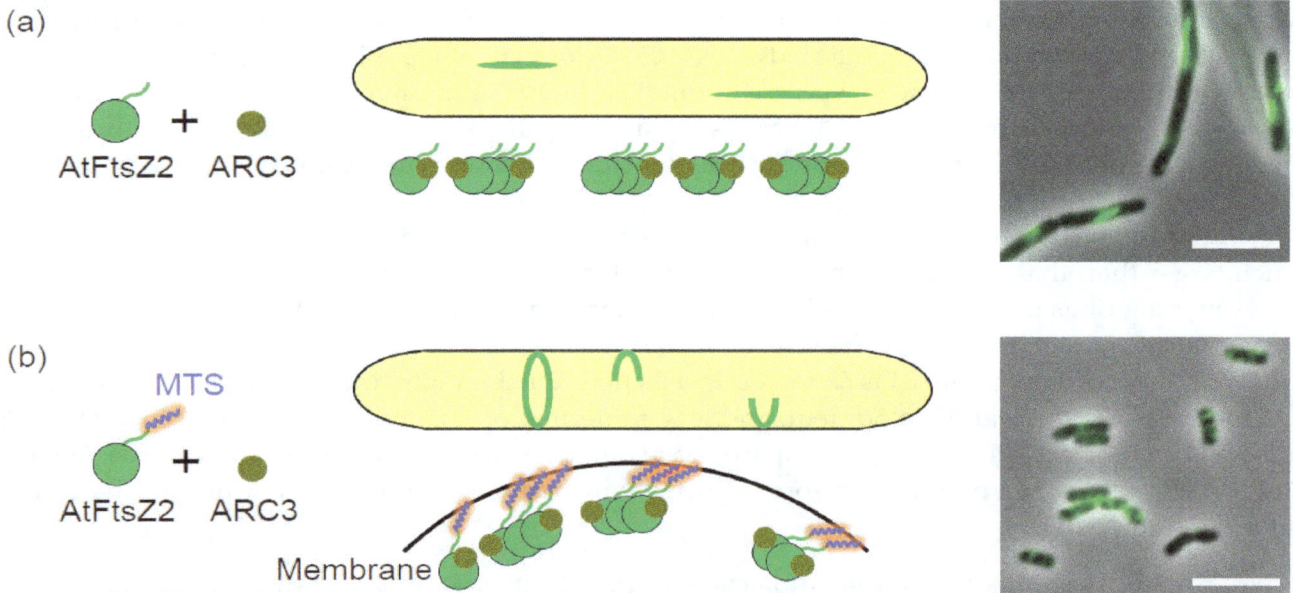

Figure 2. The effects of Accumulation and Replication of Chloroplasts 3 (ARC3) on the filaments of *Arabidopsis* chloroplast FtsZ2 in the bacterial heterologous expression system. Schematic illustration of the proposed molecular behavior of chloroplast division-related components in *E. coli* cells and its merged microscopic image of phase-contrast and GFP are shown when expressing (**a**) sfGFP-AtFtsZ2 and ARC3, and (**b**) sfGFP-AtFtsZ2-2MTS and ARC3. *E. coli* cells were grown in L broth (1% bactotryptone, 0.5% yeast extract, 0.5% NaCl) to the stationary phase at 22 °C. Scale bars: 5 µm. MTS: membrane-targeting sequence. To reduce the complexity in the diagrams, bundling of the FtsZ2 filaments was omitted.

3.2. The Positive Regulator ARC6

Using liposomes and purified EcFtsZ proteins, Osawa and Erickson (2013) demonstrated that the reconstitution of the Z ring (or Z ring-like structure) membrane-tethered by its natural partner is one goal for protein-free or heterologous expression systems in order to study the molecular mechanisms of the Z ring-centered division machinery [62]. In the case of chloroplast division machinery, a Z ring-anchoring factor has not yet been identified, but a great deal of indirect evidence implied the IEM protein ARC6 as a potential candidate [22,28,29,34,45,63,64]. In *Arabidopsis*, FP-labeled ARC6 concentrated at the chloroplast constriction sites in the shape of a ring, and ARC6 mutants exhibited a Z ring-defective phenotype, consequently leading to a small number of enlarged chloroplasts [28,45,63,64]. Using the yeast two-hybrid system, a direct interaction between ARC6 and the C-terminal conserved sequence of AtFtsZ2 was demonstrated [22,29,34]. FP-labeled ARC6 and AtFtsZ2 co-localized in the yeast cytosol, which mostly depended on the AtFtsZ2 C-terminus [34]. Collectively, all these reports revealed that ARC6 is a positive regulator of Z ring formation in chloroplasts. Therefore, the next challenge will test whether ARC6 truly anchors the chloroplast Z ring to the membrane.

We applied Osawa and Erickson's strategy to our *E. coli* heterologous expression system, where the MTS-untagged AtFtsZ2 (which itself can only form linear filaments) and ARC6 were co-expressed, and evaluated the effects of ARC6 on AtFtsZ2 filament morphology [38]. Fortunately, our challenge was a success—AtFtsZ2 polymer drastically altered its morphology from the linear filaments to Z ring-like or helical structures dependent on ARC6 (Figure 1a,c). FP-labeling of both AtFtsZ2 and ARC6 showed co-localization of these two proteins in the ring-like structures [38]. These Z ring-like structures completely depended on the ARC6-interacting sequence at the AtFtsZ2 C-terminus (here we truncated 18 amino acids, AtFtsZ2ΔC18), suggesting ARC6-mediated tethering of AtFtsZ2 filaments to the membrane (Figure 1c,d). Membrane-fractionation assays further supported the membrane attachment of AtFtsZ2 by ARC6 through their direct interaction [38]. The C-terminal region of ARC6 protrudes into the chloroplast IEM and directly interacts with the outer envelope membrane (OEM) protein Plastid Division 2 (PDV2), being able to transfer the Z ring positioning information from the stromal division machinery to the cytosolic one [76,77]. Together with previous results, our data obtained from the bacterial reconstitution system clarified that the other N-terminal side of ARC6 interacts with AtFtsZ2—a backbone protein of the Z ring through which ARC6 directly anchors the Z ring to IEM [22,28,29,34,38,45,63,64]. Bacteria such as *E. coli* and *B. subtilis* have no ARC6 ortholog, but the membrane-interacting protein FtsA interacts with FtsZ and anchors it to the membrane, hence stabilizing the Z ring [42,57,58,62,78,79]. By contrast, cyanobacteria uniquely evolved Ftn2/ZipN—an ancestor of chloroplast ARC6—as a functional analog of FtsA for Z ring-tethering [59–61]. The successful reconstitution of chloroplast FtsZ2 ring in bacterial cells by membrane-tethering through the chloroplast ARC6 indicates high stability and plasticity of the Z ring-centered division machinery that is conserved from bacteria to chloroplasts.

Additionally, in the *S. pombe* yeast expression system, it has been demonstrated that ARC6 stabilizes AtFtsZ2 filaments independent of its tethering ability [35]. On the other hand, a recent report revealed a new function of the *E. coli* FtsA in aligning FtsZ protofilaments in the unbundled state and stabilizing them, in addition to its membrane-tethering ability [80]. Thus, these functional analogs commonly work for Z ring-tethering but have additional function(s) as a positive regulator in each division system.

3.3. The Positive Regulator AtFtsZ1

In our paper, we described the unexpected function of AtFtsZ1 in positively contributing to the filament morphology of AtFtsZ2 in the bacterial expression system [38]. As mentioned above, the chloroplast Z ring is composed of AtFtsZ2 and AtFtsZ1, but the former dominantly determines the filament morphology, while the latter plays a regulatory role [15,19]. It is worth noting that the *Arabidopsis ftsZ1* mutant (like *ftsZ2* mutants) showed a small number of enlarged chloroplasts, implying the indispensable function of AtFtsZ1 in chloroplast division, although it has also been observed that chloroplasts in the *ftsZ1* mutant still exhibited a single mid-plastid constriction [22,26,30,81]. As expected, FP-labeled AtFtsZ1 and AtFtsZ2 co-localized in the *E. coli* expression system, consistent with previous observations that AtFtsZ1 and AtFtsZ2 can form a heteropolymer both in vitro and in yeast systems (Figure 3) [15,17–19,38]. However, the AtFtsZ1 surprisingly induced a morphological transformation of AtFtsZ2 filaments into ring-like and helical structures in *E. coli* cells, resembling Z ring-like structures tethered by ARC6 (Figure 3) [38]. This phenomenon was also observed in the case of AtFtsZ2ΔC18 [38]. AtFtsZ1 possesses no C-terminal sequences responsible for membrane-tethering, and independently expressed AtFtsZ1 showed only a dispersed pattern in the bacterial cytoplasm, which suggests that AtFtsZ1 itself is unlikely to interact with any of the *E. coli* endogenous components supporting membrane-tethering [38]. Thus, it still remains an open question as to how AtFtsZ2 filaments form Z ring-like structures depending on AtFtsZ1 in the bacterial expression system. In the *P. pastoris* yeast system, a ring-like structure of MTS-untagged AtFtsZ1 was observed in the absence of other related component(s), although we did not detect any AtFtsZ1 filaments or rings in *E. coli* cells [15,38]. As for membrane-tethering, these data led to the speculation that AtFtsZ1 itself may interact with the membrane. Nevertheless, it is clear that AtFtsZ1 works positively to form the chloroplast Z ring. The conservation of two distinct FtsZ proteins in green lineage indicates a unique mechanism to regulate chloroplast Z ring assembly and dynamics compared to bacterial cell division. We hope future studies will reveal unknown mechanisms of action of AtFtsZ1 apart from its ability to increase the FtsZ filament turnover rate [19].

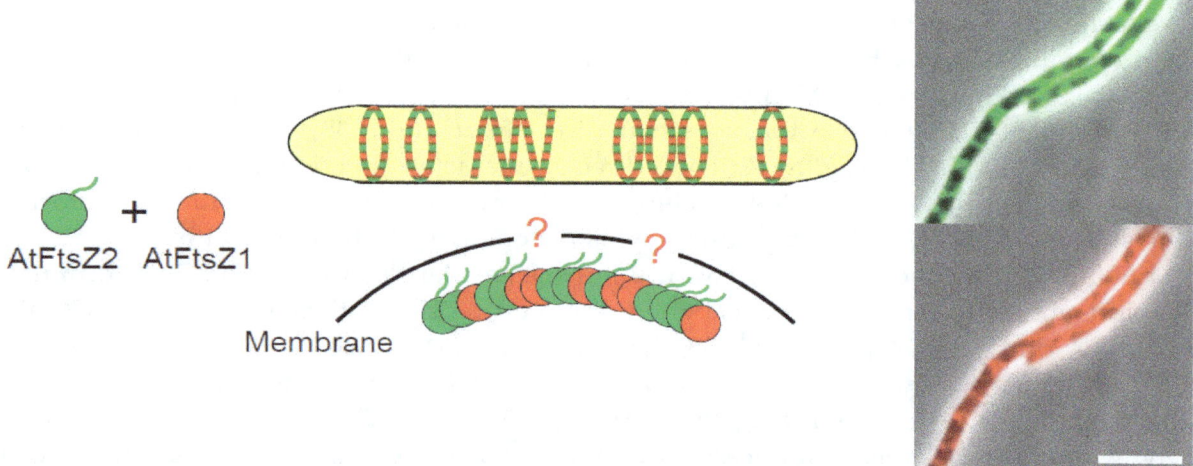

Figure 3. The effects of *Arabidopsis* chloroplast FtsZ1 on the formation of Z ring-like structures of *Arabidopsis* chloroplast FtsZ2 in the bacterial heterologous expression system. Schematic illustration of the Z ring-like structures composed of AtFtsZ1 and AtFtsZ2 heterooligomer in *E. coli* cells, although the mechanism by which the *Arabidopsis* FtsZ filaments are tethered to the membrane is unclear (indicated by red question marks in the illustration), and its merged microscopic images of phase-contrast and GFP (upper panel), and phase-contrast and mCherry (lower panel) are shown. *E. coli* cells expressing sfGFP-AtFtsZ2 and mCherry-AtFtsZ1 were grown in L broth (1% bactotryptone, 0.5% yeast extract, 0.5% NaCl) to the stationary phase at 22 °C. Scale bar: 5 μm. To reduce the complexity in the diagram, bundling of the FtsZ filaments was omitted.

4. Conclusions

The *E. coli* reconstitution system has now become one of the many useful heterologous expression systems for studying the FtsZ-centered chloroplast division machinery. Although in planta analysis is the best way to comprehensively examine the components employed in chloroplast division, heterologous expression systems lacking such native division-related contributors also provide important insights into inherent properties of the introduced component(s). In general, microbial manipulations have practical advantages, such as rapid growth, axenic culture, and genetic accessibility. Yeast systems allow the advantage of a eukaryotic environment completely lacking plant derivatives. The advantages of bacterial systems are as follows: (i) bacteria is an evolutionary progenitor of chloroplast and bacterial cytosol, wherein the Z ring-centered division system still works and is topologically equivalent to chloroplast stroma; and (ii) bacteria—especially *E. coli*—are distantly related to cyanobacteria, and partly lack the homologs of chloroplast division-related proteins. Indeed, using the *E. coli* system, we reconfirmed the negative regulation of the chloroplast FtsZ2 filaments by plant-specific ARC3 and found that this regulation was also effective for the membrane-tethered one (Z ring-like structures) [38]. We also directly demonstrated the Z ring-tethering ability of ARC6, which is unique to cyanobacteria and chloroplasts [38]. The components of the chloroplast division machinery in the stromal side (including IEM) have changed over the course of evolution [3–5]. Besides ARC3 and ARC6, the homologous proteins of PARC6 and MULTIPLE CHLOROPLAST DIVISION SITE 1 (MCD1) are absent in *E. coli* [32–34,82]; hence, they are capable of being applied to our bacterial system. Additionally, bacteria-derived chloroplast MinD and MinE can be analyzed by using the minCDE deletion mutant of *E. coli*, which creates minicells but also shows normal- and long-sized cells, as in the case of AtFtsZ2 expressed in the ftsZ-depleted *E. coli* strain [38,83]. By contrast, in the bacterial system (and in the yeast systems), the analyses of division-related components employed in the plant cytosolic side (including OEM), such as DRP5B, Plastid Division 1 (PDV1), and PDV2, appear more challenging [7,13,76,77,84,85]. Accordingly, the integration of data obtained from yeast and *E. coli* systems into in planta results is important, and will continue to accelerate the research on chloroplast division.

Acknowledgments: We thank all the members of Cellular Function Laboratory (Shiomi lab) for the helpful discussions. This work was supported by the Strategic Research Foundation Grant-aided Project for Private Universities (S1201003 to Daisuke Shiomi) from the Ministry of Education, Culture, Sports, Science and Technology, Japan.

Author Contributions: Hiroki Irieda and Daisuke Shiomi contributed to the writing of this review.

References

1. Gould, S.B.; Waller, R.F.; McFadden, G.I. Plastid evolution. *Annu. Rev. Plant Biol.* **2008**, *59*, 491–517. [CrossRef] [PubMed]

2. Keeling, P.J. The number, speed, and impact of plastid endosymbiosis in eukaryotic evolution. *Annu. Rev. Plant Biol.* **2013**, *64*, 583–607. [CrossRef] [PubMed]

3. Miyagishima, S.Y.; Nakanishi, H.; Kabeya, Y. Structure, regulation, and evolution of the plastid division machinery. *Int. Rev. Cell Mol. Biol.* **2011**, *291*, 115–153. [CrossRef] [PubMed]

4. Osteryoung, K.W.; Pyke, K.A. Division and dynamic morphology of plastids. *Annu. Rev. Plant Biol.* **2014**, *65*, 443–472. [CrossRef] [PubMed]

5. Chen, C.; MacCready, J.S.; Ducat, D.C.; Osteryoung, K.W. The molecular machinery of chloroplast division. *Plant Physiol.* **2017**. [CrossRef] [PubMed]

6. Kuroiwa, T.; Kuroiwa, H.; Sakai, A.; Takahashi, H.; Toda, K.; Itoh, R. The division apparatus of plastids and mitochondria. *Int. Rev. Cytol.* **1998**, *181*, 1–41. [CrossRef] [PubMed]

7. Yoshida, Y.; Kuroiwa, H.; Misumi, O.; Yoshida, M.; Ohnuma, M.; Fujiwara, T.; Yagisawa, F.; Hirooka, S.; Imoto, Y.; Matsushita, K.; et al. Chloroplasts divide by contraction of a bundle of nanofilaments consisting of polyglucan. *Science* **2010**, *329*, 949–953. [CrossRef] [PubMed]

8. Adams, D.W.; Errington, J. Bacterial cell division: Assembly, maintenance and disassembly of the Z ring. *Nat. Rev. Microbiol.* **2009**, *7*, 642–653. [CrossRef] [PubMed]

9. Erickson, H.P.; Anderson, D.E.; Osawa, M. FtsZ in bacterial cytokinesis: Cytoskeleton and force generator all in one. *Microbiol. Mol. Biol. Rev.* **2010**, *74*, 504–528. [CrossRef] [PubMed]

10. Mingorance, J.; Rivas, G.; Vélez, M.; Gómez-Puertas, P.; Vicente, M. Strong FtsZ is with the force: Mechanisms to constrict bacteria. *Trends Microbiol.* **2010**, *18*, 348–356. [CrossRef] [PubMed]

11. Terbush, A.D.; MacCready, J.S.; Chen, C.; Ducat, D.C.; Osteryoung, K.W. Conserved dynamics of chloroplast cytoskeletal FtsZ proteins across photosynthetic lineages. *Plant Physiol.* **2018**, *176*, 295–306. [CrossRef] [PubMed]

12. Osteryoung, K.W.; Vierling, E. Conserved cell and organelle division. *Nature* **1995**, *376*, 473–474. [CrossRef] [PubMed]

13. Yoshida, Y.; Kuroiwa, H.; Misumi, O.; Nishida, K.; Yagisawa, F.; Fujiwara, T.; Nanamiya, H.; Kawamura, F.; Kuroiwa, T. Isolated chloroplast division machinery can actively constrict after stretching. *Science* **2006**, *313*, 1435–1438. [CrossRef] [PubMed]

14. Osawa, M.; Anderson, D.E.; Erickson, H.P. Reconstitution of contractile FtsZ rings in liposomes. *Science* **2008**, *320*, 792–794. [CrossRef] [PubMed]

15. Yoshida, Y.; Mogi, Y.; TerBush, A.D.; Osteryoung, K.W. Chloroplast FtsZ assembles into a contractible ring via tubulin-like heteropolymerization. *Nat. Plants* **2016**, *2*, 16095. [CrossRef] [PubMed]

16. Osteryoung, K.W.; Stokes, K.D.; Rutherford, S.M.; Percival, A.L.; Lee, W.Y. Chloroplast division in higher plants requires members of two functionally divergent gene families with homology to bacterial FtsZ. *Plant Cell* **1998**, *10*, 1991–2004. [CrossRef] [PubMed]

17. Olson, B.J.; Wang, Q.; Osteryoung, K.W. GTP-dependent heteropolymer formation and bundling of chloroplast FtsZ1 and FtsZ2. *J. Biol. Chem.* **2010**, *285*, 20634–20643. [CrossRef] [PubMed]

18. Smith, A.G.; Johnson, C.B.; Vitha, S.; Holzenburg, A. Plant FtsZ1 and FtsZ2 expressed in a eukaryotic host: GTPase activity and self-assembly. *FEBS Lett.* **2010**, *584*, 166–172. [CrossRef] [PubMed]

19. TerBush, A.D.; Osteryoung, K.W. Distinct functions of chloroplast FtsZ1 and FtsZ2 in Z ring structure and remodeling. *J. Cell Biol.* **2012**, *199*, 623–637. [CrossRef] [PubMed]

20. Miyagishima, S.Y.; Nozaki, H.; Nishida, K.; Nishida, K.; Matsuzaki, M.; Kuroiwa, T. Two types of FtsZ proteins in mitochondria and red-lineage chloroplasts: The duplication of FtsZ is implicated in endosymbiosis. *J. Mol. Evol.* **2004**, *58*, 291–303. [CrossRef] [PubMed]

21. TerBush, A.D.; Yoshida, Y.; Osteryoung, K.W. FtsZ in chloroplast division: Structure, function and evolution. *Curr. Opin. Cell Biol.* **2013**, *25*, 461–470. [CrossRef] [PubMed]

22. Schmitz, A.J.; Glynn, J.M.; Olson, B.J.; Stokes, K.D.; Osteryoung, K.W. *Arabidopsis* FtsZ2-1 and FtsZ2-2 are functionally redundant, but FtsZ-based plastid division is not essential for chloroplast partitioning or plant growth and development. *Mol. Plant* **2009**, *2*, 1211–1222. [CrossRef] [PubMed]

23. Karamoko, M.; El-Kafafi, E.S.; Mandaron, P.; Lerbs-Mache, S.; Falconet, D. Multiple FtsZ2 isoforms involved in chloroplast division and biogenesis are developmentally associated with thylakoid membranes in *Arabidopsis*. *FEBS Lett.* **2011**, *585*, 1203–1208. [CrossRef] [PubMed]

24. Shimada, H.; Koizumi, M.; Kuroki, K.; Mochizuki, M.; Fujimoto, H.; Ohta, H.; Masuda, T.; Takamiya, K. ARC3, a chloroplast division factor, is a chimera of prokaryotic FtsZ and part of eukaryotic phosphatidylinositol-4-phosphate 5-kinase. *Plant Cell Physiol.* **2004**, *45*, 960–967. [CrossRef] [PubMed]

25. Maple, J.; Vojta, L.; Soll, J.; Møller, S.G. ARC3 is a stromal Z ring accessory protein essential for plastid division. *EMBO Rep.* **2007**, *8*, 293–299. [CrossRef] [PubMed]

26. Zhang, M.; Schmitz, A.J.; Kadirjan-Kalbach, D.K.; TerBush, A.D.; Osteryoung, K.W. Chloroplast division protein ARC3 regulates chloroplast FtsZ ring assembly and positioning in *Arabidopsis* through interaction with FtsZ2. *Plant Cell* **2013**, *25*, 1787–1802. [CrossRef] [PubMed]

27. Hu, Z.; Mukherjee, A.; Pichoff, S.; Lutkenhaus, J. The MinC component of the division site selection system in *Escherichia coli* interacts with FtsZ to prevent polymerization. *Proc. Natl. Acad. Sci. USA* **1999**, *96*, 14819–14824. [CrossRef] [PubMed]

28. Vitha, S.; Froehlich, J.E.; Koksharova, O.; Pyke, K.A.; van Erp, H.; Osteryoung, K.W. ARC6 is a J-domain plastid division protein and an evolutionary descendant of the cyanobacterial cell division protein Ftn2. *Plant Cell* **2003**, *15*, 1918–1933. [CrossRef] [PubMed]

29. Maple, J.; Aldridge, C.; Møller, S.G. Plastid division is mediated by combinatorial assembly of plastid division proteins. *Plant J.* **2005**, *43*, 811–823. [CrossRef] [PubMed]

30. McAndrew, R.S.; Olson, B.J.; Kadirjan-Kalbach, D.K.; Chi-Ham, C.L.; Vitha, S.; Froehlich, J.E.; Osteryoung, K.W. In vivo quantitative relationship between plastid division proteins FtsZ1 and FtsZ2 and identification of ARC6 and ARC3 in a native FtsZ complex. *Biochem. J.* **2008**, *412*, 367–378. [CrossRef] [PubMed]

31. Maple, J.; Chua, N.H.; Møller, S.G. The topological specificity factor AtMinE1 is essential for correct plastid division site placement in *Arabidopsis*. *Plant J.* **2002**, *31*, 269–277. [CrossRef] [PubMed]

32. Glynn, J.M.; Yang, Y.; Vitha, S.; Schmitz, A.J.; Hemmes, M.; Miyagishima, S.Y.; Osteryoung, K.W. PARC6, a novel chloroplast division factor, influences FtsZ assembly and is required for recruitment of PDV1 during chloroplast division in Arabidopsis. *Plant J.* **2009**, *59*, 700–711. [CrossRef] [PubMed]

33. Zhang, M.; Hu, Y.; Jia, J.; Li, D.; Zhang, R.; Gao, H.; He, Y. CDP1, a novel component of chloroplast division site positioning system in *Arabidopsis*. *Cell Res.* **2009**, *19*, 877–886. [CrossRef] [PubMed]

34. Zhang, M.; Chen, C.; Froehlich, J.E.; TerBush, A.D.; Osteryoung, K.W. Roles of Arabidopsis PARC6 in coordination of the chloroplast division complex and negative regulation of FtsZ assembly. *Plant Physiol.* **2016**, *170*, 250–262. [CrossRef] [PubMed]

35. TerBush, A.D.; Porzondek, C.A.; Osteryoung, K.W. Functional analysis of the chloroplast division complex using *Schizosaccharomyces pombe* as a heterologous expression system. *Microsc. Microanal.* **2016**, *22*, 275–289. [CrossRef] [PubMed]

36. Srinivasan, R.; Mishra, M.; Murata-Hori, M.; Balasubramanian, M.K. Filament formation of the Escherichia coli actin-related protein, MreB, in fission yeast. *Curr. Biol.* **2007**, *17*, 266–272. [CrossRef] [PubMed]

37. Srinivasan, R.; Mishra, M.; Wu, L.; Yin, Z.; Balasubramanian, M.K. The bacterial cell division protein FtsZ assembles into cytoplasmic rings in fission yeast. *Genes Dev.* **2008**, *22*, 1741–1746. [CrossRef] [PubMed]

38. Irieda, H.; Shiomi, D. ARC6-mediated Z ring-like structure formation of prokaryote-descended chloroplast FtsZ in *Escherichia coli*. *Sci. Rep.* **2017**, *7*, 3492. [CrossRef] [PubMed]

39. Haeusser, D.P.; Margolin, W. Splitsville: Structural and functional insights into the dynamic bacterial Z ring. *Nat. Rev. Microbiol.* **2016**, *14*, 305–319. [CrossRef] [PubMed]

40. Bi, E.F.; Lutkenhaus, J. FtsZ ring structure associated with division in *Escherichia coli*. *Nature* **1991**, *354*, 161–164. [CrossRef] [PubMed]

41. Levin, P.A.; Losick, R. Transcription factor Spo0A switches the localization of the cell division protein FtsZ from a medial to a bipolar pattern in *Bacillus subtilis*. *Genes Dev.* **1996**, *10*, 478–488. [CrossRef] [PubMed]

42. Ma, X.; Ehrhardt, D.W.; Margolin, W. Colocalization of cell division proteins FtsZ and FtsA to cytoskeletal structures in living *Escherichia coli* cells by using green fluorescent protein. *Proc. Natl. Acad. Sci. USA* **1996**, *93*, 12998–13003. [CrossRef] [PubMed]

43. Addinall, S.G.; Bi, E.; Lutkenhaus, J. FtsZ ring formation in *fts* mutants. *J. Bacteriol.* **1996**, *178*, 3877–3884. [CrossRef] [PubMed]

44. Sun, Q.; Margolin, W. FtsZ dynamics during the division cycle of live *Escherichia coli* cells. *J. Bacteriol.* **1998**, *180*, 2050–2056. [PubMed]

45. McAndrew, R.S.; Froehlich, J.E.; Vitha, S.; Stokes, K.D.; Osteryoung, K.W. Colocalization of plastid division proteins in the chloroplast stromal compartment establishes a new functional relationship between FtsZ1 and FtsZ2 in higher plants. *Plant Physiol.* **2001**, *127*, 1656–1666. [CrossRef] [PubMed]

46. Mori, T.; Kuroiwa, H.; Takahara, M.; Miyagishima, S.Y.; Kuroiwa, T. Visualization of an FtsZ ring in chloroplasts of *Lilium longiflorum* leaves. *Plant Cell Physiol.* **2001**, *42*, 555–559. [CrossRef] [PubMed]

47. Vitha, S.; McAndrew, R.S.; Osteryoung, K.W. FtsZ ring formation at the chloroplast division site in plants. *J. Cell Biol.* **2001**, *153*, 111–120. [CrossRef] [PubMed]

48. Baneyx, F.; Mujacic, M. Recombinant protein folding and misfolding in *Escherichia coli*. *Nat. Biotechnol.* **2004**, *22*, 1399–1408. [CrossRef] [PubMed]

49. Bernhardt, T.G.; de Boer, P.A. SlmA, a nucleoid-associated, FtsZ binding protein required for blocking septal ring assembly over chromosomes in *E. coli*. *Mol. Cell* **2005**, *18*, 555–564. [CrossRef] [PubMed]

50. Osawa, M.; Erickson, H.P. Probing the domain structure of FtsZ by random truncation and insertion of GFP. *Microbiology* **2005**, *151*, 4033–4043. [CrossRef] [PubMed]

51. Ma, X.; Margolin, W. Genetic and functional analyses of the conserved C-terminal core domain of *Escherichia coli* FtsZ. *J. Bacteriol.* **1999**, *181*, 7531–7544. [PubMed]

52. Johnson, C.B.; Shaik, R.; Abdallah, R.; Vitha, S.; Holzenburg, A. FtsZ1/FtsZ2 turnover in chloroplasts and the role of ARC3. *Microsc. Microanal.* **2015**, *21*, 313–323. [CrossRef] [PubMed]

53. Fujiwara, M.; Yoshida, S. Chloroplast targeting of chloroplast division FtsZ2 proteins in *Arabidopsis*. *Biochem. Biophys. Res. Commun.* **2001**, *287*, 462–467. [CrossRef] [PubMed]

54. Hale, C.A.; de Boer, P.A. Direct binding of FtsZ to ZipA, an essential component of the septal ring structure that mediates cell division in *E. coli*. *Cell* **1997**, *88*, 175–185. [CrossRef]

55. Liu, Z.; Mukherjee, A.; Lutkenhaus, J. Recruitment of ZipA to the division site by interaction with FtsZ. *Mol. Microbiol.* **1999**, *31*, 1853–1861. [CrossRef] [PubMed]

56. Mosyak, L.; Zhang, Y.; Glasfeld, E.; Haney, S.; Stahl, M.; Seehra, J.; Somers, W.S. The bacterial cell-division protein ZipA and its interaction with an FtsZ fragment revealed by X-ray crystallography. *EMBO J.* **2000**, *19*, 3179–3191. [CrossRef] [PubMed]

57. Pichoff, S.; Lutkenhaus, J. Unique and overlapping roles for ZipA and FtsA in septal ring assembly in *Escherichia coli*. *EMBO J.* **2002**, *21*, 685–693. [CrossRef] [PubMed]

58. Pichoff, S.; Lutkenhaus, J. Tethering the Z ring to the membrane through a conserved membrane targeting sequence in FtsA. *Mol. Microbiol.* **2005**, *55*, 1722–1734. [CrossRef] [PubMed]

59. Koksharova, O.A.; Wolk, C.P. A novel gene that bears a DnaJ motif influences cyanobacterial cell division. *J. Bacteriol.* **2002**, *184*, 5524–5528. [CrossRef] [PubMed]

60. Mazouni, K.; Domain, F.; Cassier-Chauvat, C.; Chauvat, F. Molecular analysis of the key cytokinetic components of cyanobacteria: FtsZ, ZipN and MinCDE. *Mol. Microbiol.* **2004**, *52*, 1145–1158. [CrossRef] [PubMed]

61. Marbouty, M.; Saguez, C.; Cassier-Chauvat, C.; Chauvat, F. ZipN, an FtsA-like orchestrator of divisome assembly in the model cyanobacterium *Synechocystis* PCC6803. *Mol. Microbiol.* **2009**, *74*, 409–420. [CrossRef] [PubMed]

62. Osawa, M.; Erickson, H.P. Liposome division by a simple bacterial division machinery. *Proc. Natl. Acad. Sci. USA* **2013**, *110*, 11000–11004. [CrossRef] [PubMed]

63. Pyke, K.A.; Rutherford, S.M.; Robertson, E.J.; Leech, R.M. *arc6*, a fertile *Arabidopsis* mutant with only two mesophyll cell chloroplasts. *Plant Physiol.* **1994**, *106*, 1169–1177. [CrossRef] [PubMed]

64. Johnson, C.B.; Tang, L.K.; Smith, A.G.; Ravichandran, A.; Luo, Z.; Vitha, S.; Holzenburg, A. Single particle tracking analysis of the chloroplast division protein FtsZ anchoring to the inner envelope membrane. *Microsc. Microanal.* **2013**, *19*, 507–512. [CrossRef] [PubMed]

65. Chen, Y.; Porter, K.; Osawa, M.; Augustus, A.M.; Milam, S.L.; Joshi, C.; Osteryoung, K.W.; Erickson, H.P. The chloroplast tubulin homologs FtsZA and FtsZB from the red alga *Galdieria sulphuraria* co-assemble into dynamic filaments. *J. Biol. Chem.* **2017**, *292*, 5207–5215. [CrossRef] [PubMed]

66. Okazaki, K.; Kabeya, Y.; Suzuki, K.; Mori, T.; Ichikawa, T.; Matsui, M.; Nakanishi, H.; Miyagishima, S. The PLASTID DIVISION1 and 2 Components of the Chloroplast Division Machinery Determine the Rate of Chloroplast Division in Land Plant Cell Differentiation. *Plant Cell* **2009**, *21*, 1769–1780. [CrossRef] [PubMed]

67. Lutkenhaus, J. Assembly dynamics of the bacterial MinCDE system and spatial regulation of the Z ring. *Annu. Rev. Biochem.* **2007**, *76*, 539–562. [CrossRef] [PubMed]

68. Rowlett, V.W.; Margolin, W. The bacterial Min system. *Curr. Biol.* **2013**, *23*, R553–R556. [CrossRef] [PubMed]

69. MacCready, J.S.; Schossau, J.; Osteryoung, K.W.; Ducat, D.C. Robust Min-system oscillation in the presence of internal photosynthetic membranes in cyanobacteria. *Mol. Microbiol.* **2017**, *103*, 483–503. [CrossRef] [PubMed]

70. Yang, Y.; Glynn, J.M.; Olson, B.J.; Schmitz, A.J.; Osteryoung, K.W. Plastid division: Across time and space. *Curr. Opin. Plant Biol.* **2008**, *11*, 577–584. [CrossRef] [PubMed]

71. Miyagishima, S.Y.; Kabeya, Y. Chloroplast division: Squeezing the photosynthetic captive. *Curr. Opin. Microbiol.* **2010**, *13*, 738–746. [CrossRef] [PubMed]

72. Pyke, K.A.; Leech, R.M. Chloroplast division and expansion is radically altered by nuclear mutations in *Arabidopsis thaliana*. *Plant Physiol.* **1992**, *99*, 1005–1008. [CrossRef] [PubMed]

73. Glynn, J.M.; Miyagishima, S.Y.; Yoder, D.W.; Osteryoung, K.W.; Vitha, S. Chloroplast division. *Traffic* **2007**, *8*, 451–461. [CrossRef] [PubMed]

74. Shaik, R.S.; Sung, M.W.; Vitha, S.; Holzenburg, A. Chloroplast division protein ARC3 acts on FtsZ2 by preventing filament bundling and enhancing GTPase activity. *Biochem. J.* **2018**, *475*, 99–115. [CrossRef] [PubMed]

75. Huang, J.; Cao, C.; Lutkenhaus, J. Interaction between FtsZ and inhibitors of cell division. *J. Bacteriol.* **1996**, *178*, 5080–5085. [CrossRef] [PubMed]

76. Glynn, J.M.; Froehlich, J.E.; Osteryoung, K.W. *Arabidopsis* ARC6 coordinates the division machineries of the inner and outer chloroplast membranes through interaction with PDV2 in the intermembrane space. *Plant Cell* **2008**, *20*, 2460–2470. [CrossRef] [PubMed]

77. Wang, W.; Li, J.; Sun, Q.; Yu, X.; Zhang, W.; Jia, N.; An, C.; Li, Y.; Dong, Y.; Han, F.; et al. Structural insights into the coordination of plastid division by the ARC6-PDV2 complex. *Nat. Plants* **2017**, *3*, 17011. [CrossRef] [PubMed]

78. Wang, X.; Huang, J.; Mukherjee, A.; Cao, C.; Lutkenhaus, J. Analysis of the interaction of FtsZ with itself, GTP, and FtsA. *J. Bacteriol.* **1997**, *179*, 5551–5559. [CrossRef] [PubMed]

79. Feucht, A.; Lucet, I.; Yudkin, M.D.; Errington, J. Cytological and biochemical characterization of the FtsA cell division protein of *Bacillus subtilis*. *Mol. Microbiol.* **2001**, *40*, 115–125. [CrossRef] [PubMed]

80. Krupka, M.; Rowlett, V.W.; Morado, D.; Vitrac, H.; Schoenemann, K.; Liu, J.; Margolin, W. *Escherichia coli* FtsA forms lipid-bound minirings that antagonize lateral interactions between FtsZ protofilaments. *Nat. Commun.* **2017**, *8*, 15957. [CrossRef] [PubMed]

81. Yoder, D.W.; Kadirjan-Kalbach, D.; Olson, B.J.; Miyagishima, S.Y.; Deblasio, S.L.; Hangarter, R.P.; Osteryoung, K.W. Effects of mutations in Arabidopsis *FtsZ1* on plastid division, FtsZ ring formation and positioning, and FtsZ filament morphology in vivo. *Plant Cell Physiol.* **2007**, *48*, 775–791. [CrossRef] [PubMed]

82. Nakanishi, H.; Suzuki, K.; Kabeya, Y.; Miyagishima, S.Y. Plant-specific protein MCD1 determines the site of chloroplast division in concert with bacteria-derived MinD. *Curr. Biol.* **2009**, *19*, 151–156. [CrossRef] [PubMed]

83. De Boer, P.A.; Crossley, R.E.; Rothfield, L. A division inhibitor and a topological specificity factor coded for by the minicell locus determine proper placement of the division septum in *E. coli*. *Cell* **1989**, *56*, 641–649. [CrossRef]

84. Gao, H.; Kadirjan-Kalbach, D.; Froehlich, J.E.; Osteryoung, K.W. ARC5, a cytosolic dynamin-like protein from plants, is part of the chloroplast division machinery. *Proc. Natl. Acad. Sci. USA* **2003**, *100*, 4328–4333. [CrossRef] [PubMed]

85. Miyagishima, S.Y.; Froehlich, J.E.; Osteryoung, K.W. PDV1 and PDV2 mediate recruitment of the dynamin-related protein ARC5 to the plastid division site. *Plant Cell* **2006**, *18*, 2517–2530. [CrossRef] [PubMed]

7

Whole-Genome Comparison Reveals Heterogeneous Divergence and Mutation Hotspots in Chloroplast Genome of *Eucommia ulmoides* Oliver

Wencai Wang [1], Siyun Chen [2] and Xianzhi Zhang [3,*]

[1]   Institute of Clinical Pharmacology, Guangzhou University of Chinese Medicine, Guangzhou 510000, China; wencaiwang@gzucm.edu.cn

[2]   Germplasm Bank of Wild Species, Kunming Institute of Botany, Chinese Academy of Sciences, Kunming 650201, China; chensiyun@mail.kib.ac.cn

[3]   College of Forestry, Northwest A&F University, Yangling 712100, China

*   Correspondence: zhangxianzhi@nwsuaf.edu.cn

Abstract: *Eucommia ulmoides* (*E. ulmoides*), the sole species of Eucommiaceae with high importance of medicinal and industrial values, is a Tertiary relic plant that is endemic to China. However, the population genetics study of *E. ulmoides* lags far behind largely due to the scarcity of genomic data. In this study, one complete chloroplast (cp) genome of *E. ulmoides* was generated via the genome skimming approach and compared to another available *E. ulmoides* cp genome comprehensively at the genome scale. We found that the structure of the cp genome in *E. ulmoides* was highly consistent with genome size variation which might result from DNA repeat variations in the two *E. ulmoides* cp genomes. Heterogeneous sequence divergence patterns were revealed in different regions of the *E. ulmoides* cp genomes, with most (59 out of 75) of the detected SNPs (single nucleotide polymorphisms) located in the gene regions, whereas most (50 out of 80) of the indels (insertions/deletions) were distributed in the intergenic spacers. In addition, we also found that all the 40 putative coding-region-located SNPs were synonymous mutations. A total of 71 polymorphic cpDNA fragments were further identified, among which 20 loci were selected as potential molecular markers for subsequent population genetics studies of *E. ulmoides*. Moreover, eight polymorphic cpSSR loci were also developed. The sister relationship between *E. ulmoides* and *Aucuba japonica* in Garryales was also confirmed based on the cp phylogenomic analyses. Overall, this study will shed new light on the conservation genomics of this endangered plant in the future.

Keywords: *Eucommia ulmoides*; chloroplast genome; heterogeneous divergence; mutation hotspots; whole-genome comparison

1. Introduction

There are profuse paleoendemics (e.g., Eucommiaceae) and/or phylogenetically primitive taxa (e.g., Cercidiphyllaceae) in China due to the glaciation refuge role played during the Quaternary period [1,2]. Unfortunately, up to *circa* 5000 flora species are currently endangered in China, some of which have already become extinct [3]. Many plant species with important medicinal values have also been threatened seriously due to the increasing demand for raw materials of medicines, over-harvesting and habitat-loss [4–6]. Conservation of medicinal plants has become one of the most urgent issues faced today in China.

Eucommia ulmoides Oliver, a dioecious woody plant endemic to China, is the sole species in the family Eucommiaceae [7]. *E. ulmoides* has been widely cultivated and used as a herbal drug to reduce blood pressure and strengthen the body in central and southern China for at least 2000 years [8,9]. *E. ulmoides* is also well-known as a "hardy rubber" tree that produces trans-polyisoprene rubber (i.e., gutta or Eu-rubber) in the leaves, bark, and pericarp [10,11]. It has been shown that *Eucommia* fossils occurred widely across the Northern Hemisphere from the Palaeocene onwards [12], which indicates that *E. ulmoides* is a representative model of Tertiary relict species, i.e., living from Tertiary to present. However, *E. ulmoides* may have been extinct in the wild and already listed in the Red List of Endangered Plant Species in China probably due to exhaustive human exploration [13,14]. Therefore, effective strategies are urgently needed to conserve this rare and endangered medicinal plant.

To date, studies on *E. ulmoides* have mainly focused on the morphological variation and the natural products [8,15]. Molecular and population genetics studies of this valuable tree lag behind largely due to limited DNA sequence resources [16,17]. Recently, nuclear microsatellites (nrSSR) were developed to investigate the genetic diversity of *E. ulmoides* [18,19]. Amplified fragment length polymorphism (AFLP) and sequence-related amplified polymorphism (SRAP) have been used to construct genetic maps of *E. ulmoides* [20,21]. The genetic markers of random amplified polymorphic DNA (RAPD), chloroplast microsatellite (cpSSR) and inter-simple sequence repeat (ISSR) have also been uncovered [22–24]. Nevertheless, the variability of these developed fingerprinting markers in *E. ulmoides* is relatively low, with limited population genetics information. A new and promising marker type i.e., Single Nucleotide Polymorphism (SNP) has gained high popularity during the last two decades [25,26]. With the on-going progress of high throughput sequencing techniques, it has become convenient to collect large-scale SNP data for genetic analyses [27,28]. Using SNP markers in conservation genetics studies of endangered plants has attracted much attention; for instance, in *Pinus ponderosa* Douglas ex Lawson [29], and *Sciadopitys verticillata* (Thunb.) Siebold and Zucc [30].

Chloroplast (cp) DNA sequences have been extensively used in the studies of plant population genetics and molecular phylogenetics [31–33]. Typically, cp genomes of land plants have a quadripartite structure with a pair of inverted repeats (IRs) separating a large single-copy (LSC) region and a small single-copy (SSC) region, ranging from 115 to 165 kilobase (kb) [34]. The cp genomes in general are inherited uniparentally, mostly maternally and are essentially recombination-free, leading to a smaller effective population size and a shorter coalescent time than the nuclear genomes [35]. Recently Wang et al. [17] reported a cp genome sequence of *E. ulmoides* with a length of 163,341 bp. Clearly, the availability of additional sequenced cp genomes from *E. ulmoides* would aid our understanding of the cp genome-wide variation at the individual level. Through comparative genomic analysis, polymorphic cpDNA loci with plentiful SNPs and indels i.e., nucleotide insertions and deletions can also be detected, which would be useful for further population genetics studies of *E. ulmoides*.

Genome skimming is currently one of the most economical techniques to obtain plastome sequences [36], through which obtaining complete cp genomes for plant phylogenomics inference becomes convenient [37]. In this study, we generated and characterized one complete cp genome of *E. ulmoides* using the genome skimming approach. By comparing the cp genome generated in this study and the one published previously [17], our main goals were to: (1) test whether the cp genomes in *E. ulmoides* show structural rearrangements; (2) reveal the divergence pattern of the cp genome in *E. ulmoides*; (3) identify highly variable cp genome-wide markers for subsequent population genetics studies of *E. ulmoides*.

2. Results

2.1. Chloroplast Genome Variation in E. ulmoides

About 20 million clean reads (4.72 Gb data) were generated from genome skimming sequencing. Two assembly methods (CLC Genomics Workbench and SPAdes software) both obtained the complete cp genome of *E. ulmoides* with high genome coverage (>180×) and there is no difference between the two assembled sequences, suggesting a high-quality cp genome map was achieved. The final cp genome size was determined to be 163,586 bp, similar to the previously published one (KU204775) (Table 1). The number of protein-coding genes, tRNA genes and rRNA genes, were the same as those in the available *E. ulmoides* cp genome (KU204775). We have deposited the newly sequenced *E. ulmoides* cp genome in GenBank with accession number MF766010. The genome skimming sequencing reads have also been deposited in the Sequence Read Archive (SRA) with the accession number PRJNA399774.

Table 1. Comparison between the newly and previously sequenced chloroplast genomes of *Eucommia ulmoides*.

Item	This Study	KU204775
Chloroplast genome size (bp)	163,586	163,341
LSC [a] length (bp)	86,764	86,592
SSC [b] length (bp)	14,166	14,149
IRa/IRb [c] length (bp)	31,328	31,300
Number of genes (unique genes)	136 (115)	136 (115)
Number of protein-coding genes (unique genes)	89 (80)	89 (80)
Number of tRNA genes (unique genes)	39 (31)	39 (31)
Number of rRNA genes (unique genes)	8 (4)	8 (4)
GC [d] content (%)	38.33%	38.34%
Protein-coding regions (%)	51.91%	51.99%

[a] LSC, large single-copy region; [b] SSC, small single-copy region; [c] IRa/IRb, two identical inverted repeat regions a/b; [d] GC, Guanine and Cytosine.

The whole genome alignments from MAFFT (Figure 1A) and MAUVE (Figure 1B) were consistent. There were no large genome rearrangements in the two cp genomes, which indicated that the cp genome structure in *E. ulmoides* is highly conserved and perfectly syntenic (Figure 1). Interestingly, small-scale nucleotide insertions and deletions were detected in the *E. ulmoides* cp genome. We found 15 insertions with more than ten nucleotides (11–111 bp) in the two cp genomes (Table 2). Five deletions in the range of 16–90 bp were also uncovered. It is worth noting that all of these insertions and deletions were involved in repeat sequence expansions and contractions (Table 2). Furthermore, across the entire cp genome of *E. ulmoides*, the sequence divergences were not uniform but highly heterogeneous (Figure 1).

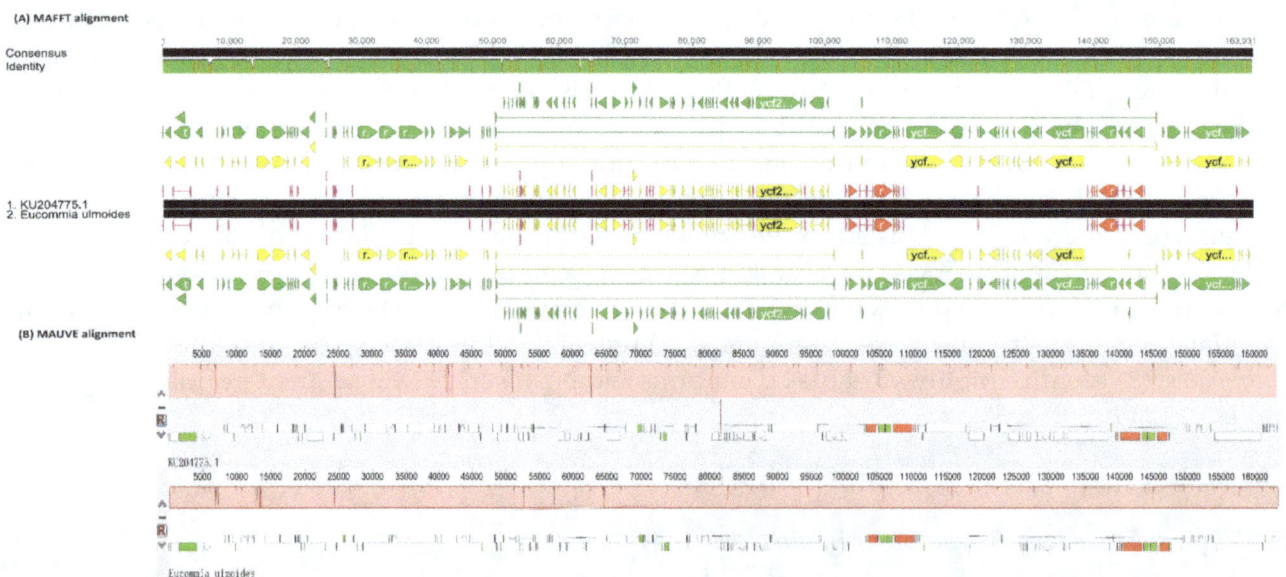

Figure 1. Conserved chloroplast genome structure in *Eucommia ulmoides*. (**A**) Pairwise chloroplast genome alignments derived from Multiple Alignment using Fast Fourier Transform (MAFFT) program. The sequence identity is indicated on the top. Label KU204775.1 represents the *E. ulmoides* chloroplast genome retrieved from GenBank, while label *E. ulmoides* indicates the newly sequenced genome in this study. (**B**) Pairwise chloroplast genome alignments derived from MAUVE software.

Table 2. DNA insertions and deletions with more than 10 nucleotides in the chloroplast genomes of *Eucommia ulmoides*.

No.	Size (bp)	Start Position	Location	Type
1	56	6851	*rps16-trnT(UGU)*	insertion
2	27	7006	*rps16-trnT(UGU)*	insertion
3	45	7196	*rps16-trnT(UGU)*	insertion
4	13	12,693	*ycf3-psaA*	insertion
5	23	12,912	*ycf3-psaA*	insertion
6	111	13,312	*ycf3-psaA*	insertion
7	12	13,471	*ycf3-psaA*	insertion
8	32	24,279	*psbD-trnT(GGU)*	insertion
9	12	26,615	*trnD(GUC)-psbM*	insertion
10	11	51,194	*rps12-rpl20*	insertion
11	17	52,547	*rps18-rpl33*	insertion
12	40	57,075	*psbJ-petA*	insertion
13	18	64,506	*accD-trnM(CAU)*	insertion
14	12	73,717	*trnL(UAA) intron*	insertion
15	14	127,486	*ndhG-ndhI*	insertion
16	16	4673	*trnK(UUU)-rps16*	deletion
17	44	24,700	*trnG(TTU)-trnE(UUC)*	deletion
18	31	41,767	*atpI-atpH*	deletion
19	44	51,109	*rps12-rpl20*	deletion
20	90	62,865	*accD-trnM(CAU)*	deletion

2.2. Molecular Marker Development

A total of 155 mutational events, including 75 nucleotide substitutions (SNPs) and 80 nucleotide indels (insertions and deletions), were detected within 71 loci of the cp genome in *E. ulmoides* (Figure 2). There were 98 mutations (51 SNPs and 47 indels) and 15 mutations (12 SNPs and 3 indels) in the LSC and SSC regions, respectively. In addition, 42 mutations (12 SNPs and 30 indels) were located in the IR region. Distribution patterns of SNPs and indels differed largely in the cp genic and intergenic regions of *E. ulmoides*. Most of the SNPs (59 out of 75) were found in the gene sequences, including 31 protein-coding genes and one tRNA gene. In contrast, indels were mainly (50 out of 80) distributed in the intergenic spacers (Figure 2). Upon further investigation of SNPs and indels in the nine intron-containing protein-coding genes (*atpF, ndhA, ndhB, rpl2, rpl16, rpoC1, rps12, rps16, ycf3*), we found that all the mutations were located in the intron regions. In all, 40 SNPs and 14 indels occurred in the plastid-coding sequence (CDS) regions.

The proportion of variability in the 71 polymorphic loci ranged from 0.03% to 1.55% with a mean value of 0.37% (Figure 3). The mutation rates in most (53 out of 71) of the loci were between 0.10 and 1.00%. Five of these DNA fragments i.e., *atpF-atpA, rps18-rpl33, psaJ, infA* and *rpl32* had variations exceeding 1.00%. Considering the relatively high percentage of variability and convenience for primer design in PCR (Polymerase Chain Reaction) and sequencing experiments, we chose 20 highly variable loci with length of 200–1500 bp as potential molecular markers for subsequent population genetic studies (Table 3). The percentage of variations in these 20 loci all exceeded 0.25%, among which 16 had a percentage of variable characters (VCs) greater than 0.30% (Table 3).

Through SSR analysis, we found a total of 31 SSR loci in the newly assembled *E. ulmoides* cp genome, among which 27 were shared by the two genomes. Further detection revealed that eight cpSSR loci were polymorphic in *E. ulmoides* (Table 4). All the polymorphic cpSSR loci were mononucleotide repeats, ranging from 10–15 bp in length. Five polymorphic cpSSR loci were located in the LSC region, with another three ones in the IR regions (Table 4).

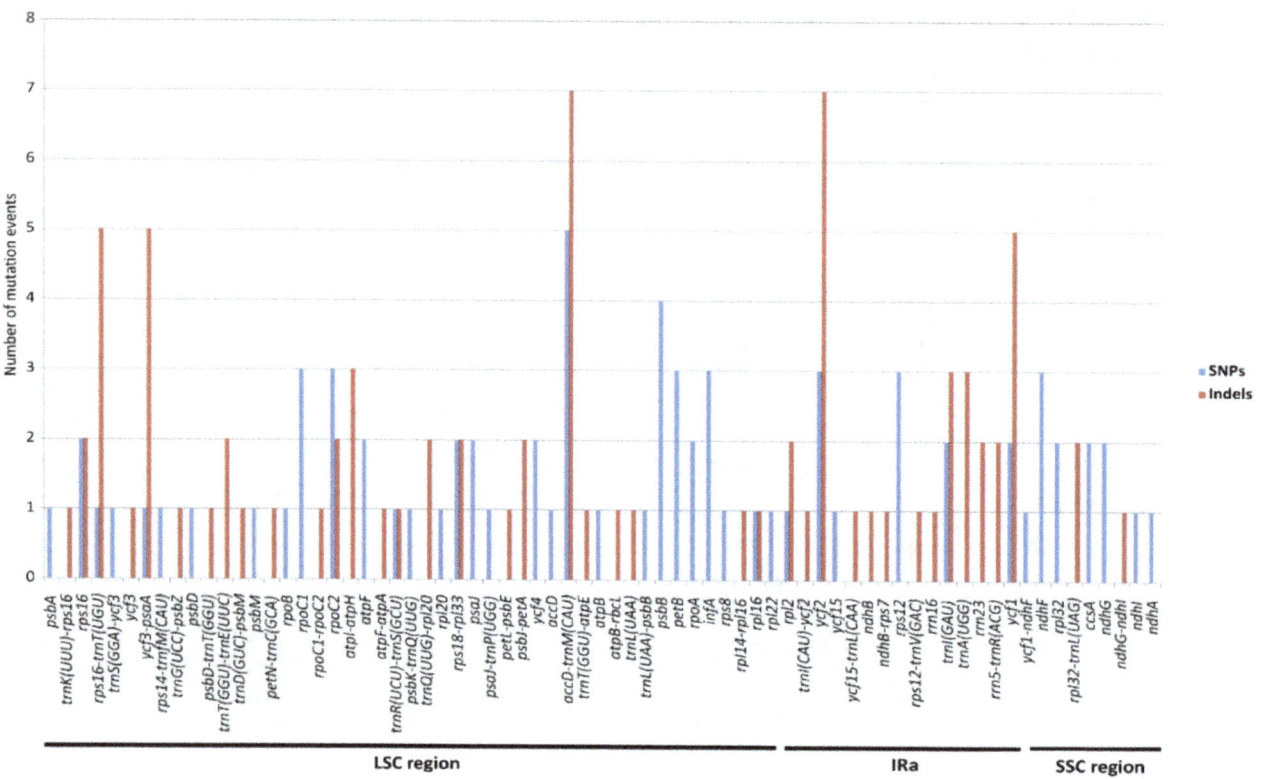

Figure 2. Mutational events (SNPs and indels) detected across the chloroplast genome of *Eucommia ulmoides*. SNPs (single nucleotide polymorphisms) indicate nucleotide substitutions and indels represent nucleotide insertions and deletions. The homologous loci are oriented according to their locations in the chloroplast genome.

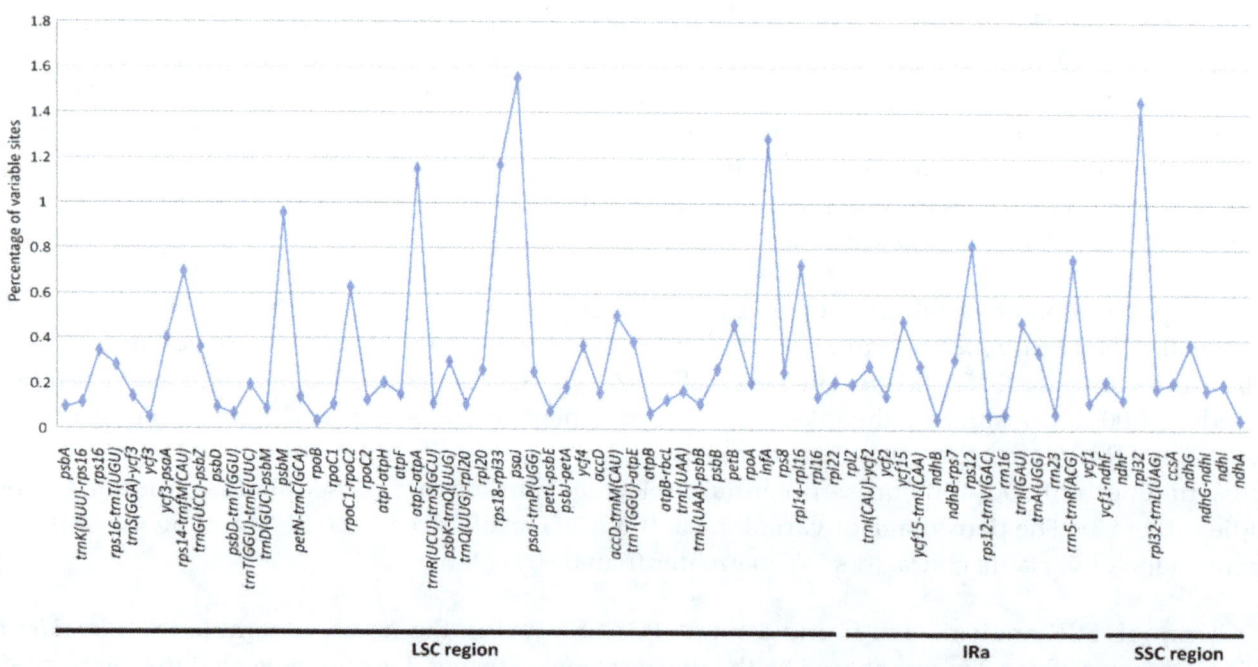

Figure 3. Percentage of variable characters (SNPs and indels) in polymorphic chloroplast loci in *Eucommia ulmoides*. The homologous loci are oriented according to their locations in the chloroplast genome.

Table 3. The 20 chloroplast DNA fragments with relative high genetic divergences identified in *Eucommia ulmoides*.

Region	Aligned Length (bp)	No. VCs [a]	Percentage of VCs (%)
infA	234	3	1.28
rps18-rpl33	343	4	1.17
rps12	369	3	0.81
rrn5-trnR(ACG)	266	2	0.75
ycf15	210	1	0.48
trnI(GAU)	1062	5	0.47
petB	651	3	0.46
ycf3-psaA	1502	6	0.40
trnT(GGU)-atpE	261	1	0.38
ndhG	531	2	0.38
ycf4	546	2	0.37
trnG(UCC)-psbZ	280	1	0.36
rps16	1170	4	0.34
trnA(UGG)	881	3	0.34
ndhB-rps7	325	1	0.31
psbK-trnQ(UUG)	338	1	0.30
trnI(CAU)-ycf2	357	1	0.28
ycf15-trnL(CAA)	359	1	0.28
psbB	1521	4	0.27
rpl20	384	1	0.26

[a] VCs: variable characters, including SNPs and indels.

Table 4. The polymorphic chloroplast SSRs identified in *Eucommia ulmoides*.

No.	SSR Repeat Motif	Length Variation (bp)	Location	Region [a]
1	(G)	10–11	*trnG(UCC)-psbZ*	LSC
2	(A)	12–15	*rpoC2*	LSC
3	(A)	12–13	*ycf1*	IRb
4	(A)	13–14	*rpl32-trnL(UAG)*	IRa
5	(T)	10–14	*psbJ-petA*	LSC
6	(T)	10–11	*trnG(GCC)-trnS(GCU)*	LSC
7	(T)	12–13	*ycf1*	IRa
8	(T)	14–15	*rpl16-rps3*	LSC

[a] LSC, large single-copy region; IRa/IRb, two identical inverted repeat regions a/b.

2.3. SNP Calling and Phylogenomic Inference

The SNPs calling analysis using the previously published *E. ulmoides* cp genome (KU204775) as reference revealed a total of 75 SNPs. This result of SNP occurrence was consistent with the aforementioned molecular marker analysis (75 SNPs, Figure 2), which indicated that the detected SNPs were really present in different individuals of *E. ulmoides*. Further examination of the 40 SNPs in the CDS regions suggested that all these SNPs were synonymous, i.e., no amino acid change at the protein level. There were 34 transitions and six transversions in the protein-coding region SNPs. The average frequency of SNPs occurrence in the *E. ulmoides* cp genome was calculated as 0.46 per kb.

The final length of the supermatrix dataset contained 96,894 unambiguously aligned nucleotide characters. Three methods produced a congruent phylogenetic tree, shown in Figure 4. Eudicots, monocots and magnoliids were all highly supported to be monophyletic (94–100/90–100/1.00). Magnoliids diverged firstly, followed by monocots, then Chloranthales and eudicots, with relatively low (60–72/74–80/0.90–0.99) support values. Two *E. ulmoides* individuals clustered together with high statistical support values (100/100/1.00), which was subsequently sister to *Aucuba japonica* Thunb.

in the Garryales clade with high statistical support values (100/100/1.00). Garryales, Gentianales and Solanales formed as the highly supported lamiids lineage (100/100/1.00), resolved as (Garryales, (Gentianales, Solanales)). Asterales and Dipsacales formed a highly supported clade, campanulids (100/100/1.00), which was sister to lamiids (100/100/1.00) in asterids. Ericales was resolved as the basal most group in the asterids lineage (100/100/1.00). Brassicales, Malvales, Myrtales and Sapindales clustered in the malvids clade (95/98/1.00) with resolution as (((Brassicales, Malvales), Sapindales), Myrtales). Fabales, Malpighiales and Rosales were located in the fabids lineage (98/100/1.00), which was a sister clade to malvids in rosdis.

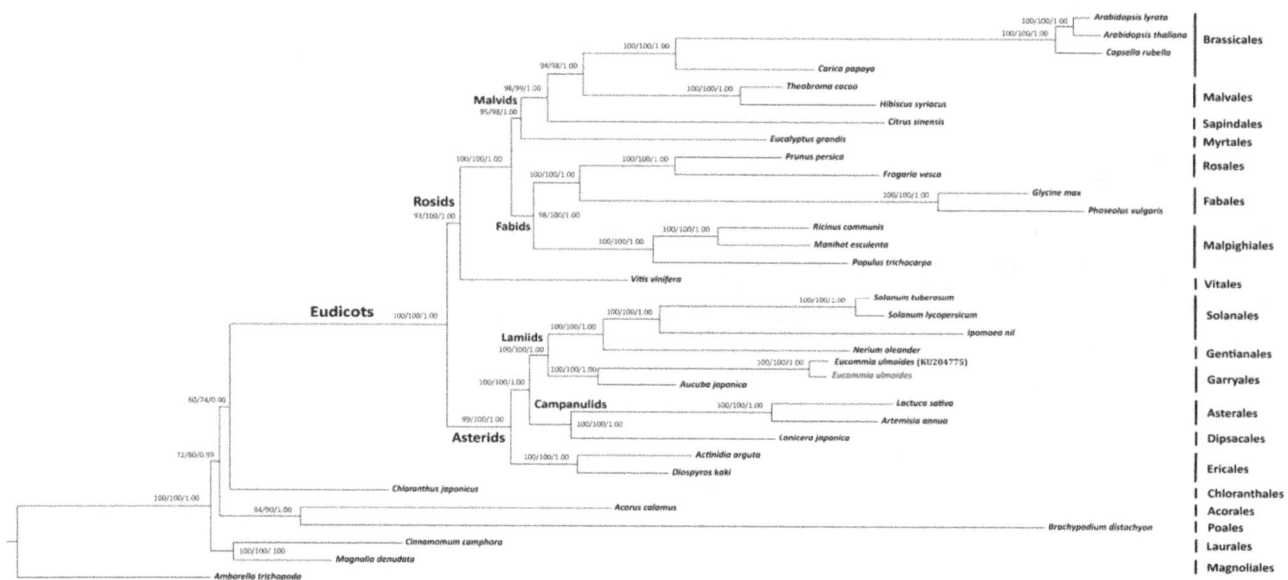

Figure 4. Maximum likelihood (ML) tree for 34 taxa based on 80 unique plastid protein-coding genes of *Eucommia ulmoides*. Values above the branches represent maximum parsimony bootstrap (MPBS)/maximum likelihood bootstrap (MLBS)/Bayesian inference posterior probability (PP). The newly sequenced *Eucommia ulmoides* chloroplast genome is indicated by red color and the previously published *E. ulmoides* chloroplast genome is followed by its GenBank accession number KU204775.

3. Discussion

3.1. Conserved Chloroplast Genome Structure in E. ulmoides

Land plant cp genomes are generally inherited as a haplotype with no recombination, providing useful genetic information to trace relationships between different species [35,38]. Within species cp genome structure was highly conserved [39]. As expected, it is the case in *E. ulmoides* in terms of the contained genes and coding regions in the cp genomes (Table 1). Further whole-genome alignments suggested that the two cp genomes of *E. ulmoides* did not show genome rearrangement having the same linear gene order (Figure 1). As such it is reasonable to use cp genomes for subsequent conservation genomics studies on *E. ulmoides*.

It is noteworthy that the newly sequenced cp genome of *E. ulmoides* (163,586 bp) in this study is 245 bp larger than that of the previously reported one (163,341 bp, [17]) (Table 1). The cp genome size variation within different individuals of the same species has been reported for several other plants, such as in *Camptotheca acuminate* Decne. with its size varied as 157,806 bp [40], 157,877 bp [41] and 162,382 bp [42]. Nuclear genome size variations in plants are mostly caused by the repeats activities (e.g., expansions/contractions) via illegitimate recombination in addition to polyploidy [43–47]. In the two *E. ulmoides* cp genomes, we detected 15 insertions and five deletions, with more than 10 nucleotides for each (Table 2). All these sequences were observed to be part of or the whole DNA repeats. For instance, the repeat sequences in *rps16-trnT(UGU)*, *psbD-trnT(GGU)*, *rps12-rpl20* and *accD-trnM(CAU)* have been detected in the study of Wang et al. [17] as well. Therefore, potentially the illegitimate recombination between repeat regions of *E. ulmoides* cp genome may contribute to the cp genome size variation.

3.2. Heterogeneous Divergence in E. ulmoides Chloroplast Genome

Heterogenous divergence patterns in cp genomes have been reported in several plant groups, such as in Actinidiaceae [37] and in Poaceae [48]. The alignment of the two available cp genomes of *E. ulmoides* revealed highly heterogeneous sequence divergences within this species (Figures 1 and 2). All the identified SNPs and indels from the intron-containing protein-coding genes were located in the intron regions. Due to natural selection CDS regions are in general more conserved than non-coding regions (i.e., intergenic sequences and introns) [49]. In addition, nucleotide substitutions likely have less destructive effect to the integrity of open reading frame (ORF) than indels [50]. We, thus, speculated that this functional constraint may lead to the contrasting occurrences of SNPs and indels in the *E. ulmoides* cp genomes.

The occurrence of synonymous SNPs was more abundant than that of the non-synonymous SNPs in CDS regions because of selection process [51,52]. As expected, all the SNPs detected in protein-coding regions were synonymous. Moreover, since the transitions rather than the transversions usually would generate more synonymous mutations in the CDS the transitions SNPs are more easily retained than the transversion ones [52,53]. In this study we found that the transition SNPs (34) were indeed more frequently detected than the transversion SNPs (6). Previous studies have reported a high level of nuclear genetic diversity at the population level of *E. ulmoides* [18,54]. In this study, we revealed that the frequency of plastid SNPs were 0.46 per kb at the whole cp genome level, lower than the average of 1.02 per kb in the nuclear genes of *E. ulmoides* [54]. The difference of SNP frequency between the cp genome and the nuclear genes could be caused by insufficient sampling in this study and/or different variation rates between the plastid and nuclear sequences.

3.3. Mutation Hotspots in E. ulmoides Chloroplast Genome

In general, protein-coding genes in the cp genome have lower sequence variation than the non-coding loci, for instance in bamboos [55] and mimosoid legume [56]. However, an accelerated variation rate of some plastid protein-coding genes has been reported, such as the *psb* in Poaceae [51], *rps* in Saxifragales [57], and *accD* and *rpl20* in Actinidiaceae [37]. In *E. ulmoides*, we found three protein-coding genes (*infA, psaJ* and *rpl32*) that varied the most quickly, all having variations exceeding 1.2% (Figure 3). The gene *infA* encodes translation initiation factor 1 and has been found missing in cp genomes of several plant lineages, e.g., in rosids [58]. The other two genes i.e., *psaJ* and *rpl32* code for photosystem I protein J and ribosomal protein L32, respectively, both of which are short in length with the former having 129 bp and the latter 138 bp (this study and [17]). The relatively high level of variation of these genes in *E. ulmoides* indicates that they are less constrained. Abnormal DNA replication, repair or recombination [59,60] may lead to the elevated divergence of these genes.

The genetic markers of SSR, AFLP and SRAP have been developed and used for the population genetics studies of *E. ulmoides* [18,20,21]. However, the above fingerprinting markers [23,24] may provide insufficient genetic information to resolve the population structure and history of *E. ulmoides*. The relationships among natural and cultivated populations of *E. ulmoides* are elusive at present [18,24]. SNP markers are ample in plant cp genomes, making them useful candidates for population genetics studies. Using cp genome SNP markers has widely received attention during the past few years with the advances of high throughput sequencing techniques [61,62]. Highly variable cpDNA fragments have been mined for phylogenetic and population genetic studies in several species using cp genomes data, such as in kiwifruit [37] and temperate woody bamboos [55]. Given that it would be easy to amplify and sequence DNA fragments with length from *circa* 200 to 1500 bp using Sanger sequencing method [62,63], we thus chose 20 cpDNA loci with relatively high genetic divergences as potential molecular markers (Table 3) for subsequent population genomics studies of *E. ulmoides*. These selected plastid genome-wide loci would genetically be informative for uncovering the genetic relationships among the natural and cultivated *E. ulmoides* populations.

Additionally, 27 cpSSR loci identified by Wang et al. [17] were also confirmed in our SSR analysis, among which eight were further mined as polymorphic cpSSR loci (Table 4). Given that polymorphic cpSSR loci could be applied as useful markers to meet certain study purposes under the circumstances of limited budget [64,65], the newly developed polymorphic cpSSR loci in *E. ulmoides* here would be potential genetic markers to facilitate subsequent population genetics studies in the future.

3.4. Phylogenomic Validation of E. ulmoides

The newly obtained *E. ulmoides* cp genome was further validated via phylogenomic analyses using 34 complete plastomes from 10 major lineages of angiosperms. The resulting phylogenomic tree highly supported the clade of two *E. ulmoides* cp genomes (Figure 4), confirming the validity of the assembled and annotated cp genome of *E. ulmoides* in this study. The sister relationship between *E. ulmoides* and *A. japonica* in the Garryales clade was highly supported, which is consistent with the results derived from five organellar genes [66] and 36 plastid genes [17], supporting the classification of *E. ulmoides* (Eucommiaceae) in the updated APG IV system [67]. *E. ulmoides* and *A. japonica* are both woody and have unisexual flowers in separate individuals, which seem to be morphological synapomorphies for the order Garryales [68]. An average of 92.6% identities between 78 common unique cp protein-coding genes in *E. ulmoides* and *A. japonica* were also detected, suggesting a high similarity between the two species at the molecular level. Garryales was shown to be closely related to the clade of (Gentianales + Solanales) in lamiids, in line with the APG IV system [67].

All 20 sampled orders were highly supported to be monophyletic separately (Figure 4), agreeing with the APG IV system [67]. Within eudicots two large sister clades i.e., asterids and rosids were uncovered, and the relationships among these two lineages were highly resolved as ((campanulids, lamiids), Ericales) and ((fabids, malvids), Vitales), respectively as stated previously [69]. The branching patterns of species within campanulids and lamiids are consistent with recent studies [66,70] as (((Gentianales, Solanales), Garryales), (Asterales, Dipsacales)). Our analyses also resolved the phylogenetic relationships within fabids and malvids as (((((Brassicales, Malvales), Sapindales), Myrtales), ((Fabales, Rosales), Malpighiales)), consistent with the results of previous studies [17,69]. It is noteworthy that the branching orders of magnoliids, monocots, Chloranthales and eudicots only obtained low-level support values here (Figure 4). Further studies with expanded taxon samples are expected to confirm the phylogeny of these lineages. Moreover, plastome is inherited uniparentally in general, which might introduce biases to species phylogeny inference [71,72]. Analyses using orthologous nuclear genes are also needed for studying the evolutionary history of *E. ulmoides* among the flowering plants [54].

4. Materials and Methods

4.1. Plant Materials and DNA Sequencing

Fresh healthy leaves were collected from an adult male individual of *E. ulmoides* growing in the Arboretum of Northwest Agriculture and Forest University in Yangling, Shanxi, China, in April 2015. After collection, the leaves were immediately immersed in liquid nitrogen and then stored at $-80\,°C$ until use. The voucher specimen of this tree was deposited at the Trees Herbarium of Northwest A and F University with accession number ZXZ15027.

Total genomic DNA was extracted by the CTAB method [73]. Paired-end (PE) libraries with insert size *circa* 500 bp were constructed from fragmented genomic DNA based on standard Illumina protocols (Illumina Inc., San Diego, CA, USA). Prepared library was then sequenced for PE 100 bp read length on the Illumina HiSeq 2000 platform at the Beijing Genomics Institute (BGI) in Shenzhen, China.

4.2. Genome Assembly and Annotation

Fastq format PE reads were supplied with adaptor sequences removed. Poor quality reads with phred scores lower than 20 for more than 10% of their bases were also removed. Two independent methods were used to assemble the *E. ulmoides* cp genome. (1) The cp genome was de novo assembled using the CLC Genomics Workbench v7.5 software (CLC Bio, Aarhus, Denmark) based on the clean reads. After discarding contigs with length <300 bp and sequences with coverage <50, the remaining contigs were searched against the available cp genome of *E. ulmoides* (GenBank accession number KU204775) that used as the reference by BLAST (http://blast.ncbi.nlm.nih.gov/) with *e*-value $<10^{-5}$. Aligned contigs with ≥90% similarity and query coverage were determined as cpDNA sequences and ordered according to the reference genome. Small gaps were filled using PE clean reads as conducted in Wang et al. [37]. (2) The clean reads were firstly mapped to the reference cp genome of *E. ulmoides* to determine the proportion of cpDNA using Bowtie v2.3.1 program [74] with a maximum of 3 mismatches. Subsequently, we applied SPAdes v3.9 software [75] with default setting to assemble the cp genome using the determined cpDNA clean reads.

DOGMA software [76] was used for initial cp genome annotation. Start/stop codons and intron/exon boundaries were checked and adjusted manually when necessary by comparing to the reference genome. tRNA genes were confirmed based on tRNAscan-SE 1.21 [77].

4.3. Genome-Wide Comparison and Divergent Hotspot Identification

The previously published cp genome of *E. ulmoides* (accession number: KU204775) was downloaded from GenBank database (https://www.ncbi.nlm.nih.gov/genbank/). This genome was aligned with the *E. ulmoides* cp genome described herein, using MAFFT program [78] and MAUVE software [79], respectively, and manually adjusted where necessary. The obtained pairwise alignment of the cp genomes was visualized in Geneious v9.0 [80]. Moreover, given the genome repeat sequences expansion and contraction may result in genome size variation [81], we examined the DNA insertions and deletions in repeat regions of the two *E. ulmoides* cp genomes.

The two *E. ulmoides* cp genomes were analyzed to identify molecular markers that can be selected in subsequent population genetic studies. We firstly extracted both the genic and intergenic DNA fragments in each cp genome using the "Extract Sequences" option in DOGMA [76]. Then the homologous loci were aligned individually by MUSCLE program (http://www.drive5.com/muscle/) [82] implemented in Geneious v9.0 [80] with default settings. Manual adjustments were made for the alignments where necessary. The proportion of mutational events for each genic and intergenic locus was calculated as follows: the proportion of variation = ((NS + ID)/L) × 100, where NS = the number of nucleotide substitutions (SNPs), ID = the number of indels (insertions and deletions), L = the aligned sequence length.

Polymorphic cpSSR loci were further mined by genome comparison. Firstly, SSRs in the newly sequenced *E. ulmoides* cp genome were detected by MISA perl script (http://pgrc.ipk-gatersleben.de/misa/). The parameter of minimum repeat unit was set as 10 for mono-, 6 for di-, and 5 for tri-, tetra-, penta-, and hexanucleotide SSRs [83]. Then, all the identified SSR loci were compared to the 29 cpSSR loci of Wang et al. [17] to develop polymorphic cpSSR markers in *E. ulmoides*.

4.4. SNPs Validation and Phylogenomic Analyses

To confirm the SNPs identified by the aforementioned cp genome alignment, we here mapped the genome skimming clean reads generated in this study to the previously published *E. ulmoides* cp genome (KU204775) [17] for SNP calling. Picard-tools v1.41 (http://broadinstitute.github.io/picard/) and samtools v0.1.18 [84] were applied to sort and remove duplicated reads and merge the bam alignment results. GATK3 software [85] was further used to perform SNPs identification. Raw vcf files were filtered with GATK standard filter method and other parameters were set as defaults. Moreover, to reveal if the coding region SNPs detected in *E. ulmoides* cp genome caused amino acid substitution on protein level, we firstly translated each protein-coding gene into amino acids in Geneious v9.0 [80]. Then the protein sequences of each gene were aligned, respectively using MUSCLE [82]. The mutational events were checked to uncover the synonymous and nonsynonymous SNPs and the nucleotide transitions and transversions.

Phylogenomic analyses were also conducted to validate the newly assembled and annotated *E. ulmoides* cp genome. 34 plastomes representing 10 major lineages of angiosperms (Table S1) were included for phylogenomic analyses. *Amborella trichopoda* from basal angiosperm lineages was defined as outgroup according to previous studies [66,67]. 80 unique plastid protein-coding genes of *E. ulmoides* (Table S2) were used for the phylogenetic inferences. Each gene was aligned individually by MUSCLE [82] in Geneious v9.0 [80], and then concatenated as a supermatrix. Gaps were not included in the dataset.

Three methods i.e., maximum parsimony (MP), maximum likelihood (ML), and Bayesian inference (BI) were used for phylogenetic reconstruction. We performed parsimony heuristic tree searches in PAUP v4.0b10 [86] with parameters set as 1000 random addition sequence replicates, tree bisection and reconnection (TBR) branch swapping, and MulTrees option in effect. 1000 bootstrap replicates [87] were calculated to evaluate the branch support (MPBS) of the MP tree. RAxML v.8.2.8 [88] and MrBayes 3.2.6 [89] in the CIPRES Science Gateway v3.3.3 [90] were applied for ML and BI analyses, respectively. Supermatrix was partitioned by genes and GTR + G model of nucleotide substitution was used. For the ML tree we conducted 1000 fast bootstrap ML reps to assess the support values (MLBS) of internal nodes. In Bayesian analysis two runs with four chains were carried out up to 50,000,000 generations, sampling one tree every 1000 generations till convergence, i.e., the average standard deviation of split frequencies <0.01. We discarded the first 25% of trees as burn-in, and used the remaining trees to estimate the majority-rule consensus BI tree and posterior probabilities (PP).

5. Conclusions

In summary, in the present study we generated one complete cp genome of *E. ulmoides* using the genome skimming approach. Through comprehensive genome-wide comparative analyses we found that the cp genomes within *E. ulmoides* were highly conserved in terms of structure and content. Nevertheless, obviously heterogeneous sequence divergences were revealed in different regions of the *E. ulmoides* cp genome. A total of 20 polymorphic DNA fragments and eight SSR loci have been identified as potential cpDNA markers for subsequent population genetics studies of this tree species. The phylogenetic placement of *E. ulmoides* in angiosperms was robustly resolved as well based on the cp genomes data, strongly supporting the sister relationship between *E. ulmoides* and *A. japonica* in the asterids lineage. The data presented here will aid further conservation genomic studies and facilitate the development of plastid genetic engineering for *E. ulmoides*.

Acknowledgments: We would like to thank Zhirong Zhang for the help in the experiment performance. This work is supported by the National Natural Science Foundation of China (31600173) and the Basic Science Fund of Northwest A&F University (2452016052).

Author Contributions: Wencai Wang and Xianzhi Zhang conceived and designed the experiments; Wencai Wang performed the experiments; Siyun Chen and Xianzhi Zhang analyzed the data; Xianzhi Zhang contributed reagents/materials/analysis tools; Wencai Wang and Xianzhi Zhang wrote the paper. All authors reviewed and approved the final manuscript.

References

1. Huang, H. Plant diversity and conservation in China: Planning a strategic bioresource for a sustainable future. *Bot. J. Linn. Soc.* **2011**, *166*, 282–300. [CrossRef] [PubMed]

2. Liu, J.; Ouyang, Z.; Pimm, S.L.; Raven, P.H.; Wang, X.; Miao, H.; Han, N. Protecting China's biodiversity. *Science* **2003**, *300*, 1240–1241. [CrossRef] [PubMed]

3. Lópezpujol, J.; Zhang, F.M.; Ge, S. Plant biodiversity in China: Richly varied, endangered, and in need of conservation. *Biodivers. Conserv.* **2006**, *15*, 3983–4026. [CrossRef]

4. Gu, J. Conservation of plant diversity in China: Achievements, prospects and concerns. *Biol. Conserv.* **1998**, *85*, 321–327. [CrossRef]

5. Chen, S.L.; Hua, Y.; Luo, H.M.; Wu, Q.; Li, C.F.; Steinmetz, A. Conservation and sustainable use of medicinal plants: Problems, progress, and prospects. *Chin. Med.* **2016**, *11*, 37. [CrossRef] [PubMed]

6. Huang, L.; Yang, B.; Wang, M.; Fu, G. An approach to some problems on utilization of medicinal plant resource in China. *China J. Chin. Mater. Med.* **1999**, *24*, 70–73.

7. Zhang, Z.Y.; Zhang, H.D.; Turland, N.J. Eucommiaceae. In *Flora of China*; Wu, Z.Y., Raven, P.H., Hong, D.Y., Eds.; Science Press and Missouri Botanical Garden: Beijing, China, 2003; p. 43.

8. Kawasaki, T.; Uezono, K.; Nakazawa, Y. Antihypertensive mechanism of food for specified health use: "*Eucommia* leaf glycoside" and its clinical application. *J. Health Sci.* **2000**, *22*, 29–36.

9. Liu, H.; Hongyan, D.U.; Tana, W. Advances in research on biotechnology breeding of *Eucommia ulmoides*. *Hunan For. Sci. Technol.* **2016**, *43*, 132–136.

10. Suzuki, N.; Uefuji, H.; Nishikawa, T.; Mukai, Y.; Yamashita, A.; Hattori, M.; Ogasawara, N.; Bamba, T.; Fukusaki, E.; Kobayashi, A. Construction and analysis of EST libraries of the trans-polyisoprene producing plant, *Eucommia ulmoides* Oliver. *Planta* **2012**, *236*, 1405–1417. [CrossRef] [PubMed]

11. Du, H.Y.; Hu, W.Z.; Yu, R. *The Report on Development of China's Eucommia Rubber Resources and Industry (2014-2015)*; Social Sciences Academic Press: Beijing, China, 2015.

12. Manchester, S.R.; Chen, Z.D.; An-Ming, L.U.; Uemura, K. Eastern Asian endemic seed plant genera and their paleogeographic history throughout the Northern Hemisphere. *J. Syst. Evol.* **2009**, *47*, 1–42. [CrossRef]

13. Mabberley, D.J. *The Plant Book*; Cambridge University Press: Cambridge, UK, 1989.

14. Fu, L.G.; Jin, J.M. *Red List of Endangered Plants in China*; Science Press: Beijing, China, 1992.

15. Du, H.Y. *China Eucommia Pictorial*; China Forestry Publishing House: Beijing, China, 2014.

16. Wang, L.; Du, H.; Wuyun, T.N. Genome-wide identification of microRNAs and their targets in the leaves and fruits of *Eucommia ulmoides* using high-throughput sequencing. *Front. Plant Sci.* **2016**, *7*, 1632. [CrossRef] [PubMed]

17. Wang, L.; Wuyun, T.N.; Du, H.; Wang, D.; Cao, D. Complete chloroplast genome sequences of *Eucommia ulmoides*: Genome structure and evolution. *Tree Genet. Genomes* **2016**, *12*, 12. [CrossRef]

18. Zhang, J.; Xing, C.; Tian, H.; Yao, X. Microsatellite genetic variation in the Chinese endemic *Eucommia ulmoides* (Eucommiaceae): Implications for conservation. *Bot. J. Linn. Soc.* **2013**, *173*, 775–785. [CrossRef]

19. Zhang, W.R.; Li, Y.; Zhao, J.; Wu, C.H.; Ye, S.; Yuan, W.J. Isolation and characterization of microsatellite markers for *Eucommia ulmoides* (Eucommiaceae), an endangered tree, using next-generation sequencing. *Genet. Mol. Res.* **2016**, *15*. [CrossRef] [PubMed]

20. Li, Y.; Wang, D.; Li, Z.; Wei, J.; Jin, C.; Liu, M. A molecular genetic linkage map of *Eucommia ulmoides* and quantitative trait loci (QTL) analysis for growth traits. *Int. J. Mol. Sci.* **2014**, *15*, 2053–2074. [CrossRef] [PubMed]

21. Wang, D.; Li, Y.; Li, L.; Wei, Y.; Li, Z. The first genetic linkage map of *Eucommia ulmoides*. *J. Genet.* **2014**, *93*, 13–20. [CrossRef] [PubMed]

22. Wang, A.Q.; Huang, L.Q.; Shao, A.J.; Cui, G.H.; Chen, M.; Tong, C.H. Genetic diversity of *Eucommia ulmoides* by RAPD analysis. *China J. Chin. Mater. Med.* **2006**, *31*, 1583–1586.

23. Yao, X.; Deng, J.; Huang, H. Genetic diversity in *Eucommia ulmoides* (Eucommiaceae), an endangered traditional Chinese medicinal plant. *Conserv. Genet.* **2012**, *13*, 1499–1507. [CrossRef]

24. Yu, J.; Wang, Y.; Peng, L.; Ru, M.; Liang, Z.S. Genetic diversity and population structure of *Eucommia ulmoides* Oliver, an endangered medicinal plant in China. *Genet. Mol. Res.* **2015**, *14*, 2471–2483. [CrossRef] [PubMed]

25. Vignal, A.; Milan, D.; Sancristobal, M.; Eggen, A. A review on SNP and other types of molecular markers and their use in animal genetics. *Genet. Sel. Evol.* **2002**, *34*, 275–305. [CrossRef] [PubMed]

26. Bernardi, J.; Mazza, R.; Caruso, P.; Reforgiato, R.G.; Marocco, A.; Licciardello, C. Use of an expressed sequence tag-based method for single nucleotide polymorphism identification and discrimination of *Citrus* species and cultivars. *Mol. Breed.* **2013**, *31*, 705–718. [CrossRef]

27. Elshire, R.J.; Glaubitz, J.C.; Sun, Q.; Poland, J.A.; Kawamoto, K.; Buckler, E.S.; Mitchell, S.E. A robust, simple genotyping-by-sequencing (GBS) approach for high diversity species. *PLoS ONE* **2011**, *6*, e19379. [CrossRef]

28. Kess, T.; Gross, J.; Harper, F.; Boulding, E.G. Low-cost ddRAD method of SNP discovery and genotyping applied to the periwinkle *Littorina saxatilis*. *J. Molluscan Stud.* **2016**, *82*, eyv042.

29. Potter, K.M.; Hipkins, V.D.; Mahalovich, M.F.; Means, R.E. Nuclear genetic variation across the range of ponderosa pine (*Pinus ponderosa*): Phylogeographic, taxonomic and conservation implications. *Tree Genet. Genomes* **2015**, *11*, 38. [CrossRef]

30. Worth, J.R.P.; Yokogawa, M.; Pérez-Figueroa, A.; Tsumura, Y.; Tomaru, N.; Janes, J.K.; Isagi, Y. Conflict in outcomes for conservation based on population genetic diversity and genetic divergence approaches: A case study in the Japanese relictual conifer *Sciadopitys verticillata* (Sciadopityaceae). *Conserv. Genet.* **2014**, *15*, 1243–1257. [CrossRef]

31. Chung, S.M.; Staub, J.E.; Lebeda, A.; Paris, H.S. Consensus chloroplast primer analysis: A molecular tool for evolutionary studies in Cucurbitaceae. In Proceedings of the Progress in Cucurbit Genetics and Breeding Research, Olomouc, Czech Republic, 12–17 July 2004.

32. Ahmed, I.; Matthews, P.J.; Biggs, P.J.; Naeem, M.; Mclenachan, P.A.; Lockhart, P.J. Identification of chloroplast genome loci suitable for high-resolution phylogeographic studies of *Colocasia esculenta* (L.) Schott (Araceae) and closely related taxa. *Mol. Ecol. Resour.* **2013**, *13*, 929–937. [CrossRef] [PubMed]

33. Zhang, Y.; Du, L.; Ao, L.; Chen, J.; Li, W.; Hu, W.; Wei, Z.; Kim, K.; Lee, S.C.; Yang, T.J. The complete chloroplast genome sequences of five *Epimedium* species: Lights into phylogenetic and taxonomic analyses. *Front. Plant Sci.* **2016**, *7*, 696. [CrossRef] [PubMed]

34. Raubeson, L.A.; Jansen, R.K. Chloroplast genomes of plants. In *Plant Diversity and Evolution: Genotypic and Phenotypic Variation in Higher Plants*; Henry, R.J., Ed.; CABI: Cambridge, MA, USA, 2005; pp. 45–68.

35. Birky, C.W., Jr. Uniparental inheritance of mitochondrial and chloroplast genes: Mechanisms and evolution. *Proc. Natl. Acad. Sci. USA* **1996**, *92*, 11331–11338. [CrossRef]

36. Straub, S.C.; Parks, M.; Weitemier, K.; Fishbein, M.; Cronn, R.C.; Liston, A. Navigating the tip of the genomic iceberg: Next-generation sequencing for plant systematics. *Am. J. Bot.* **2012**, *99*, 349–364. [CrossRef] [PubMed]

37. Wang, W.C.; Chen, S.Y.; Zhang, X.Z. Chloroplast genome evolution in Actinidiaceae: *clpP* Loss, heterogenous divergence and phylogenomic practice. *PLoS ONE* **2016**, *11*, e0162324. [CrossRef] [PubMed]

38. Wicke, S.; Schneeweiss, G.M.; Müller, K.F.; Quandt, D. The evolution of the plastid chromosome in land plants: Gene content, gene order, gene function. *Plant Mol. Biol.* **2011**, *76*, 273–297. [CrossRef] [PubMed]

39. Wicke, S.; Schneeweiss, G.M. Next-Generation Organellar Genomics: Potentials and Pitfalls of High-Throughput Technologies for Molecular Evolutionary Studies and Plant Systematics. In *Next Generation Sequencing in Plant Systematics*; International Association for Plant Taxonomy (IAPT): Bratislava, Slovakia, 2015.

40. Chen, S.Y.; Zhang, X.Z. Characterization of the complete chloroplast genome of the relict Chinese false tupelo, *Camptotheca acuminata*. *Conserv. Genet. Resour.* **2017**, 1–4. [CrossRef]

41. Yang, Z.; Ji, Y. Comparative and phylogenetic analyses of the complete chloroplast genomes of three Arcto-Tertiary relicts: *Camptotheca acuminata, Davidia involucrata*, and *Nyssa sinensis*. *Front. Plant Sci.* **2017**, *8*, 1536. [CrossRef] [PubMed]

42. Wang, W.; Liu, H.; He, Q.; Yang, W.L.; Chen, Z.; Wang, M.; Su, Y.; Ma, T. Characterization of the complete chloroplast genome of *Camptotheca acuminata*. *Conserv. Genet. Resour.* **2017**, *9*, 241–243. [CrossRef]

43. Puterova, J.; Razumova, O.; Martinek, T.; Alexandrov, O.; Divashuk, M.; Kubat, Z.; Hobza, R.; Karlov, G.; Kejnovsky, E. Satellite DNA and transposable elements in seabuckthorn (*Hippophae rhamnoides*), a dioecious plant with small Y and large X chromosomes. *Genome Biol. Evol.* **2017**, *9*, 197–212. [CrossRef] [PubMed]

44. Rocha, E.P.C. An appraisal of the potential for illegitimate recombination in bacterial genomes and its consequences: From duplications to genome reduction. *Genome Res.* **2003**, *13*, 1123–1132. [CrossRef] [PubMed]

45. Kegel, A.; Martinez, P.; Carter, S.D.; Aström, S.U. Genome wide distribution of illegitimate recombination events in *Kluyveromyces lactis*. *Nucleic Acids Res.* **2006**, *34*, 1633–1645. [CrossRef] [PubMed]

46. Dodsworth, S.; Leitch, A.R.; Leitch, I.J. Genome size diversity in angiosperms and its influence on gene space. *Curr. Opin. Genet. Dev.* **2015**, *35*, 73–78. [CrossRef] [PubMed]

47. Lisch, D. How important are transposons for plant evolution? *Nat. Rev. Genet.* **2013**, *14*, 49–61. [CrossRef] [PubMed]

48. Zhong, B.; Yonezawa, T.; Zhong, Y.; Hasegawa, M. Episodic evolution and adaptation of chloroplast genomes in ancestral grasses. *PLoS ONE* **2009**, *4*, e5297. [CrossRef] [PubMed]

49. Shaw, J.; Lickey, E.B.; Schilling, E.E.; Small, R.L. Comparison of whole chloroplast genome sequences to choose noncoding regions for phylogenetic studies in angiosperms: The tortoise and the hare III. *Am. J. Bot.* **2007**, *94*, 275–288. [CrossRef] [PubMed]

50. Zhang, Z.; Gerstein, M. Patterns of nucleotide substitution, insertion and deletion in the human genome inferred from pseudogenes. *Nucleic Acids Res.* **2003**, *31*, 5338–5348. [CrossRef] [PubMed]

51. Matsuoka, Y.; Yamazaki, Y.; Ogihara, Y.; Tsunewaki, K. Whole chloroplast genome comparison of rice, maize, and wheat: Implications for chloroplast gene diversification and phylogeny of cereals. *Mol. Biol. Evol.* **2002**, *19*, 2084–2091. [CrossRef] [PubMed]

52. Castle, J.C. SNPs occur in regions with less genomic sequence conservation. *PLoS ONE* **2011**, *6*, e20660. [CrossRef] [PubMed]

53. Allegre, M.; Argout, X.; Boccara, M.; Fouet, O.; Roguet, Y.; Bérard, A.; Thévenin, J.M.; Chauveau, A.; Rivallan, R.; Clement, D. Discovery and mapping of a new expressed sequence tag-single nucleotide polymorphism and simple sequence repeat panel for large-scale genetic studies and breeding of *Theobroma cacao* L. *DNA Res.* **2011**, *19*, 23–35. [CrossRef] [PubMed]

54. Wang, W.; Zhang, X. Identification of the sex-biased gene expression and putative sex-associated genes in *Eucommia ulmoides* Oliver using comparative transcriptome analyses. *Molecules* **2017**, *22*, 2255. [CrossRef] [PubMed]

55. Zhang, Y.-J.; Ma, P.-F.; Li, D.-Z. High-throughput sequencing of six bamboo chloroplast genomes: Phylogenetic implications for temperate woody bamboos (Poaceae: Bambusoideae). *PLoS ONE* **2011**, *6*, e20596. [CrossRef] [PubMed]

56. Magee, A.M.; Aspinall, S.; Rice, D.W.; Cusack, B.P.; Sémon, M.; Perry, A.S.; Stefanović, S.; Milbourne, D.; Barth, S.; Palmer, J.D. Localized hypermutation and associated gene losses in legume chloroplast genomes. *Genome Res.* **2010**, *20*, 1700–1710. [CrossRef] [PubMed]

57. Dong, W.; Xu, C.; Cheng, T.; Zhou, S. Complete chloroplast genome of *Sedum sarmentosum* and chloroplast genome evolution in Saxifragales. *PLoS ONE* **2013**, *8*, e77965. [CrossRef] [PubMed]

58. Millen, R.S.; Olmstead, R.G.; Adams, K.L.; Palmer, J.D.; Lao, N.T.; Heggie, L.; Kavanagh, T.A.; Hibberd, J.M.; Gray, J.C.; Morden, C.W. Many parallel losses of *infA* from chloroplast DNA during angiosperm evolution with multiple independent transfers to the nucleus. *Plant Cell* **2001**, *13*, 645–658. [CrossRef] [PubMed]

59. Guisinger, M.M.; Kuehl, J.V.; Boore, J.L.; Jansen, R.K. Genome-wide analyses of Geraniaceae plastid DNA reveal unprecedented patterns of increased nucleotide substitutions. *Proc. Natl. Acad. Sci. USA* **2008**, *105*, 18424–18429. [CrossRef] [PubMed]

60. Dugas, D.V.; Hernandez, D.; Koenen, E.J.; Schwarz, E.; Straub, S.; Hughes, C.E.; Jansen, R.K.; Nageswara-Rao, M.; Staats, M.; Trujillo, J.T. Mimosoid legume plastome evolution: IR expansion, tandem repeat expansions, and accelerated rate of evolution in *clpP*. *Sci. Rep.* **2015**, *5*, 16958. [CrossRef] [PubMed]

61. Bock, D.G.; Kane, N.C.; Ebert, D.P.; Rieseberg, L.H. Genome skimming reveals the origin of the Jerusalem Artichoke tuber crop species: Neither from Jerusalem nor an artichoke. *New Phytol.* **2014**, *201*, 1021–1030. [CrossRef] [PubMed]

62. Downie, S.R.; Jansen, R.K. A comparative analysis of whole plastid genomes from the Apiales: Expansion and contraction of the inverted repeat, mitochondrial to plastid transfer of DNA, and identification of highly divergent noncoding regions. *Syst. Bot.* **2015**, *40*, 336–351. [CrossRef]

63. Shaw, J.; Lickey, E.B.; Beck, J.T.; Farmer, S.B.; Liu, W.; Miller, J.; Siripun, K.C.; Winder, C.T.; Schilling, E.E.; Small, R.L. The tortoise and the hare II: Relative utility of 21 noncoding chloroplast DNA sequences for phylogenetic analysis. *Am. J. Bot.* **2005**, *92*, 142–166. [CrossRef] [PubMed]

64. Huang, J.; Yang, X.; Zhang, C.; Yin, X.; Liu, S.; Li, X. Development of chloroplast microsatellite markers and analysis of chloroplast diversity in Chinese jujube (*Ziziphus jujuba* Mill.) and wild jujube (*Ziziphus acidojujuba* Mill.). *PLoS ONE* **2015**, *10*, e0134519. [CrossRef] [PubMed]

65. Ren, X.; Jiang, H.; Yan, Z.; Chen, Y.; Zhou, X.; Huang, L.; Lei, Y.; Huang, J.; Yan, L.; Qi, Y. Genetic diversity and population structure of the major peanut (*Arachis hypogaea* L.) cultivars grown in China by SSR markers. *PLoS ONE* **2014**, *9*, e88091. [CrossRef] [PubMed]

66. Chen, Z.D.; Yang, T.; Lin, L.; Lu, L.M.; Li, H.L.; Sun, M.; Liu, B.; Chen, M.; Niu, Y.T.; Ye, J.F. Tree of life for the genera of Chinese vascular plants. *J. Syst. Evol.* **2016**, *54*, 277–306. [CrossRef]

67. Byng, J.W.; Chase, M.W.; Christenhusz, M.J.; Fay, M.F.; Judd, W.S.; Mabberley, D.J.; Sennikov, A.N.; Soltis, D.E.; Soltis, P.S.; Stevens, P.F. An update of the Angiosperm Phylogeny Group classification for the orders and families of flowering plants: APG IV. *Bot. J. Linn. Soc.* **2016**, *181*, 105–121.

68. Stevens, P.F. Angiosperm Phylogeny Website. Version 12, July 2012 (and More or Less Continuously Updated Since). Available online: http://www.mobot.org/MOBOT/research/APweb/ (accessed on 1 March 2018).

69. Jansen, R.K.; Cai, Z.; Raubeson, L.A.; Daniell, H.; Depamphilis, C.W.; Leebensmack, J.; Müller, K.F.; Guisingerbellian, M.; Haberle, R.C.; Hansen, A.K. Analysis of 81 genes from 64 plastid genomes resolves relationships in angiosperms and identifies genome-scale evolutionary patterns. *Proc. Natl. Acad. Sci. USA* **2007**, *104*, 19369–19374. [CrossRef] [PubMed]

70. Smith, S.A.; Beaulieu, J.M.; Donoghue, M.J. An uncorrelated relaxed-clock analysis suggests an earlier origin for flowering plants. *Proc. Natl. Acad. Sci. USA* **2010**, *107*, 5897–5902. [CrossRef]

71. Davis, C.C.; Xi, Z.; Mathews, S. Plastid phylogenomics and green plant phylogeny: Almost full circle but not quite there. *BMC Biol.* **2014**, *12*, 11. [CrossRef] [PubMed]

72. Zeng, L.; Zhang, Q.; Sun, R.; Kong, H.; Zhang, N.; Ma, H.; Zeng, L.; Zhang, Q.; Sun, R.; Kong, H. Resolution of deep angiosperm phylogeny using conserved nuclear genes and estimates of early divergence times. *Nat. Commun.* **2014**, *5*, 4956. [CrossRef] [PubMed]

73. Doyle, J.J. A rapid DNA isolation procedure for small quantities of fresh leaf tissue. *Phytochem. Bull.* **1987**, *19*, 11–15.

74. Langmead, B.; Salzberg, S.L. Fast gapped-read alignment with Bowtie 2. *Nat. Methods* **2012**, *9*, 357–359. [CrossRef] [PubMed]

75. Bankevich, A.; Nurk, S.; Antipov, D.; Gurevich, A.A.; Dvorkin, M.; Kulikov, A.S.; Lesin, V.M.; Nikolenko, S.I.; Pham, S.; Prjibelski, A.D. SPAdes: A new genome assembly algorithm and its applications to single-cell sequencing. *J. Comput. Biol.* **2012**, *19*, 455–477. [CrossRef] [PubMed]

76. Wyman, S.K.; Jansen, R.K.; Boore, J.L. Automatic annotation of organellar genomes with DOGMA. *Bioinformatics* **2004**, *20*, 3252–3255. [CrossRef] [PubMed]

77. Schattner, P.; Brooks, A.N.; Lowe, T.M. The tRNAscan-SE, snoscan and snoGPS web servers for the detection of tRNAs and snoRNAs. *Nucleic Acids Res.* **2005**, *33* (Suppl. 2), W686–W689. [CrossRef] [PubMed]

78. Katoh, K.; Standley, D.M. MAFFT multiple sequence alignment software version 7: Improvements in performance and usability. *Mol. Biol. Evol.* **2013**, *30*, 772–780. [CrossRef] [PubMed]

79. Darling, A.C.; Mau, B.; Blattner, F.R.; Perna, N.T. Mauve: Multiple alignment of conserved genomic sequence with rearrangements. *Genome Res.* **2004**, *14*, 1394–1403. [CrossRef] [PubMed]

80. Kearse, M.; Moir, R.; Wilson, A.; Stones-Havas, S.; Cheung, M.; Sturrock, S.; Buxton, S.; Cooper, A.; Markowitz, S.; Duran, C. Geneious Basic: An integrated and extendable desktop software platform for the organization and analysis of sequence data. *Bioinformatics* **2012**, *28*, 1647–1649. [CrossRef] [PubMed]

81. Wang, W.; Ma, L.; Becher, H.; Garcia, S.; Kovarikova, A.; Leitch, I.J.; Leitch, A.R.; Kovarik, A. Astonishing 35S rDNA diversity in the gymnosperm speciesCycas revolutaThunb. *Chromosoma* **2016**, *125*, 683–699. [CrossRef] [PubMed]

82. Edgar, R.C. MUSCLE: Multiple sequence alignment with high accuracy and high throughput. *Nucleic Acids Res.* **2004**, *32*, 1792–1797. [CrossRef] [PubMed]

83. Zhao, H.; Li, Y.; Peng, Z.; Sun, H.; Yue, X.; Lou, Y.; Dong, L.; Wang, L.; Gao, Z. Developing genome-wide microsatellite markers of bamboo and their applications on molecular marker assisted taxonomy for accessions in the genus *Phyllostachys*. *Sci. Rep.* **2015**, *5*, 8018. [CrossRef] [PubMed]

84. Li, H.; Handsaker, B.; Wysoker, A.; Fennell, T.; Ruan, J.; Homer, N.; Marth, G.; Abecasis, G.; Durbin, R. The sequence alignment/map (SAM) format and SAMtools. *Transpl. Proc.* **2009**, *19*, 1653–1654.

85. Van der Auwera, G.A.; Carneiro, M.O.; Hartl, C.; Poplin, R.; Del Angel, G.; Levy-Moonshine, A.; Jordan, T.; Shakir, K.; Roazen, D.; et al. From FastQ Data to High-Confidence Variant Calls: The Genome Analysis Toolkit Best Practices Pipeline. *Curr. Protoc. Bioinform.* **2013**, *43*, 1–33. [CrossRef]

86. Swofford, D.L. *PAUP*: Phylogenetic Analysis Using Parsimony, version 4.0 b10*; Sinauer Associates: Sunderland, MA, USA, 2003.

87. Felsenstein, J. Confidence limits on phylogenies: An approach using the bootstrap. *Evolution* **1985**, 783–791. [CrossRef] [PubMed]

88. Stamatakis, A. RAxML version 8: A tool for phylogenetic analysis and post-analysis of large phylogenies. *Bioinformatics* **2014**, *30*, 1312–1313. [CrossRef] [PubMed]

89. Ronquist, F.; Teslenko, M.; van der Mark, P.; Ayres, D.L.; Darling, A.; Höhna, S.; Larget, B.; Liu, L.; Suchard, M.A.; Huelsenbeck, J.P. MrBayes 3.2: Efficient Bayesian phylogenetic inference and model choice across a large model space. *Syst. Biol.* **2012**, *61*, 539–542. [CrossRef] [PubMed]

90. Miller, M.A.; Pfeiffer, W.; Schwartz, T. Creating the CIPRES Science Gateway for Inference of Large Phylogenetic Trees. In Proceedings of the Gateway Computing Environments Workshop (GCE), New Orleans, LA, USA, 14 November 2010; pp. 1–8.

8

Complete Chloroplast Genome Sequences of Four Meliaceae Species and Comparative Analyses

Malte Mader [1]**, Birte Pakull** [1]**, Céline Blanc-Jolivet** [1]**, Maike Paulini-Drewes** [1]**,**
Zoéwindé Henri-Noël Bouda [1]**, Bernd Degen** [1]**, Ian Small** [2] **and Birgit Kersten** [1,*]

[1]   Thünen Institute of Forest Genetics, Sieker Landstrasse 2, D-22927 Grosshansdorf, Germany;
      malte.mader@thuenen.de (M.M.); birte.pakull@thuenen.de (B.P.); celine.blanc-jolivet@thuenen.de (C.B.-J.);
      maike.paulini@thuenen.de (M.P.-D.); henri.bouda@thuenen.de (Z.H.-N.B.); bernd.degen@thuenen.de (B.D.)
[2]   Australian Research Centre of Excellence in Plant Energy Biology, School of Molecular Sciences,
      The University of Western Australia, 35 Stirling Highway, Crawley, WA 6009, Australia;
      ian.small@uwa.edu.au
*   Correspondence: birgit.kersten@thuenen.de

Abstract: The Meliaceae family mainly consists of trees and shrubs with a pantropical distribution. In this study, the complete chloroplast genomes of four Meliaceae species were sequenced and compared with each other and with the previously published *Azadirachta indica* plastome. The five plastomes are circular and exhibit a quadripartite structure with high conservation of gene content and order. They include 130 genes encoding 85 proteins, 37 tRNAs and 8 rRNAs. Inverted repeat expansion resulted in a duplication of *rps19* in the five Meliaceae species, which is consistent with that in many other Sapindales, but different from many other rosids. Compared to *Azadirachta indica*, the four newly sequenced Meliaceae individuals share several large deletions, which mainly contribute to the decreased genome sizes. A whole-plastome phylogeny supports previous findings that the four species form a monophyletic sister clade to *Azadirachta indica* within the Meliaceae. SNPs and indels identified in all complete Meliaceae plastomes might be suitable targets for the future development of genetic markers at different taxonomic levels. The extended analysis of SNPs in the *matK* gene led to the identification of four potential Meliaceae-specific SNPs as a basis for future validation and marker development.

Keywords: chloroplast genome; Next Generation Sequencing; genome skimming; Meliaceae; DNA marker; SNP; indel; *matK*; *rps19*

1. Introduction

The Meliaceae or mahogany family is a flowering plant family of mainly trees and shrubs (and a few mangroves and herbaceous plants) in the order Sapindales. The species of the Meliaceae family, which are contained in The Plant List [1] belong to 52 plant genera, all showing a pantropical distribution. The Plant List includes 3198 scientific plant names of species rank for the family Meliaceae. Of these, 669 are accepted species names.

Cedrela odorata L., a fast-growing deciduous tree species, is the most commercially important and widely distributed species in the genus *Cedrela*, and one of the world's most important timber species. It is found from Mexico southwards throughout central America to northern Argentina, as well as in the Caribbean. The aromatic wood is in high demand in the American tropics because it is naturally termite- and rot-resistant. It contains an aromatic and insect-repelling resin that is the source of one of its popular names, Spanish-cedar (it resembles the aroma of true cedars, *Cedrus* spp.). Other common names include Cuban cedar or cedro in Spanish. It is used for a wide range of purposes, including the production of furniture and craft items as well as different medicinal uses of the bark. It is an excellent choice for use in reforestation, because it is considered a pioneer species.

Entandrophragma cylindricum (Sprague) Sprague, commonly known as sapele or sapelli, is a large tree native to tropical Africa. The commercially important wood is reminiscent of mahogany (*Swietenia macrophylla*), a member of the same family. Demand for sapelli as a mahogany substitute, often traded as "African mahogany", has increased sharply in recent years. It is sold both in lumber and veneer form. Among other applications is its use in musical instruments.

Khaya senegalensis (Desv.) A. Juss. represents a *Khaya* species of the African riparian woodlands and savanna zone. Common names include African mahogany, dry zone mahogany, Gambia mahogany, or Senegal mahogany, among others. The wood is used for a variety of purposes, such as carpentry, interior trim, and construction. The bitter tasting bark is utilized for a variety of medical purposes.

Carapa guianensis Aubl. is a tree from the tropical regions of Southern Central America, the Amazon region, and the Caribbean. The wood resembles mahogany and is used in quality furniture. The seed oil is used in traditional medicine and as an insect repellent.

All four Meliaceae species analyzed in this study (*Cedrela odorata*, *Entandrophragma cylindricum*, *Khaya senegalensis*, *Carapa guianensis*) are listed as vulnerable on the IUCN red list of threatened species [2]. Furthermore, *Cedrela odorata* is one of the six Meliaceae species that are included in the list of CITES-protected species to ensure that international trade will not threaten the survival of this species [3]. Detecting violations of CITES regulations in the tropical timber trade requires accurate genus and species identification of wood, which is often difficult or even impossible using anatomical methods, especially if the wood is processed. Thus, there is a high demand for the development of genetic markers for these purposes.

For genus and species identification, chloroplast DNA (cpDNA) markers are often useful. In contrast to nuclear genomes, chloroplast genomes are inherited uniparentally (maternally in most seed plants) [4], show a dense gene content and a slower evolutionary rate of change (e.g., [5]). The double-stranded cpDNA is present in many copies per cell, leading to the convenient situation that cpDNA can be retrieved relatively easily from low-quantity and/or degraded DNA samples, including wood samples. These advantages of cpDNA are reflected in the recommendation of the Barcode of Life Consortia to apply molecular markers, mainly based on organelle DNA, to genetically differentiate all eukaryotic species [6]. For the genetic differentiation of vascular plants, molecular markers which are mainly based on DNA variations in two chloroplast regions (*rbcL* and *matK*) are used as a two-locus barcode [7].

Chloroplast-derived DNA sequences have been widely used for taxonomic purposes and phylogenetic studies (e.g., [8,9]). Complete cp genome sequences provide valuable data sets to resolve complex evolutionary relationships [10] and have been shown to improve resolution at lower taxonomic levels (e.g., [11]). The application of Next-Generation Sequencing (NGS) technologies [12], especially using genome-skimming strategies [13,14] has made it relatively easy to obtain complete cpDNA sequences for low cost. The high abundance of cpDNA compared to nuclear DNA allows the use of total DNA for genome skimming without prior purification of chloroplasts or cpDNA [15,16].

The comparative analysis of complete plastomes allows the extension of two-locus barcoding to next-generation barcoding (whole-plastome barcoding), gene-based phylogenetics to genome-based phylogenomics, and the development of molecular markers for taxonomic and phylogeographic purposes [15–18].

In Meliaceae, the development of cpDNA markers has so far been restricted to specific loci [19,20]. Although the Meliaceae form a large family, only the plastome sequence of one species, *Azadirachta indica*, has been previously published in GenBank (KF986530.1). In this study, we sequenced the plastomes of four additional species and compared them to that of *Azadirachta indica* as well as those of other species in the order Sapindales to analyze interspecific variation within the Meliaceae and uncover Meliaceae-specific genome features.

2. Results

2.1. The Structure of Chloroplast Genomes from Four Meliaceae Species

The complete cp genomes of *Cedrela odorata*, *Entandrophragma cylindricum*, *Khaya senegalensis* and *Carapa guianensis* were sequenced using Illumina sequencing technology in a genome-skimming approach (Table 1) and compared to the previously published plastome of *Azadirachta indica* (GenBank KF986530.1).

Table 1. Information on samples and NGS data of four Meliaceae individuals sequenced in this study.

Species	Individual (Thuenen-ID)	Origin/Location	Longitude	Latitude	Trimmed Reads	Coverage *
Cedrela odorata	CEODO_205_2	Cuba, population Guisa	−76.68	20.16	254214	101×
Entandrophragma cylindricum	c-5-ENTC-46	Cameroon, FBR, Parc National de Lobeke	15.6442	2.26286	2206300	165×
Khaya senegalensis	KS	Unknown/Green house, Thünen Institute Hamburg-Lohbrügge			2783117	346×
Carapa guianensis	CAGUI_332_1	French Guiana, region Rorota	−52.262392	4.87761	422759	135×

* The coverage is based on a final mapping of the trimmed reads to the assembled cpDNA sequence.

The plastomes of these four Meliaceae species (and of *Azadirachta indica*) are small circular DNA molecules of sizes in the range of 158,558 bp to 160,737 bp, with the typical quadripartite structure of land plant cp genomes consisting of two inverted repeats (IRa and IRb) separated by large (LSC) and small (SSC) single copy regions, respectively (Table 2).

Table 2. Summary of Meliaceae chloroplast genome features.

	Cedrela Odorata (MG724915)	*Entandrophragma Cylindricum* (KY923074.1)	*Khaya Senegalensis* (KX364458.1)	*Carapa Guianensis* (MF401522.1)	*Azadirachta Indica* * (KF986530.1)
Genome size (bp)	158,558	159,609	159,787	159,483	160,737
LSC length (bp)	86,390	87,117	87,404	87,054	88,137
SSC length (bp)	18,380	18,532	18,311	18,277	18,636
IR length (bp)	26,894	26,980	27,036	27,076	26,982
Number of genes	130	130	130	130	131

* Not sequenced in this study. Identifiers (in parenthesis) under the species name refer to GenBank accession numbers.

In each of the cp genomes of the four newly sequenced Meliaceae species (Table 1), 112 different genes were annotated (78 protein-coding, 30 tRNA and 4 rRNA genes), 18 of which were duplicated in the IR regions, giving a total of 130 genes encoding 85 proteins, 37 tRNAs and 8 rRNAs (Tables 2 and 3). Among the 112 unique genes, 18 included one or two introns. The intron sizes are ranging from 532 bp for trnL-UAA to 2535 bp for trnK-UUU when considering the plastome of *Cedrela odorata*. One gene, *rps12*, is presumed to require trans-splicing (Table 3; [21]).

Table 3. List of genes annotated in the cp genomes of four Meliaceae species sequenced in this study (Table 2).

Function	Genes
RNAs, ribosomal	*rrn23, rrn16, rrn5, rrn4.5*
RNAs, transfer	*trnA-UGC *, trnC-GCA, trnD-GUC, trnE-UUC, trnF-GAA, trnG-GCC, trnG-UCC *, trnH-GUG, trnI-CAU, trnI-GAU *, trnK-UUU *, trnL-CAA, trnL-UAA *, trnL-UAG, trnM-CAU, trnfM-CAU, trnN-GUU, trnP-UGG, trnQ-UUG, trnR-ACG, trnR-UCU, trnS-GCU, trnS-GGA, trnS-UGA, trnT-GGU, trnT-UGU, trnV-GAC, trnV-UAC *, trnW-CCA, trnY-GUA*
Transcription and splicing	*rpoA, rpoB, rpoC1 *, rpoC2, matK*
Translation, ribosomal proteins	
Small subunit	*rps2, rps3, rps4, rps7, rps8, rps11, rps12* **,T *, rps14, rps15, rps16 *, rps18, rps19*
Large subunit	*rpl2 *, rpl14, rpl16 *, rpl20, rpl22, rpl23, rpl32, rpl33, rpl36*
Photosynthesis	
ATP synthase	*atpA, atpB, atpE, atpF *, atpH, atpI*
Photosystem I	*psaA, psaB, psaC, psaI, psaJ, ycf3 **, ycf4*
Photosystem II	*psbA, psbB, psbC, psbD, psbE, psbF, psbH, psbI, psbJ, psbK, psbL, psbM, psbN, psbT, psbZ*
Calvin cycle	*rbcL*
Cytochrome complex	*petA, petB *, petD *, petG, petL, petN*
NADH dehydrogenase	**NdhA** *, **ndhB** *, *ndhC, ndhD, ndhE, ndhF, ndhG, ndhH, ndhI, ndhJ, ndhK*
Others	*clpP1 **, accD, cemA, ccsA, ycf1, ycf2*

* Genes containing one intron; ** genes containing two introns; [T] trans-splicing of the related gene. Genes in bold are located in inverted repeats (two gene copies in the genome).

The gene maps of the newly sequenced Meliaceae species are provided in Figure 1 (*Cedrela odorata*), and in Figures S1–S3 (*Entandrophragma cylindricum, Khaya senegalensis* and *Carapa guianensis*). The gene content and gene order in the cp genomes of the four species are nearly identical to the previously published plastome of *Azadirachta indica* (GenBank KF986530.1) with the exception that the *ycf1* gene at the IRa/SSC border was not annotated as a pseudogene in *Azadirachta indica* resulting in a total number of 131 genes in this species compared to 130 genes in the other four Meliaceae species (Table 2). Another exception is that the gene names of trnG-UCC and trnC-GCC are swapped due to misannotation in *Azadirachta indica*.

As mentioned above, 18 unique genes were annotated to include introns in the four newly sequenced Meliaceae species (Table 3), whereas introns are missing in the annotations of two of these genes in *Azadirachta indica* (GenBank KF986530.1), namely *petD* and *rps12* (intron in the 3′ part of the gene). The annotations of these genes are probably not correct in *Azadirachta indica*; both exons are relatively short (exon 1 of *petD*: 8 bp; exon 3 of *rps12*: 26 bp), thus making their annotation difficult.

The gene *rps19* is included in the inverted repeats (close to the IRa/LSC or IRb/LSC border, respectively; Figure 1, Figures S1–S3, Table 3) in the four Meliaceae species sequenced in this study as well as in *Azadirachta indica* (KF986530.1) thus resulting in a duplication of *rps19*.

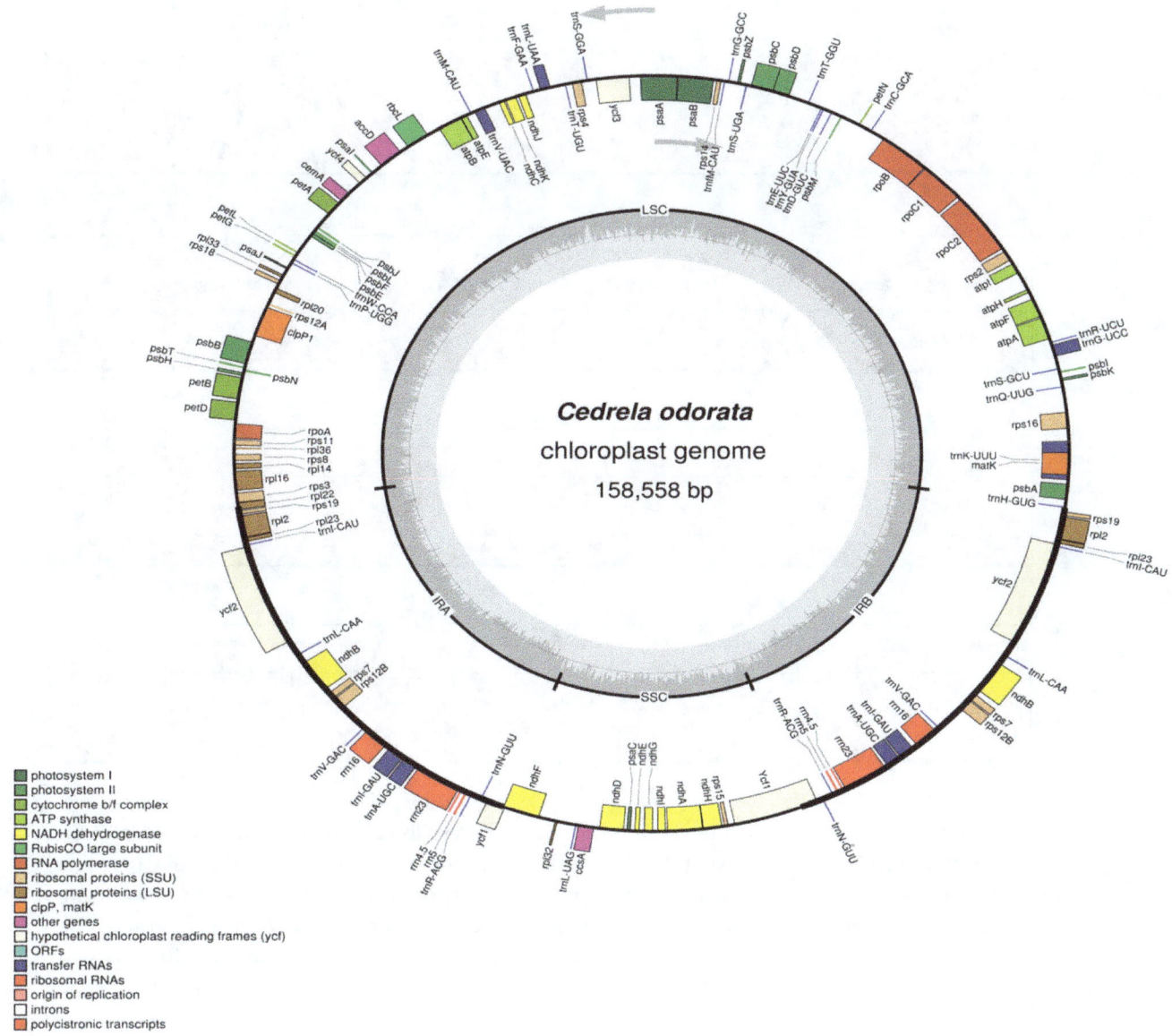

Figure 1. Gene map of the complete chloroplast genome of *Cedrela odorata* (GenBank MG724915). The grey arrows indicate the direction of transcription of the two DNA strands. A GC-content graph is depicted within the inner circle. The circle inside the GC content graph marks the 50% threshold. The maps were created using OrganellarGenomeDRAW [22].

2.2. Diversity of the Meliaceae Plastome Sequences

The newly sequenced individuals of the four different Meliaceae species share several large regions, located in intergenic regions and in the *rpl16* intron (LSC), which show low similarity to *Azadirachta indica* (Figure 2). These regions are related to large deletions which become obvious in a multiple alignment of related whole plastomes (Figure S4). The largest deletion which is in the *psbE-petL* linker is of a size of about 199 bp (Figure S4; at position 69531 bp of the *Azadirachta indica* sequence). These deletions mainly contribute to the smaller genome sizes of the four individuals compared to *Azadirachta indica* (Table 2). The *Cedrela odorata* individual, which is the individual with the smallest cp genome size (158,558 bp; Table 2), shows, compared to the four other individuals, additional large deletions in different intergenic linkers, e.g., the *ycf3-rps4* linker (Figure 2). A large exclusive deletion was also detected for the *Carapa guianensis* individual in the *psbM-psbD* linker region in the LSC (Figure 2, Figure S4).

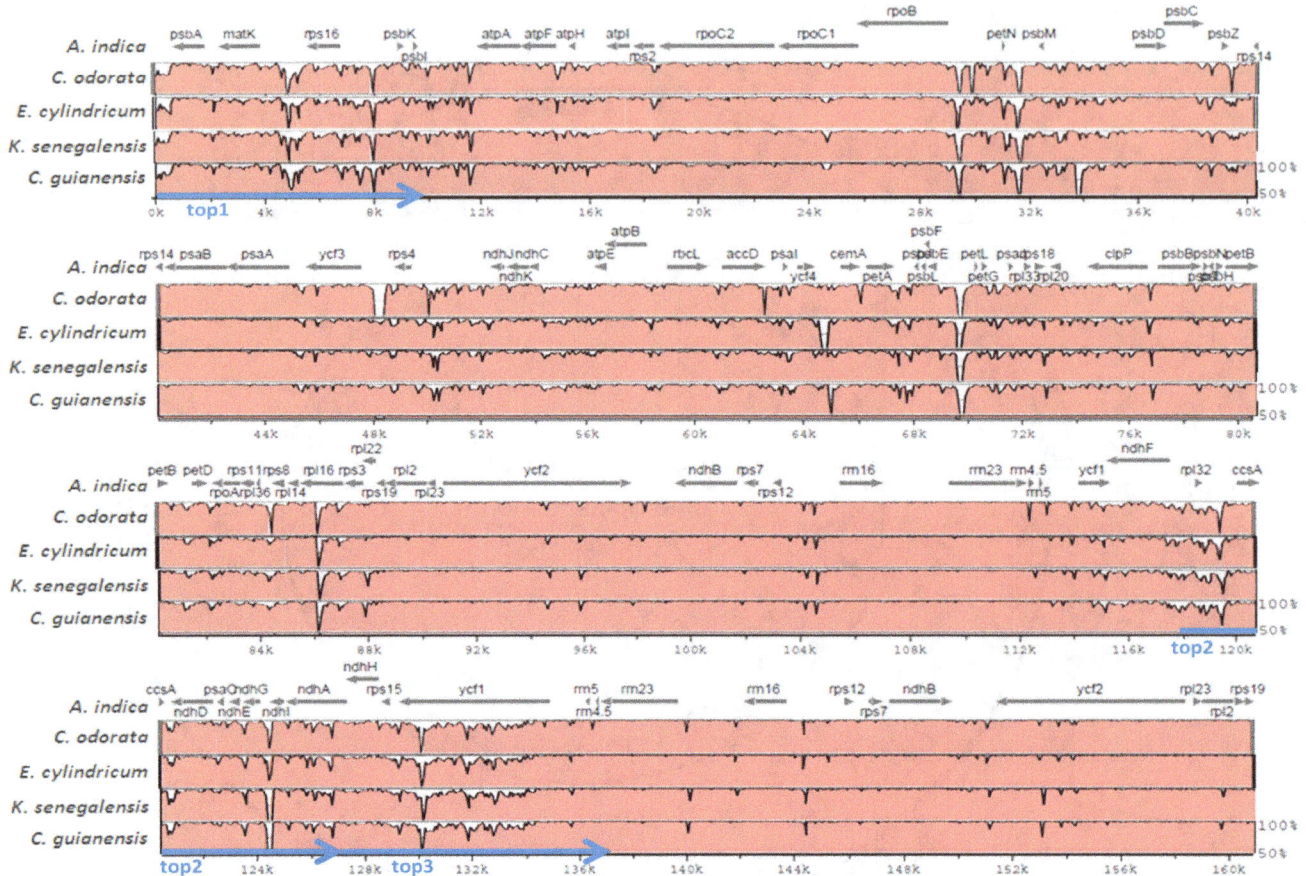

Figure 2. Visualization of pairwise alignments of complete cpDNA sequences of four Meliaceae species each with the cpDNA sequence of *Azadirachta indica* (reference). VISTA-based similarity plots portraying the sequence identity of each of the four Meliacea species with the reference *Azadirachta indica* are shown. The annotation (protein-encoding genes) is provided for *Azadirachta indica* on top (based on the related GenBank file; KF986530.1). Plastome regions with the highest diversity between the 5 Meliaceae individuals are marked by blue arrows (top1–3). Further details are provided in the text below.

Based on the multiple whole-plastome alignment of the five Meliaceae individuals (Figure S4), 7635 positions that showed DNA sequence variations (SNPs or indels) in at least one of the plastomes (compared to the consensus sequence), were called by the SNiPlay tool (SNiPlay genotyping table in Table S5). The following regions of the consensus sequence showed the highest frequencies of variations (considering intervals of 10,000 bp): 1–10,000 bp (1–9681 bp in *Azadirachta indica*) with 923 variable positions (top1); 120,000–130,000 bp (117,660–127,364 bp in *A. indica*) with 771 positions (top2); and 130,000–140,000 bp (127,365–137,199 bp in *A. indica*) with 735 positions (top3). The top1-region is located at the 5-prime part of the LSC including *psbA*, *matK*, *rps16*, *psbK* and *psbI* (Figure 2). The top2- and top3-region are connected and represent the SSC downstream of *ndhF*, the SSC/IRb border and parts of the rRNA cluster of the IRb (Figure 2).

A phylogenetic tree based on a multiple alignment of the five complete Meliaceae cpDNA sequences (Table 2) and *Acer buergerianum* (family Sapindaceae in the Sapindales; NC_034744.1) as an outgroup, shows that the analyzed Meliaceae form a monophyletic sister clade to *Azadirachta indica* within the Meliaceae. Within this clade, *Cedrela odorata* and *Entandrophragma cylindricum* group together, as do *Khaya senegalensis* and *Carapa guianensis* (Figure 3).

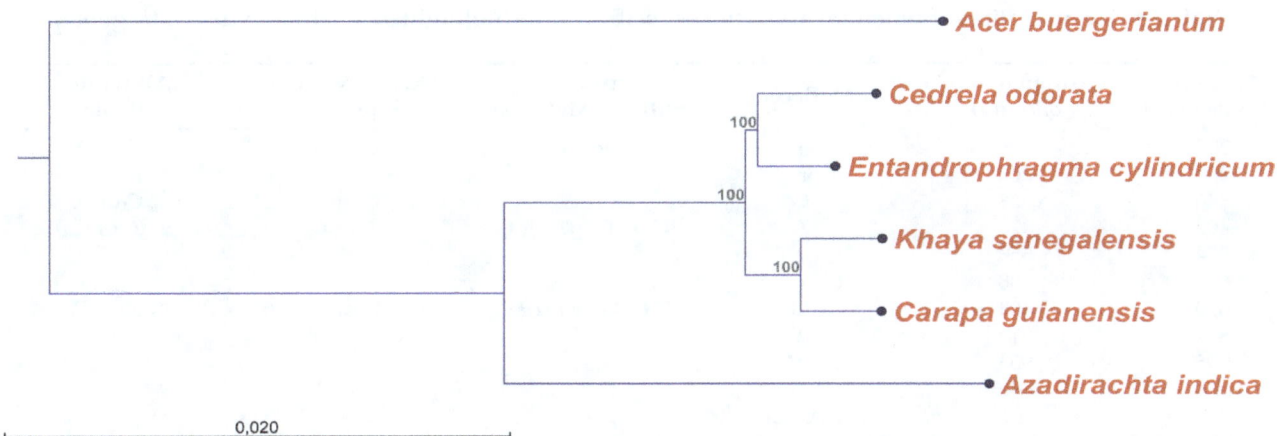

Figure 3. Phylogenetic tree (maximum likelihood) based on whole-plastome sequences of five Meliaceae species and *Acer buergerianum* (outgroup). Bootstrap values (%) are shown above branches. The bootstrap value on the branch separating *Azadirachta indica* from the other Meliaceae is below 70% and was not shown for this reason. GenBank accession numbers of the plastomes are given in Table 2.

2.3. Comparative Analyses for the Identification of Potential Meliaceae-Specific Plastome DNA Variations in One Barcoding Region

The *matK* gene, one of the cpDNA barcoding regions (see Introduction), was selected for comparative analyses aiming at the identification of potential Meliaceae-specific cpDNA variations. Based on a multiple alignment (Figure S5) of the extracted *matK* gene sequences of the five Meliaceae species listed in Table 2 together with five non-Meliaceae species of the order Sapindales (*Boswellia sacra* Flueck (Burseraceae), NC_029420.1; *Anacardium occidentale* L. (Anacardiaceae), NC_035235.1; *Leitneria floridana* Chapm. (Simaroubaceae), NC_030482.1; *Citrus sinensis* (L.) Osbeck (Rutaceae), NC_008334.1 [23]; *Acer buergerianum* Miq. (Aceraceae), NC_034744.1), 16 SNP positions were identified where the five Meliaceae individuals showed the same nucleotide, which, however, differed from the nucleotide(s) of the non-Meliaceae species at the same position (SNP positions summarized in Table 4).

These 16 SNP positions were further analyzed in a multiple alignment of *matK* gene sequences from 100 Meliaceae individuals downloaded from GenBank (Figure S6). Only SNPs where all the downloaded Meliaceae sequences showed the same nucleotide at the given position were further validated with more *matK* sequences from members of other taxonomic groups (other Sapindales, other Rosids, and other land plants; Table 4).

The SNPs at the following four positions of the *matK* gene were selected as potential Meliaceae-specific: 346 bp/328 bp (consensus sequence/*Cedrela odorata* sequence), 1318 bp/1270 bp, 1478 bp/1430 bp and 1494 bp/1446 bp (Table 4). At position 1318 bp/1270 bp, e.g., all Meliaceae individuals considered include a G (Table 4, Figure S6) in contrast to members of other Sapindales families, other Rosids, as well as other land plants, where the considered individuals show an A (Table 4). This SNP is at the first position of the codon encoding for MatK amino acid 424 in *Cedrela odorata* and will result in an amino acid exchange (Meliaceae analyzed: Gly (Figure S6); other Sapindales, Rosids and land plants analyzed: Ser or Asp).

The other barcoding region, the *rbcL* gene (see Introduction), showed less nucleotide variation compared to *matK* when analyzed in a similar way (Figure S7). In a multiple alignment of the *rbcL* sequences of five Meliaceae species and five non-Meliaceae species of the order Sapindales (Figure S7), only 2 SNP positions were identified where the five Meliaceae individuals showed another nucleotide than the non-Meliaceae species at the same position (SNP positions 225 and 582 related to the consensus sequence). However, at these two positions, some other Meliaceae individuals showed differing nucleotides at GenBank; thus, these positions were not further analyzed.

Table 4. Identification of potential Meliaceae-specific SNPs (highlighted in grey) in the *matK* gene.

Position (Consensus)	Position (*Cedrela odorata*)	Meliaceae	Sapindales without Meliaceae	Rosids without Sapindales	All without Rosids
208	208	T	A or C or T	C or A	C or A
280	262	T or C [1]			
346	328	G	C or T	C or A	C or A
402	378	A	C or T or A or G	A or C	A or C
574	550	C or T [3]			
618	588	C or A [1]			
639	609	G	T or G or A or C	T or G or A or C	G or A or T
861	819	T	C	C or T or G	C
995	953	C or T [4]			
1194	1146	T	C or T	C or T	C
1237	1189	G or A [2] or C [1]			
1239	1191	G or A [3]			
1318	1270	G	A	A	A
1389	1341	C or T [1]			
1478	1430	C	G or A	G or A or T	A or G
1494	1446	T	A or G	G or A	G or C

The SNP positions listed were validated in different multiple alignments of *matK* sequences from member individuals of different taxonomic groups downloaded from GenBank (100 sequences each). The "position (consensus)" refers to the position in the consensus sequence in the alignment in Figure S5. [1,2,3,4] Only 1/2/3/4 individual(s) show the indicated nucleotide.

3. Discussion

In this study, the complete chloroplast genomes of four Meliaceae species (order Sapindales) were sequenced by NGS (genome-skimming), annotated (Figure 1, Figures S1–S3, Table 3) and compared with each other and with the only previously published Meliaceae plastome sequence, the sequence of *Azadirachta indica* (KF986530.1; Table 2). Across eight other families within the Sapindales, there are currently 38 complete cpDNA sequences available at the Organellar Genome Resource at NCBI [24].

Gene and intron content are highly conserved among land plant plastomes, although losses have been identified in several angiosperm lineages (reviewed, e.g., in [25,26]). The following genes known to be lost in some other Angiosperm species [25] are also missing in the four newly sequenced Meliaceae plastomes (Table 3) as well as in *Azadirachta indica* (KF986530.1): *chlB, chlL, chlN, infA, trnP-GGG*. The losses of *chlB, chlL, chlN,* and *trnP-GGG* represent synapomorphies for flowering plants [25]. The most common gene loss involves *infA* [25,27]. There is evidence that *infA* has been transferred to the nucleus in some species [27], which has not yet been investigated in Meliaceae. The low depth of coverage (related to the nuclear genome) of the genome skimming data used in this study did not allow to perform such an investigation.

Intron content is also highly conserved across angiosperms with most genomes containing 18 genes with introns [25]. In the four Meliaceae species sequenced in this study, 18 unique genes with one or two introns were annotated (Table 3). The apparent lack of the *petD* intron and the intron in the 3'-*rps12* locus in *Azadirachta indica* (KF986530.1) are annotation errors in our opinion.

Hotspots for structural rearrangements within plastomes include the IRs, which are frequently subjected to expansion, contraction, or even complete loss (summarized in [21]). Inverted repeat expansion resulted in a duplication of *rps19* in the five Meliaceae plastomes analyzed so far (Figure 1, Figures S1–S3, Table 3, KF986530.1). This is consistent with research considering the plastomes of *Nitraria sibirica* (Nitrariaceae [28]) and 38 other Sapindales species [24] not belonging to the Meliaceae. However, some of these plastomes have only incomplete second *rps19* copies (*Anacardium occidentale*, NC_035235.1; *Acer miaotaiense*, NC_030343.1; *Acer buergerianum*, NC_034744.1; *Mangifera indica*, NC_035239.1) or the second copy is even completely missing, such as in *Aesculus wangii* (NC_035955.1) or *Pistacia vera* (NC_034998.1), where the first *rps19* copy is in addition incomplete. The *rps19* gene is completely absent in *Rhus chinensis* (NC_033535.1). In contrast to the Sapindales, many other rosid species include only one *rps19* gene in the LSC of the cpDNA sequence [28]. To answer the question if the duplication of *rps19* is a general plastome feature in Meliaceae, the plastomes of other member species of the 52 plant genera within the Meliaceae must be sequenced and annotated.

Complete plastome sequences are valuable for deciphering phylogenetic relationships especially between closely related taxa, or where recent divergence, rapid speciation or slow genome evolution has resulted in limited sequence variation [18,29,30]. Considering that most species-level diversity of Meliaceae in rainforests is recent [31], a whole-plastome phylogeny is highly desired for this family. The whole-plastome phylogeny that was generated in this study (Figure 3) based on the whole-plastome alignment for five Meliaceae species (Figure S4), is an initial step in this direction. The result that *Azadirachta indica* (member of the subfamily Melioidaeae) belongs to another subclade than the four other Meliaceae species (members of the subfamily Swietenioideae) as well as the further sub-grouping of the four Meliaceae species, agrees with previous studies (subfamilies according to [31–34]). In the future, more complete cpDNA sequences of member individuals of other Meliaceae species are needed to exploit the power of the whole-plastome phylogenies, especially for differentiation at lower taxonomic levels. Additional integration of complete cpDNA sequences with small amplicon datasets could further improve the phylogenetic resolution, as recently shown in *Acacia*, where the greatest support has been achieved when using a whole-plastome phylogeny as a constraint on the amplicon-derived phylogeny [30].

Whole-plastome alignments are also very useful to develop cpDNA markers for the genetic identification of species or other taxonomic categories [17]. Especially, large indels can be easily identified in whole-plastome alignments and used, e.g., for the development of robust PCR-based markers, as recently shown for a *Populus tremula*-specific marker that has been developed based on a 96 bp-indel [17,35]. In the present study, we identified and compactly visualized large indels between the analyzed Meliaceae individuals based on pairwise alignments using VISTA (Figure 2). Exclusive large deletions were identified for the *Cedrela odorata* and the *Carapa guianensis* individuals, respectively (Figure 2). Before effective markers can be developed, it must be further validated with more Meliaceae species and individuals whether these deletions are genus-, species- or individual-specific. PCR-marker development based on indels in small intergenic regions might be particularly robust, because the primers could be placed into conserved genic regions adjacent to the indel. The SNPs and indels identified between the five Meliaceae plastomes (Table S5), based on a multiple alignment, may also serve—after further validation—as targets for future marker development at different taxonomic levels.

Aiming at the identification of potential Meliaceae-specific SNPs, the *matK* gene, one of the barcoding regions [7]—was selected and the SNPs identified in this gene were analyzed in extended sets of previously published *matK* sequences. In general, nucleotide variation per site in *matK* is three times higher than in *rbcL* (the large subunit of RUBISCO; [36]), and the amino acid substitution rate is six times that of *rbcL* [37]. Here, four potential Meliaceae-specific SNPs were identified (Table 4), three of which are in the 3′-terminus of the *matK* gene encoding for the C-terminus of MatK in a region that is homolog to domain X of mitochondrial group II intron maturases [38] and has a significantly higher amount of basic amino acids compared to the N-terminal region [39]. The potential Meliaceae-specific SNPs—once further validated—should be attractive for the taxonomic differentiation of samples from wood or wood products.

4. Materials and Methods

4.1. Sampling, DNA Extraction and Sequencing

The individual trees analyzed are described in detail in Table 1. Genomic DNA was isolated from leaves or cambium according to Dumolin et al. 1995. Genomic library generation and sequencing on the Illumina MiSeq v3 (2 × 300 bp paired-end reads) was done by Eurofins Genomics (Ebersberg, Germany).

4.2. Assemblies of Chloroplast Genome Sequences and Annotation

If not otherwise stated, CLC Genomics Workbench (CLC-GWB; v8.5.1 and v9.5.3; CLC bio, A QIAGEN company; Aarhus, Denmark) was used for data processing. Initial quality control of the NGS reads was done with FastQC [40].

4.2.1. Assembly of cpDNA Sequences of Khaya Senegalensis and Entandrophragma Cylindricum

All reads were trimmed with CLC-GWB including adapter trimming, quality trimming (quality limit of 0.01), trimming of 10 nucleotides at the 5'-end and removing reads of less than 50 bp in length. All other options were set to default. Potential chloroplast reads were extracted by mapping all trimmed reads against the chloroplast sequence of *Azadirachta indica* (KF986530.1) using the tool MITObim [41]. Multiple de novo assemblies based on different word sizes were performed on extracted chloroplast reads. Overlapping contigs of one or more assemblies were used to combine the contigs to complete cpDNA sequences.

4.2.2. Assembly of cpDNA Sequences of Carapa Guianensis and Cedrela Odorata

Adapter sequences were removed by the trimming software Trimmomatic [42] using simple and palindromic trim. Further trimming was done by CLC-GWB including quality trimming (quality limit of 0.01), trimming of 10 nucleotides at the 3'- and 15 nucleotides 5'-end and removing reads shorter than 50 bp. All other options were set to default. A first set of contigs was generated by de novo assembly of all trimmed reads, using a length fraction of 0.9 and a similarity fraction of 0.95. All resulting contigs were blasted with the command line tool *blastn* [43] against the nucleotide BLAST database downloaded from GenBank [44] to identify and select chloroplast contigs. All resulting Blast hits were filtered for the keyword "chloroplast" and finally validated with the web Blast tool at NCBI [45]. Trimmed reads were mapped to the chloroplast contigs, mapped reads were extracted and stored in a separate read set as chloroplast reads. Multiple de novo assemblies based on different word sizes were performed on these chloroplast reads. Overlapping contigs of one or more assemblies were used to assemble the complete cpDNA sequence.

4.3. Annotation of the cpDNA Sequences and Preparation of GenBank Submission Files

Draft annotations were generated with the web-based software CPGAVAS [46,47] to check gene content and order. These draft annotations were corrected where necessary, guided by alignments to other well-characterized eudicot plastomes including those of *Arabidopsis thaliana* (NC_000932), *Nicotiana tabacum* (NC_001879) and *Spinacia oleracea* (NC_002202).

The file resulting from the fine annotation of the plastome of *Khaya senegalensis* (GB-format) was transferred to a draft SQN-file using the CHLOROBOX-GenBank2Sequin-tool [48] and edited using the Sequin tool (v13.05; [49]). The error-corrected SQN-file was submitted to GenBank. Because gene content and order were the same in all species analyzed (according to the fine annotation), the GenBank submission files (SQN-format) for the other species were created by updating the sqn-file of *Khaya senegalensis* with the sequences of the other three species (using the "update sequence" function in Sequin) and subsequent manual editing of shifted annotations in Sequin.

The circular gene maps of the four Meliaceae plastomes (Figure 1; Figures S1–S3) were obtained using the OrganellarGenomeDRAW software (OGDRAW v1.2; [22,50].

4.4. Alignments and Construction of a Phylogenetic Tree

Pairwise alignments of complete cpDNA sequences were performed using the VISTA tool mVISTA ("AVID" as alignment program) at VISTA [51,52]. Identity plots of each pairwise alignment were downloaded from the related VISTA-point results page.

Multiple alignments of complete cpDNA sequences were run using CLC-GWB (v. 8.5.1.; parameters: gap open cost = 10.0; gap extension cost = 1.0; end gap cost = as any other; alignment mode = very accurate; redo alignments = no; use fixpoints = no).

The phylogenetic tree was constructed based on the alignment of five Meliaceae species together with one outgroup (*Acer buergerianum*, NC_034744.1; family Sapindaceae in the order Sapindales) using the "Maximum likelihood phylogeny" tool of CLC-GWB including bootstrap analysis with 100 replicates (other parameters: construction method for the start tree = UPGMA; existing start tree = not set; nucleotide substitution model = Jukes Cantor; protein substitution model = WAG; transition/transversion ratio = 2.0; include rate variation = No; number of substitution rate categories = 4; gamma distribution parameter = 1.0; estimate substitution rate parameter(s) = Yes; estimate topology = Yes; estimate gamma distribution parameter = no).

4.5. SNP and Indel Detection in Multiple Alignments Using SNiPlay

To identify SNPs between the complete cpDNA sequences of the 5 Meliaceae species (Table 2), the alignment-FASTA of the related multiple alignment was exported from CLC-GWB and further edited (replacement of alignment spaces "-"—if present—by "?"at the 5'- or 3'- end of the sequences). The edited alignment-FASTA was used as an input for the web tool SNiPlay (pipeline v2; [53]) to run SNP/indel discovery (default parameters) [54].

4.6. NCBI-Blast Analyses of matK and Download of matK Gene Sequences of Different Taxonomic Groups for Multiple Alignments

The *matK* gene sequence of *Cedrela odorata* (MG724915) was used as a query in different BlastN searches (parameters: optimized for highly similar sequences, maximal target sequences: 100) at GenBank (NCBI; [45]): (i) restrict to Meliaceae (taxid:43707); (ii) restrict to Sapindales (taxid:41937) and exclude Meliaceae; (iii) restrict to Rosids (taxid:71275) and exclude Sapindales; and (iv) no restriction, but exclude Rosids. After each Blast analysis, all 100 hits were selected and a FASTA was downloaded with the aligned sequences. Each FASTA file was used as input for a multiple alignment together with the *matK* sequence of *Cedrela odorata* (used as reference). In the case of the Meliaceae alignment, sequences of wrong orientation in the alignment were reverse complemented and the alignment was repeated. In the case of other alignments only sequences in the right orientation (CDS orientation) were considered in the further analyses.

Supplementary Materials
Figure S1. Gene map of the complete cpDNA sequence of Entandrophragma cylindricum (KY923074). Figure S2. Gene map of the complete cpDNA sequence of Khaya senegalensis (KX364458). Figure S3. Gene map of the complete cpDNA sequence of Carapa guianensis (MF401522). Figure S4. Whole-plastome alignment of five Meliaceae species. Figure S5. Alignment of the matK gene sequences of 5 Meliaceae species (Table 2) and 5 non-Meliaceae members of the order Sapindales (Boswellia sacra, NC_029420.1; Anacardium occidentale, NC_035235.1; Leitneria floridana, NC_030482.1; Citrus sinensis, NC_008334.1; Acer buergerianum, NC_034744.1). Figure S6. Alignment of the matK gene sequences of 100 Meliaceae individuals (GenBank) and Cedrela odorata (MG724915; reference). Figure S7. Alignment of the rbcL gene sequences of 5 Meliaceae species (Table 2) and 5 non-Meliaceae members of the order Sapindales (Boswellia sacra, NC_029420.1; Anacardium occidentale, NC_035235.1; Leitneria floridana, NC_030482.1; Citrus sinensis, NC_008334.1; Acer buergerianum, NC_034744.1). Table S1. SNPs and Indels identified by SNIplay in the whole-plastome alignment of 5 Meliaceae species.

Acknowledgments: This work was funded by the Federal Ministry of Food and Agriculture (BMEL), Germany in the scope of the "Large scale project on genetic timber verification" and by the International Tropical Timber Organization (project PD 620/11 Rev.1). We would like to thank Susanne Bein for technical assistance in the laboratory, Stephen Cavers and Niklas Tysklind for providing samples from South America. Open access costs were covered by the Thuenen Institute, Braunschweig, Germany.

Author Contributions: B.D. initiated the project; B.P., M.M., C.B.-J., B.D. and B.K. conceived and designed the experiments; B.P. and M.P.-D. performed the experiments; M.M., B.K. and I.S. analyzed the data; Z.H.-N.B. collected and contributed sample material; B.K., M.M. and I.S. wrote the manuscript; and all authors edited and approved the final manuscript.

Abbreviations

cp   chloroplast

cpDNA   chloroplast DNA

CITES   Convention on International Trade in Endangered Species of Wild Fauna and Flora

IRa   Inverted Repeat a

IRb   Inverted Repeat b

IUCN   International Union for Conservation of Nature

LSC   Large Single-Copy Region

NGS   Next-Generation Sequencing

SSC   Small Single-Copy Region

References

1.   The Plant List. Available online: http://www.theplantlist.org/ (accessed on 15 January 2018).
2.   The IUCN Red List of Threatened Species. Available online: http://www.iucnredlist.org/ (accessed on 11 January 2018).
3.   CITES—Convention on International Trade in Endangered Species of Wild Fauna and Flora/the Cites Species. Available online: https://cites.org/eng/disc/species.php (accessed on 11 January 2018).
4.   Birky, C.W. Uniparental inheritance of mitochondrial and chloroplast genes—Mechanisms and evolution. *Proc. Natl. Acad. Sci. USA* **1995**, *92*, 11331–11338. [CrossRef] [PubMed]
5.   Drouin, G.; Daoud, H.; Xia, J. Relative rates of synonymous substitutions in the mitochondrial, chloroplast and nuclear genomes of seed plants. *Mol. Phylogenet. Evol.* **2008**, *49*, 827–831. [CrossRef] [PubMed]
6.   Barcode of Life. Available online: http://www.barcodeoflife.org/ (accessed on 1 January 2018).
7.   Hollingsworth, P.M.; Forrest, L.L.; Spouge, J.L.; Hajibabaei, M.; Ratnasingham, S.; van der Bank, M.; Chase, M.W.; Cowan, R.S.; Erickson, D.L.; Fazekas, A.J.; et al. A DNA barcode for land plants. *Proc. Natl. Acad. Sci. USA* **2009**, *106*, 12794–12797. [CrossRef]
8.   Shaw, J.; Shafer, H.L.; Leonard, O.R.; Kovach, M.J.; Schorr, M.; Morris, A.B. Chloroplast DNA sequence utility for the lowest phylogenetic and phylogeographic inferences in angiosperms: The tortoise and the hare IV. *Am. J. Bot.* **2014**, *101*, 1987–2004. [CrossRef] [PubMed]
9.   Wang, W.; Li, H.L.; Chen, Z.D. Analysis of plastid and nuclear DNA data in plant phylogenetics-evaluation and improvement. *Sci. China Life Sci.* **2014**, *57*, 280–286. [CrossRef] [PubMed]
10.   Moore, M.J.; Soltis, P.S.; Bell, C.D.; Burleigh, J.G.; Soltis, D.E. Phylogenetic analysis of 83 plastid genes further resolves the early diversification of eudicots. *Proc. Natl. Acad. Sci. USA* **2010**, *107*, 4623–4628. [CrossRef] [PubMed]
11.   Carbonell-Caballero, J.; Alonso, R.; Ibanez, V.; Terol, J.; Talon, M.; Dopazo, J. A phylogenetic analysis of 34 chloroplast genomes elucidates the relationships between wild and domestic species within the genus Citrus. *Mol. Biol. Evol.* **2015**, *32*, 2015–2035. [CrossRef] [PubMed]
12.   Goodwin, S.; McPherson, J.D.; McCombie, W.R. Coming of age: Ten years of next-generation sequencing technologies. *Nat. Rev. Genet.* **2016**, *17*, 333–351. [CrossRef] [PubMed]
13.   Dodsworth, S. Genome skimming for next-generation biodiversity analysis. *Trends Plant Sci.* **2015**, *20*, 525–527. [CrossRef] [PubMed]
14.   Straub, S.C.K.; Parks, M.; Weitemier, K.; Fishbein, M.; Cronn, R.C.; Liston, A. Navigating the tip of the genomic iceberg: Next-generation sequencing for plant systematics. *Am. J. Bot.* **2012**, *99*, 349–364. [CrossRef] [PubMed]
15.   Pakull, B.; Mader, M.; Kersten, B.; Ekue, M.R.M.; Dipelet, U.G.B.; Paulini, M.; Bouda, Z.H.N.; Degen, B. Development of nuclear, chloroplast and mitochondrial SNP markers for *Khaya* sp. *Conserv. Genet. Resour.* **2016**, *8*, 283–297. [CrossRef]
16.   Schroeder, H.; Cronn, R.; Yanbaev, Y.; Jennings, T.; Mader, M.; Degen, B.; Kersten, B. Development of molecular markers for determining continental origin of wood from white oaks (*Quercus* l. Sect. Quercus). *PLoS ONE* **2016**, *11*, e0158221. [CrossRef] [PubMed]

17. Kersten, B.; Rampant, P.F.; Mader, M.; Le Paslier, M.C.; Bounon, R.; Berard, A.; Vettori, C.; Schroeder, H.; Leple, J.C.; Fladung, M. Genome sequences of *Populus tremula* chloroplast and mitochondrion: Implications for holistic poplar breeding. *PLoS ONE* **2016**, *11*, e0147209. [CrossRef] [PubMed]

18. Tonti-Filippini, J.; Nevill, P.G.; Dixon, K.; Small, I. What can we do with 1000 plastid genomes? *Plant J.* **2017**, *90*, 808–818. [CrossRef] [PubMed]

19. Duminil, J.; Kenfack, D.; Viscosi, V.; Grumiau, L.; Hardy, O.J. Testing species delimitation in sympatric species complexes: The case of an african tropical tree, *Carapa* spp. (Meliaceae). *Mol. Phylogenet. Evol.* **2012**, *62*, 275–285. [CrossRef] [PubMed]

20. Holtken, A.M.; Schroder, H.; Wischnewski, N.; Degen, B.; Magel, E.; Fladung, M. Development of DNA-based methods to identify cites-protected timber species: A case study in the Meliaceae family. *Holzforschung* **2012**, *66*, 97–104. [CrossRef]

21. Wicke, S.; Schneeweiss, G.M.; dePamphilis, C.W.; Muller, K.F.; Quandt, D. The evolution of the plastid chromosome in land plants: Gene content, gene order, gene function. *Plant Mol. Biol.* **2011**, *76*, 273–297. [CrossRef] [PubMed]

22. Lohse, M.; Drechsel, O.; Kahlau, S.; Bock, R. Organellargenomedraw-a suite of tools for generating physical maps of plastid and mitochondrial genomes and visualizing expression data sets. *Nucleic Acids Res.* **2013**, *41*, W575–W581. [CrossRef] [PubMed]

23. Bausher, M.G.; Singh, N.D.; Lee, S.B.; Jansen, R.K.; Daniell, H. The complete chloroplast genome sequence of *Citrus sinensis* (L.) Osbeck var 'ridge pineapple': Organization and phylogenetic relationships to other angiosperms. *BMC Plant Biol.* **2006**, *6*, 21. [CrossRef] [PubMed]

24. Organelle Resources at NCBI. Available online: http://www.ncbi.nlm.nih.gov/genome/organelle/ (accessed on 1 February 2018).

25. Jansen, R.K.; Cai, Z.; Raubeson, L.A.; Daniell, H.; Depamphilis, C.W.; Leebens-Mack, J.; Muller, K.F.; Guisinger-Bellian, M.; Haberle, R.C.; Hansen, A.K.; et al. Analysis of 81 genes from 64 plastid genomes resolves relationships in angiosperms and identifies genome-scale evolutionary patterns. *Proc. Natl. Acad. Sci. USA* **2007**, *104*, 19369–19374. [CrossRef] [PubMed]

26. Green, B.R. Chloroplast genomes of photosynthetic eukaryotes. *Plant J.* **2011**, *66*, 34–44. [CrossRef] [PubMed]

27. Millen, R.S.; Olmstead, R.G.; Adams, K.L.; Palmer, J.D.; Lao, N.T.; Heggie, L.; Kavanagh, T.A.; Hibberd, J.M.; Giray, J.C.; Morden, C.W.; et al. Many parallel losses of infa from chloroplast DNA during angiosperm evolution with multiple independent transfers to the nucleus. *Plant Cell* **2001**, *13*, 645–658. [CrossRef] [PubMed]

28. Lu, L.; Li, X.; Hao, Z.; Yang, L.; Zhang, J.; Peng, Y.; Xu, H.; Lu, Y.; Zhang, J.; Shi, J.; et al. Phylogenetic studies and comparative chloroplast genome analyses elucidate the basal position of halophyte *Nitraria sibirica* (Nitrariaceae) in the sapindales. *Mitochondrial DNA Part A DNA Mapp. Seq. Anal.* **2017**, 1–11. [CrossRef] [PubMed]

29. Daniell, H.; Lin, C.S.; Yu, M.; Chang, W.J. Chloroplast genomes: Diversity, evolution, and applications in genetic engineering. *Genome Biol.* **2016**, *17*, 134. [CrossRef] [PubMed]

30. Williams, A.V.; Miller, J.T.; Small, I.; Nevill, P.G.; Boykin, L.M. Integration of complete chloroplast genome sequences with small amplicon datasets improves phylogenetic resolution in Acacia. *Mol. Phylogenet. Evol.* **2016**, *96*, 1–8. [CrossRef] [PubMed]

31. Koenen, E.J.M.; Clarkson, J.J.; Pennington, T.D.; Chatrou, L.W. Recently evolved diversity and convergent radiations of rainforest mahoganies (Meliaceae) shed new light on the origins of rainforest hyperdiversity. *New Phytol.* **2015**, *207*, 327–339. [CrossRef] [PubMed]

32. Pennington, T.D.; Styles, B.T. A generic monograph of the meliaceae. *Blumea* **1975**, *22*, 419–540.

33. Muellner, A.N.; Samuel, R.; Johnson, S.A.; Cheek, M.; Pennington, T.D.; Chase, M.W. Molecular phylogenetics of Meliaceae (Sapindales) based on nuclear and plastid DNA sequences. *Am. J. Bot.* **2003**, *90*, 471–480. [CrossRef] [PubMed]

34. Muellner, A.N.; Pennington, T.D.; Chase, M.W. Molecular phylogenetics of neotropical Cedreleae (mahogany family, Meliaceae) based on nuclear and plastid DNA sequences reveal multiple origins of "cedrela odorata". *Mol. Phylogenet. Evol.* **2009**, *52*, 461–469. [CrossRef] [PubMed]

35. Schroeder, H.; Kersten, B.; Fladung, M. Development of multiplexed marker sets to identify the most relevant poplar species for breeding. *Forests* **2017**, *8*, 492. [CrossRef]

36. Soltis, D.E.; Soltis, P.S. Choosing an approach and an appropriate gene for phylogenetic analysis. In *Molecular Systematics of Plants II*; Soltis, D.E., Soltis, P.S., Doyle, J.J., Eds.; Kluwer Academic Publishers: Boston, MA, USA, 1998; pp. 1–42.

37. Olmstead, R.G.; Palmer, J.D. Chloroplast DNA systematics—A review of methods and data-analysis. *Am. J. Bot.* **1994**, *81*, 1205–1224. [CrossRef]

38. Neuhaus, H.; Link, G. The chloroplast tRNALys(UUU) gene from mustard (*Sinapis alba*) contains a class-II intron potentially coding for a maturase-related polypeptide. *Curr. Genet.* **1987**, *11*, 251–257. [CrossRef] [PubMed]

39. Barthet, M.M.; Hilu, K.W. Evaluating evolutionary constraint on the rapidly evolving gene *matK* using protein composition. *J. Mol. Evol.* **2008**, *66*, 85–97. [CrossRef] [PubMed]

40. Braham Bioinformatics/Fastqc. Available online: http://www.bioinformatics.babraham.ac.uk/projects/fastqc/ (accessed on 1 December 2017).

41. Hahn, C.; Bachmann, L.; Chevreux, B. Reconstructing mitochondrial genomes directly from genomic next-generation sequencing reads-a baiting and iterative mapping approach. *Nucleic Acids Res.* **2013**, *41*. [CrossRef] [PubMed]

42. Bolger, A.M.; Lohse, M.; Usadel, B. Trimmomatic: A flexible trimmer for illumina sequence data. *Bioinformatics* **2014**, *30*, 2114–2120. [CrossRef] [PubMed]

43. Camacho, C.; Coulouris, G.; Avagyan, V.; Ma, N.; Papadopoulos, J.; Bealer, K.; Madden, T.L. Blast+: Architecture and applications. *BMC Bioinf.* **2009**, *10*, 421. [CrossRef] [PubMed]

44. Genbank. Available online: http://www.ncbi.nlm.nih.gov/genbank/ (accessed on 1 January 2018).

45. Ncbi/Blast. Available online: https://blast.ncbi.nlm.nih.gov/Blast.cgi?PROGRAM=blastn&PAGE_TYPE=BlastSearch&LINK_LOC=blasthome (accessed on 19 January 2018).

46. Liu, C.; Shi, L.C.; Zhu, Y.J.; Chen, H.M.; Zhang, J.H.; Lin, X.H.; Guan, X.J. Cpgavas, an integrated web server for the annotation, visualization, analysis, and genbank submission of completely sequenced chloroplast genome sequences. *BMC Genom.* **2012**, *13*, 715. [CrossRef] [PubMed]

47. CpGAVAS: Chloroplast Genome Annotation, Visualization, Analysis and Genbank. Available online: http://www.herbalgenomics.org/0506/cpgavas (accessed on 1 February 2018).

48. Chlorobox/genbank2sequin. Available online: https://chlorobox.mpimp-golm.mpg.de/GenBank2Sequin.html (accessed on 1 February 2018).

49. NCBI/sequin. Available online: https://www.ncbi.nlm.nih.gov/Sequin/ (accessed on 1 February 2018).

50. OrganellarGenomeDRAW. Available online: http://ogdraw.mpimp-golm.mpg.de/ (accessed on 1 February 2018).

51. VISTA. Available online: http://genome.lbl.gov/vista/index.shtml (accessed on 1 February 2018).

52. Mayor, C.; Brudno, M.; Schwartz, J.R.; Poliakov, A.; Rubin, E.M.; Frazer, K.A.; Pachter, L.S.; Dubchak, I. VISTA: Visualizing global DNA sequence alignments of arbitrary length. *Bioinformatics* **2000**, *16*, 1046–1047. [CrossRef] [PubMed]

53. SNiPlay Pipeline v2. Available online: http://sniplay.cirad.fr/cgi-bin/analysis.cgi (accessed on 1 February 2018).

54. Dereeper, A.; Homa, F.; Andres, G.; Sempere, G.; Sarah, G.; Hueber, Y.; Dufayard, J.F.; Ruiz, M. SNiPlay3: A web-based application for exploration and large scale analyses of genomic variations. *Nucleic Acids Res.* **2015**, *43*, W295–W300. [CrossRef] [PubMed]

9

Mutational Biases and GC-Biased Gene Conversion Affect GC Content in the Plastomes of *Dendrobium* Genus

Zhitao Niu, Qingyun Xue, Hui Wang, Xuezhu Xie, Shuying Zhu, Wei Liu and Xiaoyu Ding *

College of Life Sciences, Nanjing Normal University, Nanjing 210023, China; niuzhitaonj@163.com (Z.N.);
qyxue1981@126.com (Q.X.); wanghui201711@163.com (H.W.); naive0312@126.com (X.X.);
zhushuy@126.com (S.Z.); liuwei4@njnu.edu.cn (W.L.)
* Correspondence: dingxynj@263.net

Abstract: The variation of GC content is a key genome feature because it is associated with fundamental elements of genome organization. However, the reason for this variation is still an open question. Different kinds of hypotheses have been proposed to explain the variation of GC content during genome evolution. However, these hypotheses have not been explicitly investigated in whole plastome sequences. *Dendrobium* is one of the largest genera in the orchid species. Evolutionary studies of the plastomic organization and base composition are limited in this genus. In this study, we obtained the high-quality plastome sequences of *D. loddigesii* and *D. devonianum*. The comparison results showed a nearly identical organization in *Dendrobium* plastomes, indicating that the plastomic organization is highly conserved in *Dendrobium* genus. Furthermore, the impact of three evolutionary forces—selection, mutational biases, and GC-biased gene conversion (gBGC)—on the variation of GC content in *Dendrobium* plastomes was evaluated. Our results revealed: (1) consistent GC content evolution trends and mutational biases in single-copy (SC) and inverted repeats (IRs) regions; and (2) that gBGC has influenced the plastome-wide GC content evolution. These results suggest that both mutational biases and gBGC affect GC content in the plastomes of *Dendrobium* genus.

Keywords: *Dendrobium*; plastome assembly; selection; mutational biases; GC-biased gene conversion (gBGC); GC$_{eq}$

---

1. Introduction

Chloroplasts, responsible for photosynthesis and other biosynthesis processes in plants, have essential effects on plant growth and development. Their own genomes (plastomes) are usually uniparentally inherited and highly conserved in their quadripartite structure, which consists of a pair of inverted repeats (IRs) regions and two single-copy (SC) regions [1]. Comparative studies of the plastome sequence have revealed: (1) a relatively higher GC content in IR regions than that of SC regions and (2) varied GC content in different gene and non-coding loci e.g., [2,3]. The variation of GC content is a key genome feature because it is associated with fundamental elements of genome organization [4,5]. For instance, GC-rich regions exhibit higher gene density, more conserved mutation rates, and higher recombination rates, relative to GC-poor regions. Therefore, resolving the origin and causes of the variation in base composition has practical significance for a better understanding of the plastome organization.

Three major kinds of hypotheses have been proposed to explain the variation of GC content in genome evolution. The first hypothesis, "natural selection hypothesis", has suggested that high GC content can be selected for by their thermal stability [4,6]. Natural selection also affects the probability of fixation of a mutation based on the mutation fitness advantage or disadvantage of the organism [7,8].

The second hypothesis, the so-called "mutational biases hypothesis", is that the GC content is driven by the heterogeneous mutational biases along genomes [9]. The third hypothesis involves GC-biased gene conversion (gBGC), a process that takes place during meiotic recombination. The gBGC process prefers repairing DNA mismatches with GC bases and tends to increase the GC content of recombining DNA over evolutionary time [10,11]. This was recently confirmed by a broad range of genome comparison studies in eukaryotes and prokaryotes [12–14]. For example, in yeast genomes, gBGC occurred to repair the mismatches located at the extremities of the conversion regions [15]. The gBGC was also demonstrated to affect the GC content of third codon position and intron regions in grasses genomes [16]. Recently, Wu and Chaw, 2015 measured the gene conversion events of non-coding regions in cycads plastomes and reported the first case of plastome GC-biased gene conversion [17].

However, these hypotheses have not been explicitly investigated in whole plastome sequences. Orchids (Orchidaceae) are the largest family in the monocots, including about 25,000 species in 880 genera and five subfamilies [18]. Recent studies showed a diversified evolution of the plastome sequence among different orchid genera [19,20]. Moreover, orchids also present a peculiar plastomic structure: they exhibit variable IRs that caused the drastic reductions in small single-copy (SSC) regions (e.g., the expansion/contraction of IRs has led to the length of the SSC region in *Vanilla* being only about one-eighth of that in *Goodyera*) [20]. These features result in a variable GC content among different plastome sequences. However, there is still little known about the GC content evolution of the orchid plastome sequences.

Recently, plastome sequences have been made available for more than 20 orchid genera. However, in this study, we chose to focus on the *Dendrobium* genus because it is the only genus of orchids for which more than 30 plastome sequences have been sequenced [21,22]. *Dendrobium* is one of the largest genera in the orchid species. In China, there are about 80 *Dendrobium* species, some of which are well known for their high horticultural and medicinal value [23,24]. Although many plastome sequences have been published, the research of plastomic structure and base composition remains very limited in the *Dendrobium* species. In this study, we surveyed 10 plastomes of *Dendrobium* species, including two newly sequenced ones, and we addressed two key questions: (1) Is the plastome structure conserved among *Dendrobium* species or variable due to the expansion/contraction of IRs? (2) Which evolutionary forces—selection, mutational biases, or gBGC—have a significant impact on GC content in orchid plastomes?

To address the questions mentioned above, the complete plastome of two more *Dendrobium* species (*D. loddigesii* and *D. devonianum*) were sequenced and assembled by two different methods. The plastomic structures among *Dendrobium* plastomes were compared. Moreover, we evaluated the selection forces among the plastid protein-coding genes of 10 *Dendrobium* species. Meanwhile, biased mutations of protein-coding genes and non-coding regions were also measured on the basis of the estimated equilibrium GC content (GC_{eq}). Our results suggest that both mutational biases and gBGC affect GC content in the plastomes of the *Dendrobium* genus.

2. Results

2.1. Plastome Assembly of D. loddigesii and D. devonianum

A total of approximately 3.84 Gb of 150 bp pair-end reads each for *D. loddigesii* and *D. devonianum* was obtained from the Illumina paired-end sequencing. Reads with error probability >0.05 were discarded. After that, the plastomes of *D. loddigesii* and *D. devonianum* were assembled by two different ways: (1) de novo assembly by using SOAPdenovo version 1.12 and (2) using the reference-guided mapping method with CLC Genomics Workbench 6.0.1. For the de novo assembly analysis, 43,509 contigs of *D. loddigesii* and 32,667 contigs of *D. devonianum* were included in the initial assembly. After comparison with plant plastomes, 88 contigs and 67 contigs were obtained with E-values <10^{-10} and average coverage depth >30× for *D. loddigesii* and *D. devonianum*, respectively. Six of these contigs (length >17 kb and coverage depth >117×) resulted in a nearly complete draft genome for

D. loddigesii. Four contigs (length >25 kb and coverage depth >172×) were employed for the plastome assembly of *D. devonianum*. After assembly and gap closure, the plastome sequences of *D. loddigesii* and *D. devonianum* were 152,874 bp and 152,215 bp in length, respectively. For the reference-guided mapping analysis, the trimmed reads were mapped to the plastome sequence of *D. moniliforme* (AB893950), which served as a reference sequence. After gap closure, we obtained the plastome sequences of *D. loddigesii* and *D. devonianum* with 149,674 bp and 150,973 bp in length, respectively.

The assembled plastomes of *D. loddigesii* and *D. devonianum* were compared and plotted using mVISTA, with *D. moniliforme* as the reference (Figure S1). The comparison results showed that the differences—including single nucleotide polymorphism (SNP), insertion-deletion (InDel), and sequence repeat—between the plastomes were mainly distributed in the non-coding regions and *ndh* pseudo genes. Each type of these errors was corrected and validated by PCR amplification and Sanger sequencing. After that, the complete plastomes of *D. loddigesii* and *D. devonianum* were obtained with 152,498 bp and 151,715 bp respectively.

2.2. Highly Conserved Plastomic Structure and Organization in Dendrobium Genus

Genome maps of the newly sequenced two *Dendrobium* plastomes, including *D. loddigesii* and *D. devonianum*, were circular, as shown in Figure 1. The plastome contained a pair of inverted repeat (IR) regions, which separated the single-copy (SC) region into large SC (LSC) and small SC (SSC) regions. The LSC, SSC, and IR regions of *D. loddigesii* and *D. devonianum* ranged from 84,089 to 84,897 bp, 14,311 to 16,932 bp, and 25,736 to 25,800 bp, respectively. The overall GC content (37.38–37.56%) was similar with other *Dendrobium* plastomes [20,22], with 35.12–35.36% and 30.54–30.61% in LSC and SSC regions, respectively, and a higher content of 43.51–43.57% in the IR regions (Table 1). The two *Dendrobium* plastomes contained 102 unique functional coding genes, including 30 tRNA genes, four rRNA genes, and 68 protein-coding genes.

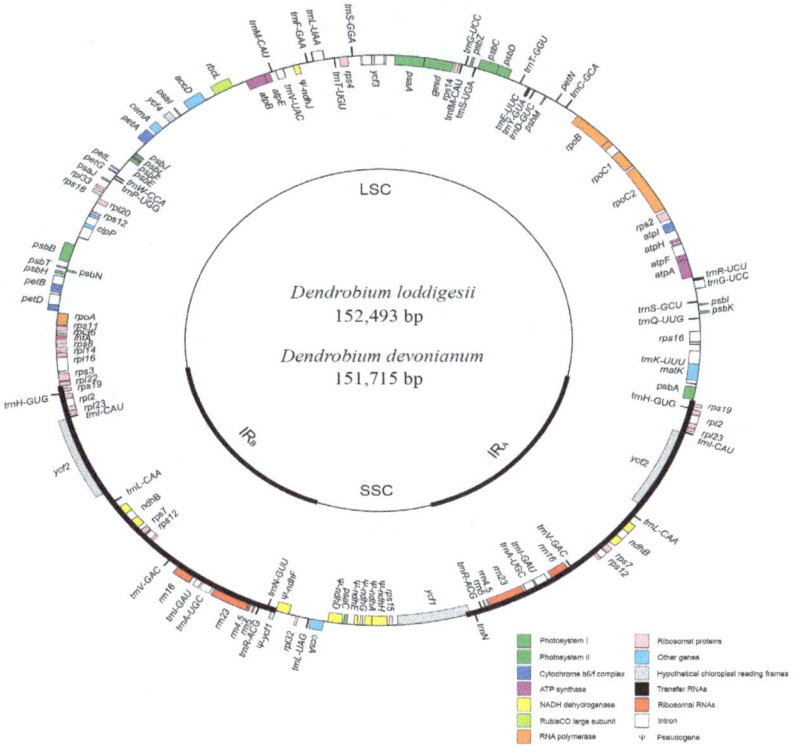

Figure 1. Genome map of two newly sequenced *Dendrobium* plastomes. Only the plastome of *D. loddigesii* is shown because it has an identical structure with *D. devonianum*. Genes outside and inside the circle are transcribed clockwise and counterclockwise, respectively. LSC: large single-copy; SSC: small single-copy; IR$_A$ and IR$_B$: two identical inverted repeats.

Table 1. Genome feature of the two newly sequenced *Dendrobium* plastomes.

Species	Plastome Length (bp)	LSC Length (bp)	SSC Length (bp)	IR Length (bp)	GC Content (%)	GC Content of LSC (%)	GC Content of SSC (%)	GC Content of IR (%)	Accession
Dendrobium loddigesii	152,493	84,089	16,932	25,736	37.38	35.63	30.54	43.57	LC317044
Dendrobium devonianum	151,715	84,897	14,311	25,800	37.56	35.12	30.61	43.51	LC317045

Abbreviations: LSC, large single-copy; SSC, small single-copy; IR, inverted repeats.

The plastome sequences of *D. loddigesii*, *D. devonianum*, *D. officinale*, and *D. moniliforme* were used for the plastome comparison (Figure 2). Comparative plastomes of these four *Dendrobium* species revealed distinct loss or retention of *ndh* genes, which indicated that the *ndh* genes have experienced independent loss during their evolution. Compared to the variable expansion/contraction of IRs in different orchid genera, e.g., [20,21], the IRs of plastomes in the *Dendrobium* genus were conserved. Overall, these *Dendrobium* plastomes appeared to have a nearly identical organization reflecting the highly conserved plastomic organization in the *Dendrobium* genus.

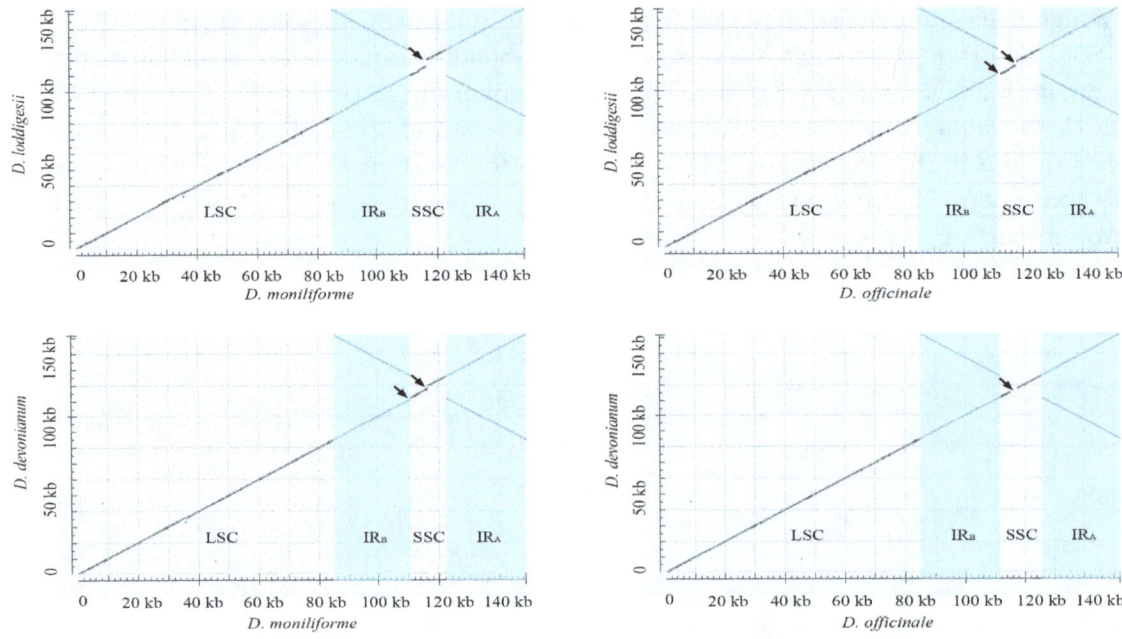

Figure 2. Dot-plot analysis of the four *Dendrobium* plastomes. The *Dendrobium* plastomes appear to have a nearly identical organization, which indicates that their plastomic organization is highly conserved. The black arrows indicate the different loss/retention of *ndh* genes.

2.3. Mutational Biases of Non-Coding Loci Are Associated with the Plastomic Structure

The phylogenetic relationship of these 10 *Dendrobium* species was inferred from the whole plastome sequence (Figure 3). The phylogenetic tree was highly resolved with the support value of all nodes = 100%. This tree was utilized to construct the ancestral sequences of non-coding loci, including intergenic and intronic loci for the 10 *Dendrobium* species. Then, the point mutations between ancestral and current sequences were calculated.

In all examined *Dendrobium* species, the nucleotide mutations were mainly distributed in the non-coding loci of SC regions. The frequencies of transversions are higher than that of transitions in both SC and IR regions. Figure 4 shows the relative rates of the six nucleotide pair mutations. The most common mutation is G/C to A/T mutations in SC regions. Therefore, we divided the nucleotide mutations into two groups: AT-rich (G/C to A/T) and GC-rich (A/T to G/C) mutations

(Table 2). The number of AT-rich mutations was larger than that of GC-rich mutations in the SC regions, while the two types of mutations did not display significant difference in the IR regions. These results indicated that the mutational biases in *Dendrobium* plastomes are associated with the plastomic structure. Moreover, the contrasting biased mutations between the SC and IR regions were also evident by the GC_{eq} values estimated from the non-coding loci of the SC and IR regions. In line with the counting results shown in Table 2, the GC_{eq} values for the SC regions were remarkably smaller than equilibrium ($GC_{eq} = 50\%$) (Figure 5). However, in the IR regions, GC_{eq} values of all *Dendrobium* species were greater than 50%, except for *D. loddigesii*, *D. lohohense*, and *D. salaccense* (with GC_{eq} values 44.83%, 42.86% and 33.97%, respectively). This result disagreed with the data shown in Table 2, that the IR regions for *Dendrobium* species have no significant difference between GC-rich and AT-rich mutations. Because only few mutations were detected in the IR regions, the data shown in Table 1 might not precisely reflect the mutational biases of non-coding loci. Thus, based on the estimated GC_{eq} values, the mutational biases in the IR region were toward GC-richness. Considering the different mutational trends in *Dendrobium* plastomes: (1) toward AT-richness in SC regions and (2) toward GC-richness in IR regions, we proposed that the mutational biases of non-coding loci are associated with the plastomic structure in *Dendrobium* species.

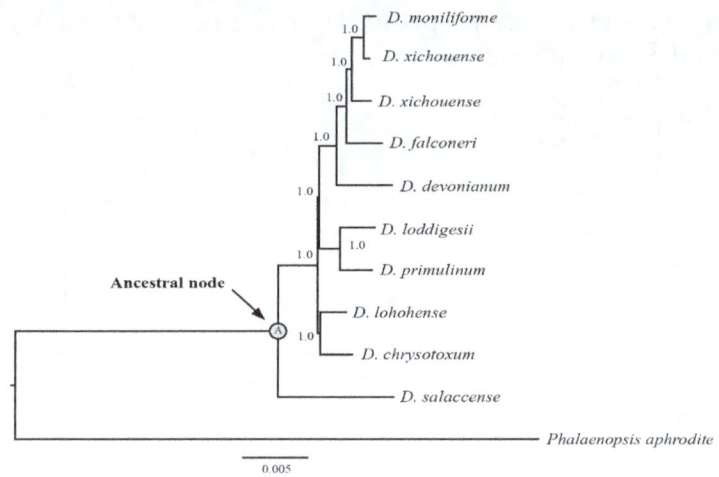

Figure 3. The BI tree inferred from whole plastome sequences of *Dendrobium* species. Tree node labeled with "A" denotes to the ancestor for each *Dendrobium* species. The values of posterior probabilities are showed for each node.

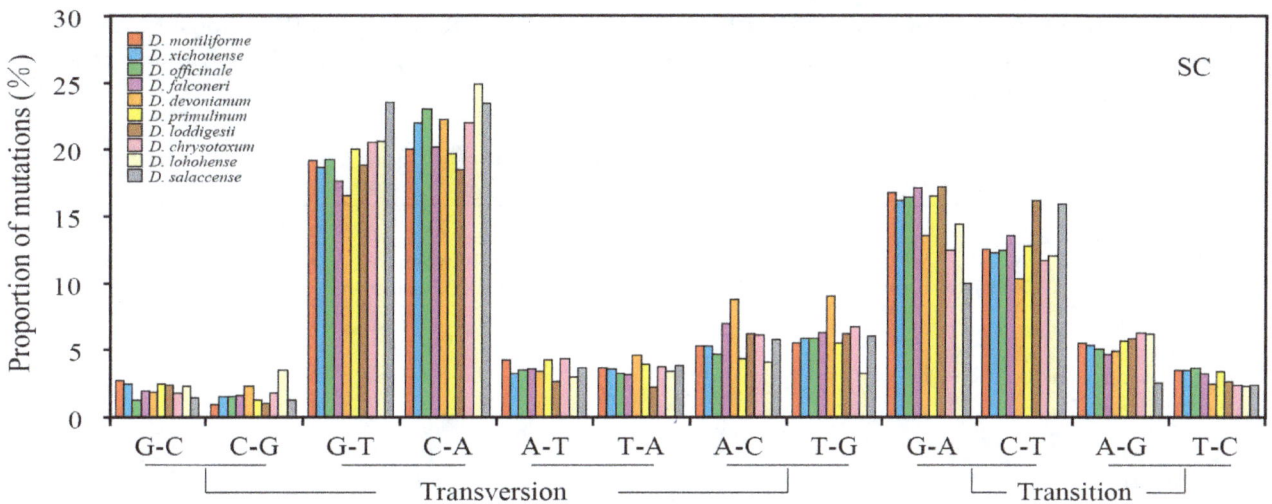

Figure 4. *Cont.*

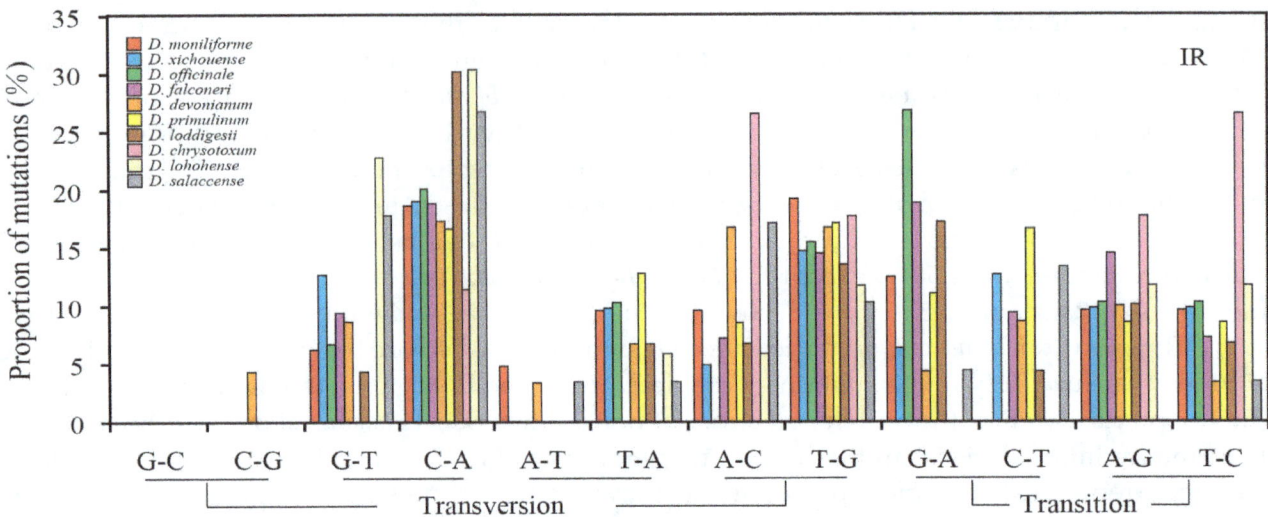

Figure 4. Proportion of the six nucleotide-pair mutations estimated from non-coding loci of the SC and IR regions of 10 *Dendrobium* species. The numbers of A to G mutations are normalized for the unequal nucleotide content of *Dendrobium* species. The frequencies of transversions are higher than that of transitions in both SC and IR regions.

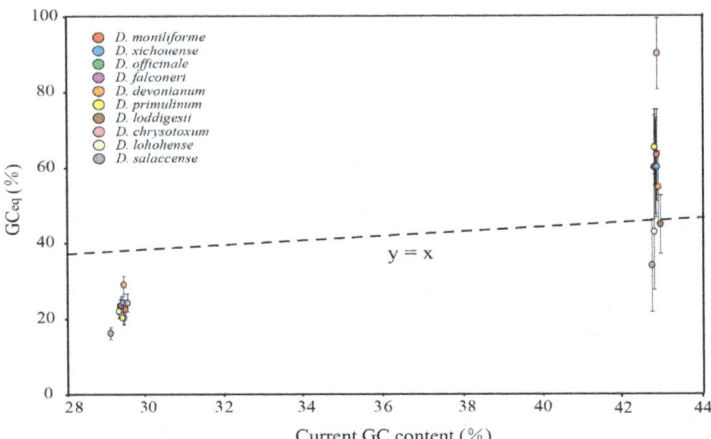

Figure 5. Comparison of equilibrium GC content (GC$_{eq}$) values between the SC and IR regions for the non-coding regions. Error bars depict 95% confidence intervals for GC$_{eq}$. Note that, the non-coding regions showed contrast GC$_{eq}$ values (<50% in SC and >50% in IR) in SC and IR regions. Moreover, the estimated GC$_{eq}$ values are lower than current GC content in SC regions, but higher than current GC content in IR regions.

Table 2. Summary of mutations in non-coding loci.

Species	SC		IR	
	Numbers of GC-AT Mutations	Numbers of Normalized AT-GC Mutations	Numbers of GC-AT Mutations	Numbers of Normalized AT-GC Mutations
D. moniliforme	229	66.65	6	7.74
D. xichouense	226	65.58	8	6.18
D. officinale	229	61.93	8	5.39
D. falconeri	248	76.54	6	4.63
D. devonianum	273	109.66	9	10.84
D. primulinum	221	60.78	8	7.71
D. loddigesii	210	62.25	13	8.62
D. chrysotoxum	188	61.10	1	7.74
D. lohohense	185	40.77	7	5.40
D. salaccense	452	104.40	14	6.92

2.4. Contrasting Mutational Biases of Protein-Coding Genes between SC and IR Regions

Three pairs of site models (M0 vs. M3, M1a vs. M2a, and M7 vs. M8) were used to test whether the evolution of plastid protein-coding genes was driven by positive selection. Among 68 plastid protein-coding genes, twelve genes (*accD*, *ccsA*, *matK*, *psaB*, *rbcL*, *rpl20*, *rpoC1*, *rpoC2*, *rps3*, *rps16*, *ycf1* and *ycf2*) were detected under positive selection (Table S1). The plastid protein-coding genes were classified into two categories, positive selected and non-positive selected data sets. The mutational biases were counted. However, the two data sets showed the same mutational trends: (1) the frequencies of transitions were higher than that of transversions (Figure S2); (2) counts of AT-rich mutations were larger than that of GC-rich mutations (Table S2); and (3) the GC_{eq} values were smaller than equilibrium (GC_{eq} = 50%) (Figure S3). Similar results were also observed in each codon position, which suggests that positive selection has no effect on gene conversions. Therefore, the plastid protein-coding genes were re-divided into SC and IR data sets based on their locations. The genes of *rpl22*, *rps12*, and *ycf1* were discarded because they were distributed in both SC and IR regions. Consistent with the counting results of non-coding loci, protein-coding genes also showed a contrast mutational bias between SC and IR regions (Table 3 and Figures 6 and 7).

Table 3. Summary of mutations in protein-coding loci.

Species	SC		IR	
	Numbers of GC-AT Mutations	Numbers of Normalized AT-GC Mutations	Numbers of GC-AT Mutations	Numbers of Normalized AT-GC Mutations
D. moniliforme	79	44.73	1	7.73
D. xichouense	115	45.95	8	5.78
D. officinale	100	51.09	8	10.29
D. falconeri	94	49.20	2	9.66
D. devonianum	89	53.71	5	5.79
D. primulinum	94	49.86	7	6.43
D. loddigesii	86	46.03	5	5.79
D. chrysotoxum	152	36.24	13	1.28
D. lohohense	59	48.04	3	5.79
D. salaccense	57	42.26	3	5.79

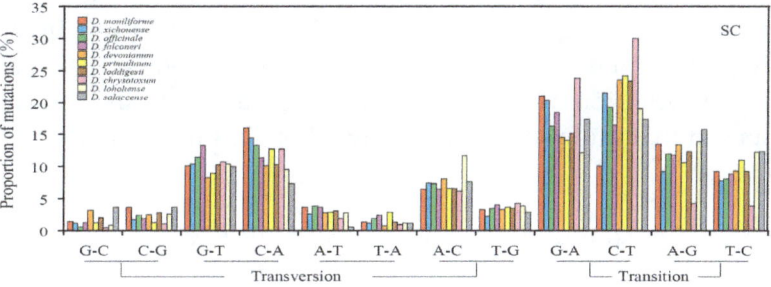

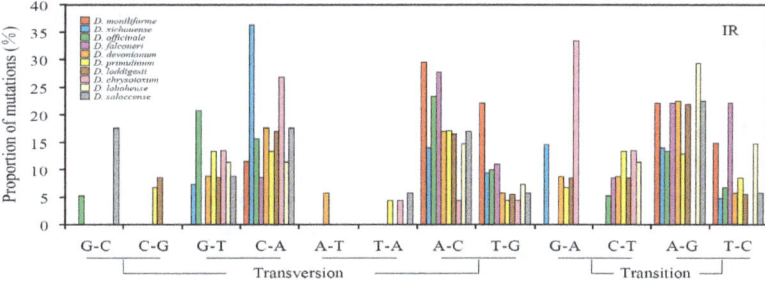

Figure 6. Proportion of the six nucleotide-pair mutations estimated from protein-coding genes. In contrast to the counting results of non-coding loci, the frequencies of transitions are higher than that of transversions in both SC and IR regions.

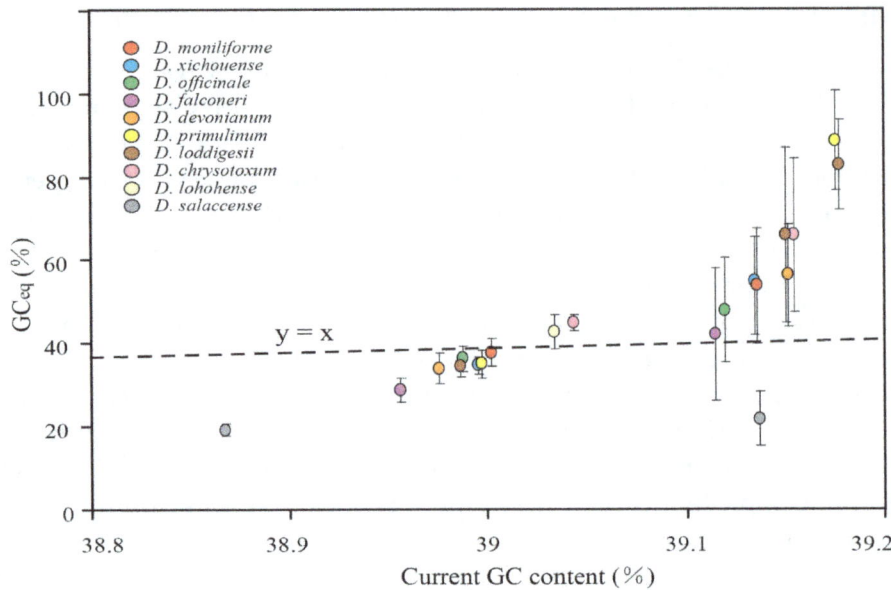

Figure 7. Comparison of GC_{eq} values between the SC and IR regions for the protein-coding genes. Error bars depict 95% confidence intervals for GC_{eq}. The protein-coding genes showed the same evolution trends with the non-coding loci.

2.5. Mutational Biases Have Directly Impact on the Evolution of GC Content in Dendrobium Plastomes

As mentioned above, our analysis revealed contrast mutational biases between SC and IR regions in both protein-coding genes and non-coding loci. To assess the effect of mutational biases on the evolution of GC content in *Dendrobium* plastomes, the GC_{eq} values were compared to current GC content. As shown in Figures 5 and 7, the GC_{eq} values of protein-coding genes and non-coding loci in SC regions were estimated to be 19.25–44.88% and 16.05–28.92%, respectively. The estimated GC_{eq} values were lower than current GC content in SC regions indicating that the GC content of SC regions was evolved toward AT-richness. On the contrary, the results show that GC_{eq} values of IRs (42.86–90.02% for protein-coding genes and 21.72–88.54% for non-coding loci) were higher than current GC content, suggesting that the evolution trends of GC content in IRs were toward GC-richness. Considering the consistent evolution trends between mutational biases and the evolution of GC content in SC and IRs, we proposed that mutational biases have directly impact on the evolution of GC content in *Dendrobium* plastomes.

2.6. gBGC Has an Effect on Maintaining Higher GC Content in IR Regions

Figure S4 illustrated the non-synonymous (dn) and synonymous (ds) substitution rates of protein-coding genes and mutation rates estimated from the syntenic non-coding loci in the SC and IR regions. The two-sided Mann–Whitney test indicated that the substitution rates were higher in SC regions than that in IR regions of *Dendrobium* plastomes ($p < 0.01$). The reduced substitution rates of IR regions most likely resulted from gene conversion. The intraplastomic recombination via IR regions frequently occurred in the plastid because of the identical sequences in the two IR regions (Figure S5A). Therefore, the GC-biased mutation and reduced substitution rates in IRs could be caused by gBGC. To determine whether the gBGC has an effect on the evolution of GC content, we compared the GC content of IR regions between the two groups: (1) the one-IR group, including five plastomes from Pinaceae and cupressophytes, which contain only one copy of IR region; and (2) the two-IR group, including 10 *Dendrobium* plastomes, which contain two copies of IR regions. Our comparison revealed a significantly higher GC content in the IR regions of the two-IR group than that of the one-IR group (Mann–Whitney 2-sides, $p < 0.01$), which suggested that the gBGC has influenced the GC content in IR regions (Figure S5B). Furthermore, the significant correlation between the GC_{eq} values and current GC content in protein-coding genes and non-coding regions (Spearman's $r = 0.851, 0.841$, both $p < 0.01$) indicated that gBGC has a long-term effect on the GC content in *Dendrobium* plastomes.

3. Discussion

3.1. The High-Quality Complete Plastome Sequences That Could Represent for Dendrobium Species

Plastome sequences of seed plants, in general, have relatively small sizes, conserved gene contents, dense coding regions, and slower evolutionary rates as compared to nuclear and mitochondrial genomes [1]. These unique features have attracted intense attention from researchers, leading to numerous plastome sequences published, e.g., [20,22,25,26]. The first record of a *Dendrobium* plastome, *Dendrobium officinale*, was obtained by sequencing from the isolated chloroplast DNA [21]. However, it is extremely hard to extract the chloroplast from *Dendrobium* species due to their high percentage of polysaccharides, e.g., [27,28]. The plant whole genome sequencing data always contains high copy numbers of plastome sequence reads, which provide an opportunity for the complete plastome sequence assembly. Thus, a lot of methods were established to assemble the plastome sequence from the genome sequencing resources [29–31]. Among them, the reference-guided mapping method is a good approach for studies of related species with known reference sequences. Recently, more than 30 *Dendrobium* plastomes have been sequenced and assembled by using this approach [22]. However, tentative errors, such as SNP, InDel, and repeat sequences, may generate by using different assembly tools or mapping data to different references [31]. As shown in Figure S1, the sequences in the non-coding regions and *ndh* genes of *D. loddigesii* and *D. devonianum* were distinct from different assembly methods. Therefore, the high-quality plastome sequence that could be representing for *Dendrobium* species was urgently needed. In this study, two different methods were employed for the plastome assembly of *Dendrobium* species. According to the mVISTA analysis, the mis-assembled error sequences were detected, which indicated that at least two different methods should be used in the plastome assembly analysis to avoid assembling errors. After correcting and validating these errors by PCR amplification and Sanger sequencing, the complete plastomes of *D. loddigesii* and *D. devonianum* were obtained. Therefore, we proposed that the high-quality complete plastome sequences of *D. loddigesii* and *D. devonianum* were obtained in this study, which could represent for *Dendrobium* species for the future studies.

3.2. Both Mutational Biases and gBGC Affect GC Content in the Plastomes of Dendrobium Genus

The causes of GC content variation in plastome sequences were less clearly established than in nuclear genome sequences, and the role of gene conversion has only been investigated in non-coding regions more recently [17]. Previously, more than 30 plastomes of *Dendrobium* species have been published, which provides a huge resource for exploring the evolutionary mechanism of the variation of GC content in *Dendrobium* plastomes. Niu et al., (2017) proposed that evolutionary stasis should use at least 10 plastome sequences [22]. Here, we examined and discussed the three major hypotheses based on our newly sequenced data coupled with other eight *Dendrobium* plastomes.

3.2.1. Natural Selection

Natural selection has great impact on the variations in base composition. For example, in plastomes, the codon-usage of the *psbA* gene was proposed to be associated with positive selection for maintaining the translation efficiency [32]. Moreover, several studies have suggested a correlation between GC content and the function and expression of genes e.g., [5,33,34], which enhanced the hypothesis that natural selection has a direct impact on GC content. The "natural selection hypothesis" was also supported by the study of Shi et al., (2007), who found that the selection is the primary cause of the GC content variation in rice genes [35]. In this study, positive selection analysis was performed for the 68 plastid protein-coding genes. The higher GC content in the IRs than that in SC regions leads us to expect that genes located in IRs have a higher proportion under positive selection. However, only one protein-coding gene (*ycf2*) was identified under positive selection in the IR regions. Moreover, the same mutational trends were observed in both positive selected and non-positive selected data sets. These results suggested that natural selection does not appear to contribute significantly to the varied GC content of plastid protein-coding genes. Therefore, we ruled out the effect of natural selection, which leaves the mutational bias and gBGC as potential determinants for the evolution of GC content in the *Dendrobium* plastomes.

3.2.2. Mutational Bias

In this study, our analysis revealed a contrast mutational bias in *Dendrobium* plastomes that: (1) toward AT-richness in SC and (2) toward GC-richness in IR regions. The contrast mutational biases would lead to a lower GC content in SC regions. Indeed, the GC contents of LSC and SSC regions are lower than that of IRs. Moreover, the result estimated GC_{eq} values lower than current GC content in SC regions and higher than that of IRs, indicating that the GC content variation in SC regions is primarily determined by the mutational biases. Therefore, we proposed that mutational biases have direct impact on the variation of GC content in *Dendrobium* plastomes. Mutational biases toward AT-richness have been reported in gnetophytes [3]. Their plastomes were discovered to be compacting with enriched AT content, which was thought to have many advantages for resource consumption. For example, an A/T nucleotide contains seven atoms of nitrogen, one less than the G/C nucleotide. Moreover, AT-richness in base composition would benefit the rapid replication of plastomes [36]. Therefore, the sequence feature that mutated toward AT-richness would be favored in the plastome.

3.2.3. gBGC

The variation in base composition is proven to be affected by gene conversion, e.g., [37,38]. The gene conversion of the plastome sequences is typically modeled as an independent process, in which changes at any one site are independent of changes at any other site [39]. However, gene conversion is not an entirely stochastic process, as it acts according to certain deterministic biases that affect the GC content evolution. Over the past 10 years, GC-biased gene conversion (gBGC) has been clearly established as the main process affecting GC content evolution in the nuclear genome [40–43]. However, whether the gBGC affects the GC content in plastome sequences is poorly understood. In this study, although no direct evidence is detected, we cannot ignore the impact of gBGC to the GC content evolution in *Dendrobium* genus. Based on the comparison results of 10 *Dendrobium* plastomes, we proposed that the gBGC has influenced the plastome wide GC content evolution due to three reasons.

Firstly, the higher GC content in IR regions could be explained by gBGC. In this study, the reduced substitution rates and GC-biased mutation indicated that gBGC occurred in IRs. The gBGC would affect GC/AT heterozygous sites yields more frequently to GC than to AT alleles, which lead to increasing the GC content over evolutionary time [13]. Secondly, the recombination of IRs provides an opportunity for gBGC to occur. IRs are the recombination hotspots in plastome sequence, which favors the gBGC [44–46]. Moreover, a gBGC model during recombination has been put forth to interpret the GC-biased mutations observed in the IR regions [17]. Through that model, GC to AT mutations were unfavored, whereas AT to GC ones were fixed after gBGC. After loss of one IR region, the GC content of IR regions in Pinaceae and cupressophytes species was evolved toward AT-richness. Thirdly, the significant correlation between the GC_{eq} values and GC content values in protein-coding genes and non-coding regions (Spearman's $r = 0.851, 0.841$, both $p < 0.01$) indicated that in *Dendrobium*, gBGC has shaped plastome-wide mutations and GC content. Therefore, we proposed that both mutational biases and gBGC affect GC content in the plastomes of the *Dendrobium* genus.

4. Materials and Methods

4.1. DNA Extraction and Plastome Sequencing

Two grams each of fresh leaves were harvested from individuals of *D. loddigesii* (voucher specimen: Niu14007) and *D. devonianum* (voucher specimen: Niu15008). They were cultivated in the greenhouse of Nanjing Normal University. The total DNA of each species was extracted using the Qiagen DNeasy Plant Mini Kit (Qiagen, Hilden, Germany), according to the manufacturer's instructions. DNA samples that met the quality of concentrations >300 ng/mL, A260/A280 ratio = 1.8–2.0, and A260/A230 ratio >1.7, were used for next-generation sequencing. The sequencing depth was 3.84 GB of 150 bp paired-end reads for each species.

4.2. Plastome Assembly, Annotation and Comparison

To obtain high-quality complete plastome sequences of *D. loddigesii* and *D. devonianum*, two different assembly methods: (1) de novo assembly methods and (2) reference-guided mapping method, were employed in this study. For the de novo assembly, raw reads were first trimmed with error probability <0.05, then de novo assembled using SOAPdenovo version 1.12 with the default parameters [47]. The de Bruijn graph approach was applied to assembly with an optimal K-mer size of 79. Contigs with length <200 bp were discarded. The remaining contigs were compared with the plastome sequences of *D. moniliforme* (AB893950) using BLAST searches. Matched contigs with E-values of <10^{-10} were designated as plastomic contigs. The gaps between plastomic contigs were closed by PCR amplification with specific primers and Sanger sequencing. For reference-guided assembly, the trimmed reads were mapped to the reference plastome sequence of *D. moniliforme* (AB893950) using CLC Genomics Workbench 6.0.1 (CLC Bio, Aarhus, Denmark). The four junctions between LSC/SSC and IRs and the regions with uncertain nucleotide "N" were validated with PCR amplification.

Protein-coding genes and ribosomal RNA genes were annotated using DOGMA [48]. The boundaries of each annotated gene were confirmed by their alignment with their orthologous genes from other *Dendrobium* plastomes. Genes of tRNA were predicted using tRNAscan-SE 1.21 [49]. The differences between plastome sequences of *D. loddigesii* and *D. devonianum* that assembled from different methods were compared by using the mVISTA software [50]. The plastome sequence of *D. moniliforme* (AB893950) was used as an outgroup.

4.3. Phylogenetic Analysis

The plastome sequences of 10 *Dendrobium* species and *Phalaenopsis aphrodite* were aligned based on the MAFFT method [51]. The gaps within alignment were excluded. The phylogenetic tree was constructed using MrBayes 3.2 [52]. For the Bayesian inference (BI) analysis, two simultaneous runs were conducted, each consisting of four chains. The parameters were set as "lset nst = 6 rates = γ". In total, chains were run for 2,000,000 generations, with topologies sampled every 100 generations. The first 25% of our sampled trees were discarded. The remaining trees were used to construct a majority-rule consensus tree and calculate posterior probabilities (PPs) for each node.

4.4. Ancestral Non-Coding Sequences Reconstruction and Calculation of the Relative Rates of the Six Nucleotide Pair Mutations

The non-coding loci, including intergenic and intronic loci, were manually retrieved from the 10 *Dendrobium* plastome sequences (Table S3). The loci were aligned by using MUSCLE [53] with the "Refining" option implemented in Mega 5.2 [54]. Gaps located at the 5'- and 3'-ends of alignments were excluded, and then concatenated using SequenceMatrix 1.8 [55]. The alignment of non-coding loci was divided into SC and IR data sets based on their locations. The ancestral non-coding sequences of 10 *Dendrobium* species were reconstructed using the maximum likelihood (ML) method with GTR + G model in Mega 5.2. The BI tree inferred from whole plastome sequences was designated as the "user tree". DAMBE 5 [56] was used to count the nucleotide changes between ancestral and current sequences. After that, we calculated the relative rates of the six nucleotide pair mutations. Firstly, we normalized the counts of the mutations from A/T to G/C, C/G, or T/A, and followed Hershberg and Petrov's method (2010) by multiplying them with current genome wide number of GC sites/AT sites [57]. For example, the normalized number of A to G mutations in *D. loddigesii* = the number of A to G changes between ancestral sequence and the current sequence of *D. loddigesii* × the current number of GC sites/AT sites. In this way, we could determine the expected number of mutations under equal GC and AT contents. Then, the relative rates of each nucleotide mutation were calculated with the formula = the number of mutations from G/C or the normalized number of mutations from A/T/ (the total number of mutations from G/C + the normalized total number of mutations from A/T) × 100.

4.5. Estimation of GC_{eq}

The GC_{eq} values were estimated according to the method of Hershberg and Petrov (2010) [57] and Wu et al., 2015 [17]. The GC_{eq} values were calculated as: rate of AT to GC/(rate of AT to GC + rate of GC to AT). The rate of AT to GC = the total number of AT to GC changes/the current number of AT sites, while the rate of GC to AT = the total number of GC to AT changes/the current number of GC sites. The SC and IR data sets including ancestral and current sequences were bootstrapped with 100 replications using PHYLIP 3.695 [58]. Then, the GC_{eq} values were recalculated 100 times and the resulting values were used to estimate the 95% confidence intervals for GC_{eq}.

4.6. Natural Selection Analysis and Evolutionary Stasis of Protein-Coding Genes

Except for the 11 *ndh* genes, positive selection analysis was performed for the other 68 protein-coding genes using the codeml program in PAML vs. 4.9 [59]. The natural selection at the codon level was detected using the three pairs of site models (M0 vs. M3, M1a vs. M2a, and M7 vs. M8) as implemented in codeml. The likelihood ratio tests (LRTs) were used to compare the site models. The relative rates of the six nucleotide pair mutations and GC_{eq} values of protein-coding genes were also counted.

4.7. Estimation of Substitution Rates

The synonymous (ds) and non-synonymous (dn) substitution rates were estimated with the codeml program of PAML [59]. The parameters were set to the following: seqtype = 1, runmodel = −2, and CodonFreq = 3. The protein-coding genes of *Phalaenopsis aphrodite* were used as the reference. The non-coding loci that flanked by the same genes/exons were identified as syntenic. The loci with sequence lengths less than 150 bp were discarded. The mutation rates between *P. aphrodite* and the 10 *Dendrobium* species were estimated with the baselml program of PAML using the REV model [59].

4.8. Statistical Analysis

Statistical analyses were performed using SPSS Statistics 20.0.

5. Conclusions

In conclusion, this study is the first to observe the impact of evolutionary forces, selection, mutational biases, and GC-biased gene conversion (gBGC) on the variation of GC content in the whole plastome sequences. The high-quality complete plastomes of *D. loddigesii* and *D. devonianum* were obtained and compared with eight other *Dendrobium* plastomes. The results indicate that the plastomes of *Dendrobium* species are highly conserved in plastomic organization. Because tentative errors generated by different assembly methods can lead to low quality of plastome sequence, we thus strongly suggest using two different assembly methods in the plastome assembly analysis. Furthermore, we examined three major hypotheses based on the plastome sequences of 10 *Dendrobium* species. Our results demonstrated that both mutational biases and gBGC affect GC content in the plastomes of the *Dendrobium* genus.

Acknowledgments: This work was supported by the National Natural Science Foundation of China (Grant No. 31170300 and No. 31670330) and the Priority Academic Program Development of Jiangsu Higher Education Institutions to Xiaoyu Ding.

Author Contributions: Xiaoyu Ding and Zhitao Niu designed the study topic. Zhitao Niu, Hui Wang, Xuezhu Xie, and Shuying Zhu performed the experiments. Zhitao Niu, Qingyun Xue, and Wei Liu analyzed data. Zhitao Niu wrote the manuscript. Xiaoyu Ding, Zhitao Niu, and Qingyun Xue revised the final version of manuscript. All authors read and approved the final manuscript.

References

1. Raubeson, L.A.; Jansen, R.K. Chloroplast genomes of plants. In *Plant Diversity and Evolution: Genotypic and Phenotypic Variation in Higher Plants*; Henry, R.J., Ed.; CAB International: London, UK, 2005; pp. 45–68.

2. Wolfe, K.H.; Li, W.H.; Sharp, P.M. Rates of nucleotide substitution vary greatly among plant mitochondrial, chloroplast, and nuclear DNAs. *Proc. Natl. Acad. Sci. USA* **1987**, *84*, 9054–9058. [CrossRef] [PubMed]

3. Wu, C.S.; Lai, Y.T.; Lin, C.P.; Wang, Y.N.; Chaw, S.M. Evolution of reduced and compact chloroplast genomes (cpDNAs) in gnetophytes: Selection toward a lower-cost strategy. *Mol. Phylogenet. Evol.* **2009**, *52*, 115–124. [CrossRef] [PubMed]

4. Eyre-Walker, A.; Hurst, L.D. The evolution of isochores. *Nat. Rev. Genet.* **2001**, *2*, 549–555. [CrossRef] [PubMed]

5. Mukhopadhyay, P.; Basak, S.; Ghosh, T.C. Nature of selective constraints on synonymous codon usage of rice differs in GC-poor and GC-rich genes. *Gene* **2007**, *400*, 71–81. [CrossRef] [PubMed]

6. Bernardi, G. Isochores and the evolutionary genomics of vertebrates. *Gene* **2000**, *241*, 3–17. [CrossRef]

7. Wang, H.C.; Singer, G.A.C.; Hickey, D.A. Mutational bias affects protein evolution in flowering plants. *Mol. Biol. Evol.* **2004**, *21*, 90–96. [CrossRef] [PubMed]

8. Günther, T.; Lampei, C.; Schmid, K.J. Mutational bias and gene conversion affect the intraspecific nitrogen stoichiometry of the *Arabidopsis thaliana* transcriptome. *Mol. Biol. Evol.* **2013**, *30*, 561–568. [CrossRef] [PubMed]

9. Fryxell, K.J.; Zuckerkandl, E. Cytosine deamination plays a primary role in the evolution of mammalian isochores. *Mol. Biol. Evol.* **2000**, *17*, 1371–1383. [CrossRef] [PubMed]

10. Marais, G. Biased gene conversion: Implications for genome and sex evolution. *Trends Genet.* **2003**, *19*, 330–338. [CrossRef]

11. Duret, L.; Galtier, N. Biased gene conversion and the evolution of mammalian genomic landscapes. *Annu. Rev. Genom. Hum. Genet.* **2009**, *10*, 285–311. [CrossRef] [PubMed]

12. Smith, D.R.; Lee, R.W. Mitochondrial genome of the colorless green alga *Polytomella capuana*: A linear molecule with an unprecedented GC content. *Mol. Biol. Evol.* **2008**, *25*, 487–496. [CrossRef] [PubMed]

13. Pessia, E.; Popa, A.; Mousset, S.; Rezvoy, C.; Duret, L.; Marais, G.A.B. Evidence for widespread GC-biased gene conversion in eukaryotes. *Genome Biol. Evol.* **2012**, *4*, 675–682. [CrossRef] [PubMed]

14. Lassalle, F.; Périan, S.; Bataillon, T.; Nesme, X.; Duret, L.; Daubin, V. GC-content evolution in bacterial genomes: The biased gene conversion hypothesis expands. *PLoS Genet.* **2015**, *11*, e1004941. [CrossRef] [PubMed]

15. Lesecque, Y.; Mouchiroud, D.; Duret, L. GC-biased gene conversion in yeast is specifically associated with crossovers: Molecular mechanisms and evolutionary significance. *Mol. Biol. Evol.* **2013**, *30*, 1409–1419. [CrossRef] [PubMed]

16. Muyle, A.; Serres-Giardi, L.; Ressayre, A.; Escobar, J.; Glémin, S. GC-biased gene conversion and selection affect GC content in the *oryza* genus (rice). *Mol. Biol. Evol.* **2011**, *28*, 2695. [CrossRef] [PubMed]

17. Wu, C.S.; Chaw, S.M. Evolutionary stasis in cycad plastomes and the first case of plastome GC-biased gene conversion. *Genome Biol. Evol.* **2015**, *7*, 2000–2009. [CrossRef] [PubMed]

18. Givnish, T.J.; Spalink, D.; Ames, M.; Lyon, S.P.; Hunter, S.J.; Zuluaga, A.; Iles, W.J.D.; Clements, M.A.; Arroyo, M.T.K.; Leebens-Mack, J.; et al. Orchid phylogenomics and multiple drivers of their extraordinary diversification. *Proc. Biol. Sci. B* **2015**, *282*, 2108–2111. [CrossRef] [PubMed]

19. Shaw, J.; Shafer, H.L.; Leonard, O.R.; Kovach, M.J.; Schorr, M.; Morris, A.B. Chloroplast DNA sequence utility for the lowest phylogenetic and phylogeographic inferences in angiosperms: The tortoise and the hare IV. *Am. J. Bot.* **2014**, *101*, 1987–2004. [CrossRef] [PubMed]

20. Niu, Z.; Xue, Q.; Zhu, S.; Sun, J.; Liu, W.; Ding, X. The complete plastome sequences of four orchid species: Insights into the evolution of the Orchidaceae and the utility of plastomic mutational hotspots. *Front. Plant. Sci.* **2017**, *8*, 715. [CrossRef] [PubMed]

21. Luo, J.; Hou, B.W.; Niu, Z.T.; Liu, W.; Xue, Q.Y.; Ding, X.Y. Comparative chloroplast genomes of photosynthetic orchids: Insights into evolution of the Orchidaceae and development of molecular markers for phylogenetic applications. *PLoS ONE* **2014**, *9*, e99016. [CrossRef] [PubMed]

22. Niu, Z.; Zhu, S.; Pan, J.; Li, L.; Sun, J.; Ding, X. Comparative analysis of *Dendrobium* plastomes and utility of plastomic mutational hotspots. *Sci. Rep.* **2017**, *7*, 2073.

23. Wood, H.P. *The Dendrobiums*; Timber Press: Portland, OR, USA, 2006.

24. Feng, S.; Jiang, Y.; Wang, S.; Jiang, M.; Chen, Z.; Ying, Q.; Wang, H. Molecular identification of *Dendrobium* species (Orchidaceae) based on the DNA barcode ITS2 region and its application for phylogenetic study. *Int. J. Mol. Sci.* **2014**, *16*, 21975–21988. [CrossRef] [PubMed]

25. Niu, Z.; Pan, J.; Zhu, S.; Li, L.; Xue, Q.; Liu, W.; Ding, X. Comparative analysis of the complete plastomes of *Apostasia wallichii* and *Neuwiedia singapureana* (Apostasioideae) reveals different evolutionary dynamics of IR/SSC boundary among photosynthetic orchids. *Front. Plant. Sci.* **2017**, *8*, 1713. [CrossRef] [PubMed]

26. Wu, C.S.; Chaw, S.M. Large-scale comparative analysis reveals the mechanisms driving plastomic compaction, reduction, and inversions in conifers II (cupressophytes). *Genome Biol. Evol.* **2016**, *8*, 3740–3750. [CrossRef] [PubMed]

27. Xu, H.; Hou, B.; Zhang, J.; Min, T.; Yuan, Y.; Niu, Z.; Ding, X. Detecting adulteration of *Dendrobium officinale* by real-time PCR coupled with ARMS. *Int. J. Food Sci. Technol.* **2012**, *47*, 1695–1700. [CrossRef]

28. Yan, W.J.; Zhang, J.Z.; Zheng, R.; Sun, Y.L.; Ren, J.; Ding, X.Y. Combination of SYBR Green II and TaqMan Probe in the adulteration detection of *Dendrobium devonianum* by fluorescent quantitative PCR. *Int. J. Food Sci. Technol.* **2016**, *50*, 2572–2578. [CrossRef]

29. Straub, S.C.; Fishbein, M.; Livshultz, T.; Foster, Z.; Parks, M.; Weitemier, K.; Cronn, R.C.; Liston, A. Building a model: Developing genomic resources for common milkweed (*Asclepias syriaca*) with low coverage genome sequencing. *BMC Genom.* **2011**, *12*, 211. [CrossRef] [PubMed]

30. Wysocki, W.P.; Clark, L.G.; Kelchner, S.A.; Burke, S.V.; Pires, J.C.; Edger, P.P.; Mayfield, D.R.; Triplett, J.K.; Columbus, J.T.; Ingram, A.L.; et al. A multi-step comparison of short-read full plastome sequence assembly methods in grasses. *Taxon* **2014**, *63*, 899–910.

31. Kim, K.; Lee, S.C.; Lee, J.; Yu, Y.; Yang, T.J.; Choi, B.S.; Koh, H.; Waminal, N.E.; Choi, H.; Kim, N.; et al. Complete chloroplast and ribosomal sequences for 30 accessions elucidate evolution of *Oryza* AA genome species. *Sci. Rep.* **2015**, *5*, 15655. [CrossRef] [PubMed]

32. Morton, B.R. Chloroplast DNA codon use: Evidence for selection at the *psbA* locus based on tRNA availability. *J. Mol. Evol.* **1993**, *3*, 273–280. [CrossRef]

33. Tatarinova, T.V.; Alexandrov, N.N.; Bouck, J.B.; Feldmann, K.A. GC3 biology in corn, rice, sorghum and other grasses. *BMC Genom.* **2010**, *11*, 308. [CrossRef] [PubMed]

34. Tatarinova, T.; Elhaik, E.; Pellegrini, M. Cross-species analysis of genic GC3 content and DNA methylation patterns. *Genome Biol. Evol.* **2013**, *5*, 1443–1456. [CrossRef] [PubMed]

35. Shi, X.; Wang, X.; Li, Z.; Zhu, Q.; Yang, J.; Ge, S. Evidence that natural selection is the primary cause of the guanine-cytosine content variation in rice genes. *J. Integr. Plant Biol.* **2007**, *49*, 1393–1399. [CrossRef]

36. McCoy, S.R.; Kuehl, J.V.; Boore, J.L.; Raubeson, L.A. The complete plastid genome sequence of *Welwitschia mirabilis*: An unusually compact plastome with accelerated divergence rates. *BMC Evol. Biol.* **2008**, *8*, 130. [CrossRef] [PubMed]

37. Glémin, S.; Clément, Y.; David, J.; Ressayre, A. GC content evolution in coding regions of angiosperm genomes: A unifying hypothesis. *Trends Genet.* **2014**, *30*, 263. [CrossRef] [PubMed]

38. Zhu, A.; Guo, W.; Gupta, S.; Fan, W.; Mower, J.P. Evolutionary dynamics of the plastid inverted repeat: The effects of expansion, contraction, and loss on substitution rates. *New Phytol.* **2016**, *209*, 1747–1756. [CrossRef] [PubMed]

39. Drouin, G.; Daoud, H.; Xia, J. Relative rates of synonymous substitutions in the mitochondrial, chloroplast, and nuclear genomes of seed plants. *Mol. Phylogenet. Evol.* **2008**, *49*, 827–831. [CrossRef] [PubMed]

40. Galtier, N.; Duret, L.; Glémin, S.; Ranwez, V. GC-biased gene conversion promotes the fixation of deleterious amino acid changes in primates. *Trends Genet.* **2009**, *25*, 1–5. [CrossRef] [PubMed]

41. Escobar, J.S.; Glémin, S.; Galtier, N. GC-biased gene conversion impacts ribosomal DNA evolution in vertebrates, angiosperms, and other eukaryotes. *Mol. Biol. Evol.* **2011**, *28*, 2561–2575. [CrossRef] [PubMed]

42. Gotea, V.; Elnitski, L. Ascertaining regions affected by GC-biased gene conversion through weak-to-strong mutational hotspots. *Genomics* **2014**, *103*, 349–356. [CrossRef] [PubMed]

43. Goubert, C.; Modolo, L.; Vieira, C.; Valientemoro, C.; Mavingui, P.; Boulesteix, M. De novo assembly and annotation of the Asian tiger mosquito (*Aedes albopictus*) repeatome with dnaPipeTE from raw genomic reads and comparative analysis with the yellow fever mosquito (*Aedes aegypti*). *Genome Biol. Evol.* **2015**, *7*, 1192–1205. [CrossRef] [PubMed]

44. Glémin, S.; Arndt, P.F.; Messer, P.W.; Petrov, D.; Galtier, N.; Duret, L. Quantification of GC-biased gene conversion in the human genome. *Genome Res.* **2015**, *25*, 1215–1228. [CrossRef] [PubMed]

45. Palmer, J.D. Chloroplast DNA exists in two orientations. *Nature* **1983**, *301*, 92–93. [CrossRef]

46. Khakhlova, O.; Bock, R. Elimination of deleterious mutations in plastid genomes by gene conversion. *Plant J.* **2006**, *46*, 85–94. [CrossRef] [PubMed]

47. Li, R.; Fan, W.; Tian, G.; Zhu, H.; He, L. The sequence and de novo assembly of the giant panda genome. *Nature* **2010**, *463*, 311–317. [CrossRef] [PubMed]

48. Wyman, S.K.; Jansen, R.K.; Boore, J.L. Automatic annotation of organellar genomes with DOGMA. *Bioinformatics* **2004**, *20*, 3252–3255. [CrossRef] [PubMed]

49. Schattner, P.; Brooks, A.N.; Lowe, T.M. The tRNAscan-SE, snoscan and snoGPS web servers for the detection of tRNAs and snoRNAs. *Nucleic Acids Res.* **2005**, *33*, W686–W689. [CrossRef] [PubMed]

50. Frazer, K.A.; Pachter, L.; Poliakov, A.; Rubin, E.M.; Dubchak, I. VISTA: Computational tools for comparative genomics. *Nucleic Acids Res.* **2004**, *32*, W273–W279. [CrossRef] [PubMed]

51. Katoh, K.; Kuma, K.; Toh, H.; Miyata, T. MAFFT version 5: Improvement in accuracy of multiple sequence alignment. *Nucleic Acids Res.* **2005**, *33*, 511–518. [CrossRef] [PubMed]

52. Ronquist, F.; Teslenko, M.; Mark, P.V.D.; Ayres, D.L.; Darling, A.; Höhna, S.; Larget, B.; Liu, L.; Suchard, M.A.; Huelsenbeck, J.P. MrBayes 3.2: Efficient Bayesian phylogenetic inference and model choice across a large model space. *Syst. Biol.* **2012**, *61*, 539–542. [CrossRef] [PubMed]

53. Edgar, R.C. MUSCLE: Multiple sequence alignment with high accuracy and high throughput. *Nucleic Acids Res.* **2004**, *32*, 1792–1797. [CrossRef] [PubMed]

54. Tamura, K.; Peterson, D.; Peterson, N.; Stecher, G.; Nei, M.; Kumar, S. MEGA5: Molecular evolutionary genetics analysis using maximum likelihood, evolutionary distance, and maximum parsimony methods. *Mol. Biol. Evol.* **2011**, *28*, 2731–2739. [CrossRef] [PubMed]

55. Vaidya, G.; Lohman, D.J.; Meier, R. SequenceMatrix: Concatenation software for the fast assembly of multi-gene datasets with character set and codon information. *Cladistics* **2011**, *27*, 171–180. [CrossRef]

56. Xia, X. DAMBE5: A comprehensive software package for data analysis in molecular biology and evolution. *Mol. Biol. Evol.* **2013**, *30*, 1720–1728. [CrossRef] [PubMed]

57. Hershberg, R.; Petrov, D.A. Evidence that mutation is universally biased towards AT in bacteria. *PLoS Genet.* **2010**, *6*, e1001115. [CrossRef] [PubMed]

58. Felsenstein, J. *PHYLIP (Phylogeny Inference Package) Version 3.6*; Department of Genome Sciences, University of Washington: Seattle, DC, USA, 2005.

59. Yang, Z. PAML 4: Phylogenetic analysis by maximum likelihood. *Mol. Biol. Evol.* **2007**, *24*, 1586–1591. [CrossRef] [PubMed]

10

The Complete Chloroplast Genome of *Catha edulis*: A Comparative Analysis of Genome Features with Related Species

Cuihua Gu [1,2], **Luke R. Tembrock** [2], **Shaoyu Zheng** [1] **and Zhiqiang Wu** [3,*]

[1] School of Landscape and Architecture, Zhejiang Agriculture and Forestry University, Hangzhou 311300, China; gu_cuihua@126.com (C.G.); aggies.collins@gmail.com (S.Z.)

[2] Department of Biology, Colorado State University, Fort Collins, CO 80523, USA; Luke.R.Tembrock@aphis.usda.gov

[3] Department of Ecology, Evolution, and Organismal Biology, Ames, IA 50011, USA

* Correspondence: wu.zhiqiang.1020@gmail.com

Abstract: Qat (*Catha edulis*, Celastraceae) is a woody evergreen species with great economic and cultural importance. It is cultivated for its stimulant alkaloids cathine and cathinone in East Africa and southwest Arabia. However, genome information, especially DNA sequence resources, for *C. edulis* are limited, hindering studies regarding interspecific and intraspecific relationships. Herein, the complete chloroplast (cp) genome of *Catha edulis* is reported. This genome is 157,960 bp in length with 37% GC content and is structurally arranged into two 26,577 bp inverted repeats and two single-copy areas. The size of the small single-copy and the large single-copy regions were 18,491 bp and 86,315 bp, respectively. The *C. edulis* cp genome consists of 129 coding genes including 37 transfer RNA (tRNA) genes, 8 ribosomal RNA (rRNA) genes, and 84 protein coding genes. For those genes, 112 are single copy genes and 17 genes are duplicated in two inverted regions with seven tRNAs, four rRNAs, and six protein coding genes. The phylogenetic relationships resolved from the cp genome of qat and 32 other species confirms the monophyly of Celastraceae. The cp genomes of *C. edulis*, *Euonymus japonicus* and seven Celastraceae species lack the *rps16* intron, which indicates an intron loss took place among an ancestor of this family. The cp genome of *C. edulis* provides a highly valuable genetic resource for further phylogenomic research, barcoding and cp transformation in Celastraceae.

Keywords: chloroplast (cp) genome; *Catha edulis*; next generation sequencing; phylogeny; repeat sequence

1. Introduction

Qat (Celastraceae: *Catha edulis* (Vahl) Forssk. ex Endl.) is a woody evergreen species of major cultural and economic importance in southwest Arabia and East Africa, which is cultivated for its stimulant alkaloids cathine and cathinone. An estimated 20 million people consume qat on a daily basis in eastern Africa [1], and its use and cultivation has been expanding in recent years [2]. Qat is the only species in Celastraceae that is cultivated on a large scale. The cultivation and/or collection (in some instances illegally from wild sources in protected areas) of qat takes place primarily in Israel, Ethiopia, Kenya, Madagascar, Rwanda, Tanzania, Somalia, Uganda, and Yemen [2–4].

The cultivation and sale of qat has become an important driver in the local and regional economies of East Africa and Yemen. In Yemen, 6% of the gross domestic product is generated from qat cultivation and sales [5]. Ethiopia has become the number one producer of qat in the world with exports in 1946 equaling only 26 tons valued at $5645, while 15,684 tons were exported in 2000 valued at $72 million [6]. A similar expansion in qat cultivation and sales has occurred in Kenya with the current trade from

Kenya to Somalia estimated at $100 million per year. Trade of qat has become international in scale with, for example, 2.26 million kilograms of qat imported into England from Ethiopian and Kenya in 2013 [7]. The biosynthesis of cathinone and similar stimulant alkaloids is rare among green plants, known only in *Catha edulis* and several Asian species of *Ephedra* [8]. In addition, Celastraceae species produce numerous unique phytochemicals of potential pharmaceutical value [9]. Chloroplast transformations of qat and related species may prove useful for the production of cathinone related alkaloids and/or novel drugs.

The phylogenetic placement of qat within the Celastraceae has been inferred from 18S, 26S, *atpB*, ITS (as Nuclear ribosomal internal transcribed spacer), *matK*, *phyB*, and *rbcL* [10]. Phylogeographic work using SSR (as simple sequence repeats) loci has been done for wild and cultivated qat in the historic areas of production—Ethiopia, Kenya, and Yemen [7,11]. Beyond these studies, no genetic resources of which we are aware have been developed for qat. In addition, no chloroplast (cp) genome has been fully sequenced and published in the genus *Catha*. Therefore, our completed cp genome will be an important genetic resource for further evolutionary studies both within the Celastrales generally and economically important qat specifically.

The cp genome in plants is noted as being highly conserved in gene content [12]. Despite the consistency between cp genomes in plants, the differences in the size of cp genomes appear to be driven by intron and gene loss, and structural changes such as loss or gain of repeat units in different types of repetitive DNA [13]. In particular, genes that straddle inversion junctions such as *ycf1* appear to be undergoing rapid evolution [14].

Contrary to the structure of most nuclear plant genomes, the cp genome is typically comprised of a highly conserved quadripartite structure which is 115 to 165 kb in length, uniparentally inherited [12,15], and with similar gene content and order shared among most land plants [16]. From the advancements made by next-generation sequencing (NGS), complete, high quality cp genomes are becoming increasingly common [17]. At present, more than 2000 completed cp genomes of angiosperm species can be downloaded in the public database of the National Center for Biotechnology Information (NCBI; [18], Available online: https://www.ncbi.nlm.nih.gov/genomes/ GenomesGroup.cgi?taxid=2759&opt=plastid). Large databases of complete cp genomes provide an indispensable resource for researchers identifying species [19], designing molecular markers for plant population studies, and for research concerning cp genome transformation [20–22]. The essentially non-recombinant structures of cp genomes make them particularly useful for the above applications. For example, cp genomes maintain a positive homologous recombination system [23–26]. Thus, in the transformation process, genes can be precisely transferred to specific genomic regions. A variety of homologous cp sites have proven useful at multiple levels of classification, including inter-specific and intra-specific [27]. In more recent years, systematic studies have employed entire cp genomes to attain high resolution phylogenies [28].

In this paper, we report the completely sequenced cp genome in the Celastrales and discuss the technical aspects of sequencing and assembly. In addition, we conduct phylogenetic analysis using other fully sequenced cp genomes from species in the closely related orders Malpighiales and Rosales. These analyses were conducted to find the top twenty loci for phylogenetic analysis and find which structural changes have taken place across cp genomes between the orders Rosales, Malpighiales, and Celastrales. The completed cp genome is a valuable resource for studying evolution and population genetics of both wild and cultivated populations of qat as well as genetic transformations related to the production of pharmaceuticals in qat or related Celastraceae species.

2. Results and Discussion

2.1. Chloroplast Assembly and Genome Features

The *C. edulis* cp genome was completely assembled into a single molecule of 157,960 bp, by combining Illumina and Sanger sequencing results. By mapping the completed genome using the paired reads, we confirm the size of our assembly for the completed cp genome with 497,848 (representing 5% of all reads) mapped pair-end reads evenly spanning the entire genome with mean read depth of 785× coverage (Figure S1). Given these quality controls and processing steps, the cp genome for qat is high quality.

Although the genome structure is highly conserved in the cp genome, several features such as the presence or lack of introns, the size of the intergenic region, gene duplication, and the length, type and number of repeat regions can vary [29]. The complete *C. edulis* cp genome has the conserved quadripartite structure and size that resembles most land plant cp genomes which are normally 115–165 kb in size including two inverted repeats (IRs) and two single-copy regions as large single copy and small single copy (LSC and SSC).

The cp genome of *C. edulis* consists of two single-copy regions isolated by two identical IRs of 26,577 bp each, one SSC region of 18,491 bp and one LSC region of 86,315 bp. The proportion of LSC, SSC, and IRs size in the entire cp genome is 54.6%, 11.7% and 33.6%, respectively (Figure 1 and Table 1). The GC contents of the LSC, IR, SSC, and the whole cp genome are 35.1%, 42.7%, 31.8%, and 37.3%, respectively, which are consistent with the published Rosid cp genomes [30].

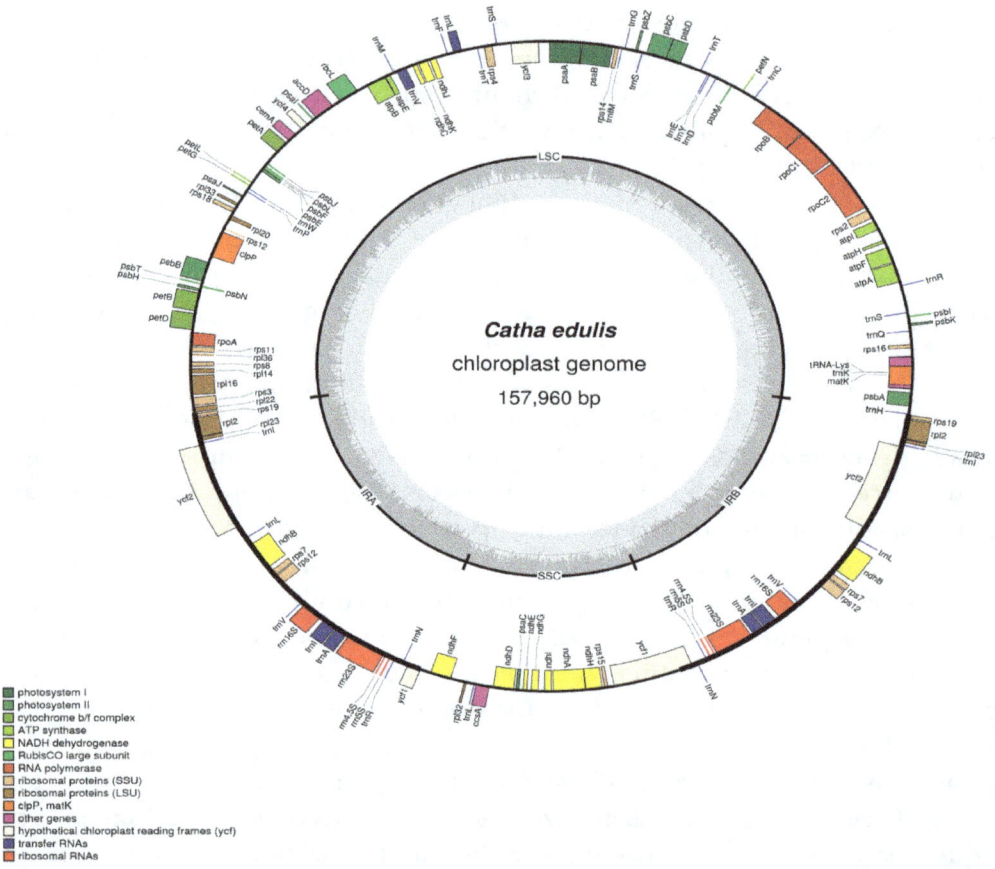

Figure 1. Circular map of the *C. edulis* cp genome. Genes shown inside and outside of the outer circle are transcribed clockwise and counterclockwise, respectively. The innermost shaded area inside the inner circle corresponds to GC content in the cp genome. Genes in different functional groups are color coded. IR, inverted repeat; LSC, large single copy region; SSC, small single copy region. The map is drawn using OGDRAW (V 1.2, Max Planck Institute of Molecular Plant Physiology, Am Mühlenberg, Germany).

The *C. edulis* cp genome is composed of tRNAs, protein coding genes and rRNAs, intergenic and intronic regions (Table 2). Non-coding DNA accounts for 67,633 bp (42.8%) of the whole *C. edulis* cp genome, protein-coding genes account for 78,471 bp (49.7%), tRNA accounts for 2806 bp (1.8%), and rRNA accounts for 9050 bp (5.7%). By comparison with seven other species, gene order, gene content, the coding genes, and non-coding region proportions are similar among these cp genomes (Table 2).

Table 1. Comparison of plastid genome size among eight species.

Region	Features	C. edulis	E. japonicus	H. brasiliensis	M. esculenta	P. euphratica	R. communis	S. purpurea	V. seoulensis
LSC	Length (bp)	86,315	85,941	89,209	89,295	84,888	89,651	84,452	85,691
	GC Content (%)	35.1	35.1	33.2	33.3	34.5	33.3	34.4	33.8
	Length Percentage (%)	54.6	54.5	55.3	55.3	54.1	54.9	54.3	54.8
SSC	Length (bp)	18,491	18,340	18,362	18,250	16,586	18,816	16,220	18,008
	GC Content (%)	31.8	31.8	29.5	29.6	30.6	29.5	31	29.6
	Length Percentage (%)	11.7	11.6	11.4	11.3	10.6	11.5	10.4	11.5
IR	Length (bp)	26,577	26,678	26,810	26,954	27,646	27,347	27,459	26,404
	GC Content (%)	42.7	42.7	42.2	42.3	41.9	41.9	41.9	42.6
	Length Percentage (%)	16.8	16.9	16.6	16.7	17.6	16.8	17.6	16.9
Total	Length (bp)	157,960	157,637	161,191	161,453	156,766	163,161	155,590	156,507
	GC Content (%)	37.3	37.3	35.7	35.9	36.7	35.7	36.7	36.3

LSC, large single copy region; SSC, small single copy region; IR, inverted repeat.

Table 2. Comparison of coding and non-coding region size among eight species.

Region	Species	C. edulis	E. japonicus	H. brasiliensis	M. esculenta	P. euphratica	R. communis	S. purpurea	V. seoulensis
Protein coding	length (bp)	78,471	77,331	78,852	79,089	78,728	78,119	77,898	78,310
	Length Percentage (%)	49.7	49.1	48.9	49.0	50.2	47.9	50.1	50.0
	GC Content (%)	38	38.2	37.1	37.2	37.6	37.5	37.6	37.2
tRNA	length (bp)	2806	2806	2798	2742	2796	2802	2792	2810
	Length Percentage (%)	1.8	1.8	1.7	1.7	1.8	1.7	1.8	1.8
	GC Content (%)	52.6	53.3	53.2	53.3	53	53.2	52.9	53
rRNA	length (bp)	9,050	9050	9050	9050	9050	9050	9,050	9050
	Length Percentage (%)	5.7	5.7	5.6	5.6	5.8	5.5	5.8	5.8
	GC Content (%)	55.2	55.4	55.4	55.5	55.5	55.5	55.4	55.4
Intron	length (bp)	18,474	19,287	18,538	18,479	18,210	18,278	17,321	18,348
	Length Percentage (%)	11.7	12.2	11.5	11.4	11.6	11.2	11.1	11.7
	GC Content (%)	37.1	36.6	36.6	36.9	36.9	37.1	37.3	36.7
Intergenic	length (bp)	49,159	49,163	51,953	52,093	47,982	54,912	48,529	47,989
	Length Percentage (%)	31.1	31.2	32.2	32.3	30.6	33.7	31.2	30.7
	GC Content (%)	31.9	31.7	29	29	31	28.7	30.7	30.1

2.2. Gene Content and Structure

The cp genome of *C. edulis* consisted of 129 coding regions made up of 37 tRNAs, 84 protein-coding genes, and eight rRNAs, of which 112 genes are unique and 17 genes were repeated in two inverted regions consisting of seven tRNAs, six protein coding genes, and four rRNAs (Figure 1 and Table 3). Among these 112 unique genes, three genes crossed different cp boundaries: $trnH^{GUG}$ crossed the IR_B and LSC regions, *ycf1* crossed the IR_B and SSC regions, *rps12* crossed two IR regions and the LSC region (two 3′ end exons repeated in IRs and 5′ end exon situated in LSC) (Figure 1). Of the remaining 109 genes, 80 are situated in LSC including 59 protein coding genes and 21 tRNAs, 17 in two inverted repeats (six coding genes, seven tRNAs, and four rRNAs), and 12 in the SSC including 11 coding genes and one tRNA.

Table 3. List of genes in the *C. edulis* plastid genome.

Gene Category	Groups of Genes	Name of Genes
Self-replication	Transfer RNA genes	$trnA^{UGC}$ [a,b] $trnC^{GCA}$ $trnD^{GUC}$ $trnE^{UUC}$ $trnF^{GAA}$ $trnfM^{CAU}$ $trnG^{UCC}$ $trnG^{GCC}$ $trnH^{GUG}$ $trnI^{CAU}$ [b] $trnI^{GAU}$ [a,b] $trnK^{UUU}$ [a] $trnL^{CAA}$ [b] $trnL^{UAA}$ [a] $trnL^{UAG}$ $trnM^{CAU}$ $trnN^{GUU}$ [b] $trnP^{UGG}$ $trnQ^{UUG}$ $trnR^{ACG}$ [b] $trnR^{UCU}$ $trnS^{GCU}$ $trnS^{GGA}$ $trnS^{UGA}$ $trnT^{GGU}$ $trnT^{UGU}$ $trnV^{GAC}$ [b] $trnV^{UAC}$ [a] $trnW^{CCA}$ $trnY^{GUA}$
	Small subunit of ribosome	*rps2 rps3 rps4 rps7b rps8 rps11 rps12* [a,b] *rps14 rps15 rps16 rps18 rps19*
	Ribosomal RNA genes	*rrn16* [b] *rrn23* [b] *rrn4.5* [b] *rrn5* [b]
	Large subunit of ribosome	*rpl2* [b] *rpl14 rpl16* [a] *rpl20 rpl22 rpl23* [b] *rpl32 rpl33 rpl36*
	DNA dependent RNA polymerase	*rpoA rpoB rpoC1* [a] *rpoC2*
Photosynthesis	Subunits of photosystem I	*psaA psaB psaC psaI psaJ*
	Subunits of photosystem II	*psbA psbB psbC psbD psbE psbF psbH psbI psbJ psbK psbL psbM psbN psbT psbZ*
	Subunits of cytochrome	*petA petB* [a] *petD* [a] *petG petL petN*
	Subunits of ATP synthase	*atpA atpB atpE atpF* [a] *atpH atpI*
	ATP-dependent protease subunit p gene	*clpP* [a]
	Large subunit of Rubisco	*rbcL*
	Subunits of NADH dehydrogenase	*ndhA* [a] *ndhB* [a,b] *ndhC ndhD ndhE ndhF ndhG ndhH ndhI ndhJ ndhK*
Other genes	Maturase	*matK*
	Envelop membrane protein	*cemA*
	Subunit of acetyl-CoA-carboxylase	*accD*
	c-type cytochrome synthesis gene	*ccsA*
Genes of unknown function	Conserved open reading frames	*ycf1 ycf2* [b] *ycf3* [a] *ycf4*

[a] Genes containing introns; [b] Duplicated gene (Genes present in the IR regions).

Most of the protein-coding genes contain only one exon, while 17 genes contain one intron, of which four occur in both IRs, 12 genes are distributed in LSC, and one in the SSC (Table 4), among them three genes (*rps12*, *clpP* and *ycf3*) contain two introns, while 14 genes ($trnA^{GUC}$, $trnI^{GAU}$, $trnG^{UCC}$, $trnL^{UAA}$, $trnK^{UUU}$, and $trnV^{UAC}$, *rpoC1*, *atpF*, *rpl16*, *rpl2*, *petB*, *petD*, *ndhA*, and *ndhB*) contain one intron. The longest intron of $trnK^{UUU}$ is 2495 bp including the 1533 bp encoding the *matK* gene [13]. The *rps12* gene was predicted to be trans-spliced with a repeated 3′ end duplicated in two IRs and a single 5′ end exon in LSC [31].

Table 4. Genes with intron and their length of exons and introns in plastid genome of *C. edulis*.

Gene Name	Location	Exon I (bp)	Intron I (bp)	Exon II (bp)	Intron II (bp)	Exon III (bp)
rpoC1	LSC	1632	817	441		
atpF	LSC	396	699	159		
petB	LSC	6	773	642		
petD	LSC	8	784	475		
ndhB	IR	756	687	777		
ndhA	SSC	540	1178	573		
rpl16	LSC	399	1119	9		
rpl2	IR	471	648	393		
rps12	LSC	114		27	546	231
ycf3	LSC	153	727	228	731	126
clpP	LSC	231	676	291	849	69
trnK-UUU	LSC	29	2495	37		
trnL-UAA	LSC	37	540	50		
trnV-UAC	LSC	37	663	39		
trnI-GAU	IR	42	939	35		
trnA-UGC	IR	38	801	35		
trnG-UCC	LSC	23	761	48		

2.3. Comparison of the cp Genomes

The cp genome of *C. edulis* (Celastraceae) was compared to species from 14 genera, including *Populus, Salix, Viola, Hevea, Manihot, Ricinus, Euonymus* and seven out-group species using dot-plot analysis. Besides a unique rearrangement of one 30-kb inversion in the *H. brasiliensis* cp genome [32], no other large structural differences (inversions) were detected among all compared species in the dot-plot analysis. This is consistent with the extremely conserved cp genomes in land plants [16]. The limited structural differences across the 14 species cp genomes demonstrate that gene order, gene content, and entire genome structure are conserved (Figure S3).

Based on the limited structural variation of cp genomes, we focused on seven closely related species of *C. edulis* to examine finer scale structural differences in genome length. Among these seven cp genomes, the length of genomes ranged from 155,590 bp (*S. purpurea*) to 163,161 bp (*R. communis*). The length of the LSC region varied from 84,452 bp (*S. purpurea*) to 89,651 bp (*R. communis*), and from 16,220 bp (*S. purpurea*) to 18,816 bp (*R. communis*) in SSC, and from 26,404 bp (*V. seoulensis*) to 27,646 bp (*P. euphratica*) in the IR regions (Table 2).

The entire GC content of the complete *C. edulis* cp genome is 37.3%, with 33.6% GC content in IRs, 35.1% in LSC, and 31.8% in SSC. These GC contents are consistent with other published cp genomes [33]. The whole GC content in the two Celastrales and six cp genomes of Malpighiales species ranged from 35.7% to 37.3% of the total genome, with *R. communis* having the lowest and *C. edulis* and *E. japonicus* having the highest GC content (Table 1).

These eight species have similar genetic composition at the IR-SSC and IR-LSC boundaries except *rps19*, which is not present from the border of LSC and IR$_A$ in *P. euphratica* and *R. communis* in which *rpl22* crosses the border of IR$_A$ and LSC (Figure 2).

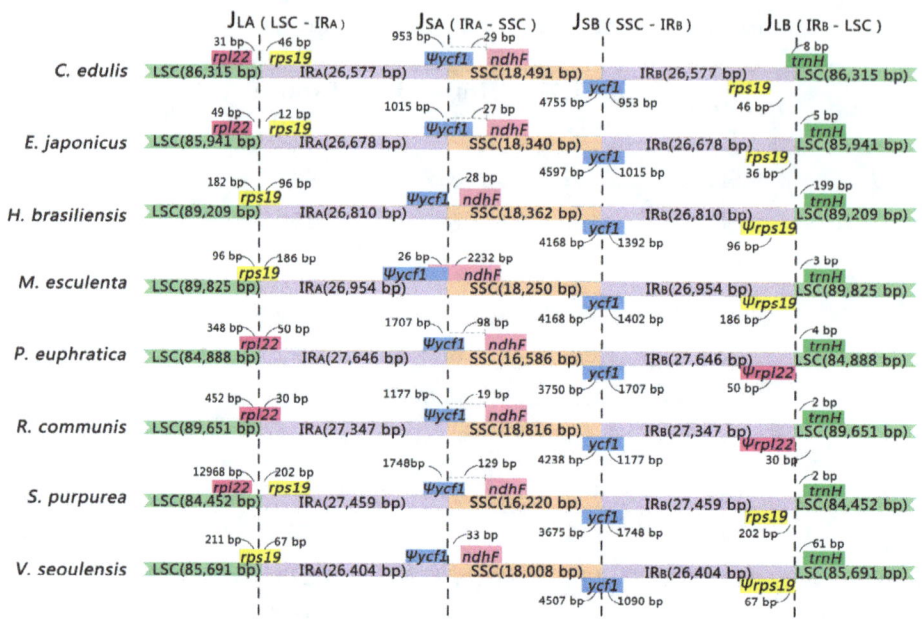

Figure 2. Comparison of junctions between the LSC, SSC, and IRs among eight species. Number above indicates the distance in bp between the ends of genes and the borders sites (distances are not to scale in this figure). The ψ symbol represents pseudogenes.

2.4. Contraction and Expansion in the Four Junction Regions

Although genomic structure including gene composition and genome size are highly conserved, expansion and contraction of IRs are common differences between plant cp genomes. Kim [34] proposed that the IRs size differ within plant cp genomes mainly results from the contraction or expansion at the junctions. Comparison of the inverted repeat-single copy (IR-SC) boundary regions of the two Celastrales and six Malpighiales species genomes showed very small differences in boundaries (Figure 2). We inspected the four boundaries (J_{LA}, J_{LB}, J_{SA}, and J_{SB}) across the two Celastrales and six Malpighiales species to detect the detailed boundary variation between the two SC regions and IRs using the methods described in [18].

The size of the IRs varied from 26,404 to 27,646 bp. The IR$_A$-LSC junction (J_{LA}) was situated in the *rps19* gene in *H. brasiliensis*, *M. esculenta*, and *V. seoulensis* which crossed inside the IR$_A$ region 96 bp, 186 bp, and 67 bp, respectively, and as a result duplicated pseudogene *rps19* (ψ*rps19*) was nested within IR$_B$ for these three species. However, in *C. edulis*, *E. japonicus* and *S. purpurea*, J_{LA} is situated in the intergenic regions between *rpl22* and *rps19* in which the distances from *rps19* to the J_{LA} were 46 bp, 12 bp and 202 bp. In two other species, *P. euphratica* and *R. communis*, J_{LA} is situated in the coding region of *rpl22* which spread into IR$_A$ 50 bp and 30 bp, respectively, and resulted in the generation of pseudogene *rpl22* (ψ*rpl22*) in IR$_B$.

The IR$_A$-SSC junction (J_{SA}) was situated in or adjoined pseudogene *ycf1* (ψ*ycf1*) for all eight species; J_{SA} of three species (*H. brasiliensis*, *M. esculenta*, and *V. seoulensis*) were all situated just adjacent to the end of ψ*ycf1*. Overlap between *ndhF* and ψ*ycf1* was found in *M. esculenta*, in which *ndhF* expanded into the IR$_A$ region for 26 bp. For the other five species, J_{SA} was located near ψ*ycf1*. In the other six species (*C. edulis*, *E. japonicus*, *H. brasiliensis*, *P. euphratica*, *R. communis*, *S. purpurea* and *V. seoulensis*), the distances between *ndhF* and J_{SA} were 29 bp, 27 bp, 28 bp, 98 bp, 19 bp, 129 bp and 33 bp, respectively.

The IR$_B$-SSC junction (J_{SB}) is situated in the *ycf1* coding region which spans into the IR$_B$ region in all eight species. However, the length of *ycf1* in the IR region varied among the eight species from 953 bp to 1748 bp highlighting the dynamic variation of the junction regions.

The IR$_B$-LSC junctions (J_{LB}) were located between *rps19* and *trnH* in *E. japonicus* and *S. purpurea*; situated at the end of ψ*rps19* in *H. brasiliensis*, *M. esculenta*; and *V. seoulensis*; and at the end of ψ*rpl22* in *P. euphratica* and *R. communis*. In the J$_{LB}$ junction, the *trnH* gene is 8 bp into IR$_B$ region in *C. edulis*. In the other seven species, 2–199 bp distance is found between the *trnH* gene and the IR$_B$-SSC junction.

The variation in the IR-SC boundary area is due to the contraction or expansion of the IR observed in the IR-SSC boundaries. These expansions/contractions are likely to be mediated by molecular recombination within the two short, straight repeating sequences that occur frequently in the genes within the boundary [34].

2.5. Verification of the rps16 Intron Loss from Catha and Seven Other Celastraceae Species

The gene composition in the *C. edulis* cp genome is similar to the other angiosperm species analyzed in this study. However, we found that the *rps16* gene had no intron in the *C. edulis* cp genome. The structure and the intron size for *rps16* are conserved in the model species *Arabidopsis thaliana* and in our sampled species (NC_000932). However, it has been reported that *rps16* gene or the intron of *rps16* has been lost multiple times in numerous lineages [35,36].

To test whether the loss of the *rps16* intron is common throughout the Celastraceae family or just in certain species, two primers were designed in the flanking exons to amplify and then sequence the intron region (or lack thereof) for eight species in the Celastraceae family. Based on the PCR amplification (Figure S2), the length of this *rps16* amplicon is about 550 bp in all eight sampled Celastraceae species indicating that the intron has been lost throughout the Celastraceae family. We also conducted Sanger sequencing to verify the alignment of the *rps16* gene (Figure 3). From this alignment, all species sampled from the Celastraceae family do not contain the *rps16* intron (Figure 3A). The Sanger sequencing data provide additional evidence that all eight-species do not have this intron (Figure 3B).

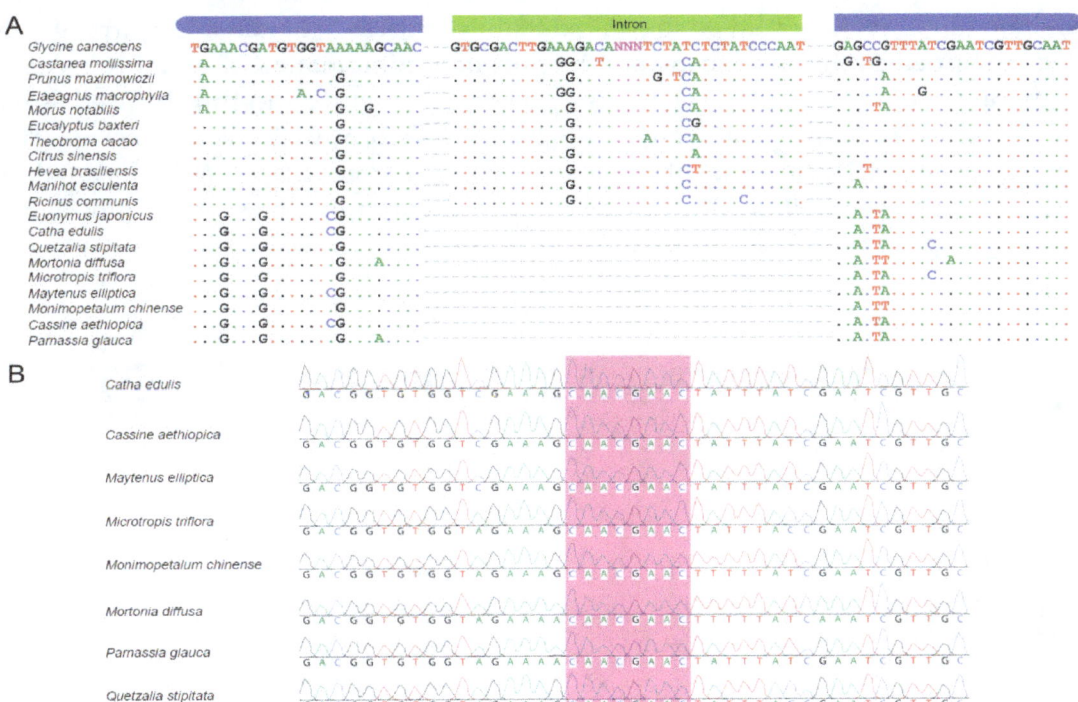

Figure 3. The sequence variation for *rps16* gene with and without intron: (**A**) The structural components of *rps16* gene in 20 species. All Species outside of Celastraceae family contained the *rps16* intron. (**B**) The purple area in all eight species from different genera of the Celastraceae family showed the connection of two exons indicating the lost intron.

Intron loss in cp genomes have been reported multiple times in different species, such as species in Desmodieae (Fabaceae) [37] and reported in both dicots and monocots. Loss of the *rps16* intron could probably be best explained by a homologous recombination and the reverse-transcriptase mediated mechanism [35]. However, intron loss from DNA fragment deletions or gene transfer between introns could be due to yet unexplained processes [37]. By increasing the sampling density within Celastraceae and its closest relatives, the timing of the *rps16* intron loss was inferred to occur between the Celastrales and Oxalidales + Malpighiales approximately 80 million years ago [38].

2.6. Identification of Long Repetitive Sequences

Long repetitive sequences play key functions in cp genome evolution, genome rearrangements and can be informative in phylogenetic studies [39]. Comparison of forward, complement, reverse, and palindromic repeats (≥30 bp) (with a sequence identity of ≥90% per repeat unit) were conducted across *C. edulis* and seven related species using REPuter (Available online: https://bibiserv.cebitec. uni-bielefeld.de/reputer/; (University of Bielefeld, Bielefeld, Germany)). *Catha edulis* had the fewest (8) repeats while its cp genome was not the shortest among those examined (157,960 bp) which is inconsistent with the general trend of shorter genomes possessing fewer repetitive regions [40].

A total of 175 unique repeats consisting of forward, reverse, complementary and palindromic were found from the eight-species examined (Figure 4A). The species *E. japonicus* included the most repeats consisting of: 14 palindromic repeats, 19 forward repeats, and eight reverse repeats, for a total of 41 repeats (Figure 4A and Table S3). In *H. brasiliensis, M. esculenta, P. euphratica, R. communis, S. purpurea* and *V. seoulensis* cp genomes, 29, 35, 20, 22, 10, and 10 total repeat pairs were found respectively (Figure 4A). Among them, 19 forward repeats were most commonly found in *E. japonicus* and *M. esculenta* and in all species the most common repeat type was forward (Figure 4A). Forward repeats are often the result of transposon activity [41], which can increase under cellular stress [42]. However, the origins and multiplication of long repetitive repeats is not fully understood [43]. Previous studies suggested that the existence of genome rearrangement could be attributed to slipped-strand mispairing and inapposite recombination of repetitive sequences [43]. Moreover, forward repeats can lead to changes in genomic structure and thus be used as markers in phylogenetic studies. The length of repeats is variable in this study, with the shortest at 30 bp and the longest at 95 bp (Table S3). The majority of repeats (82%) varied from 30 bp to 40 bp in length (Figure 4B and Table S3). Given the variability of these repeats between lineages, they can be informative regions for developing genomic markers for population and phylogenetic studies [44].

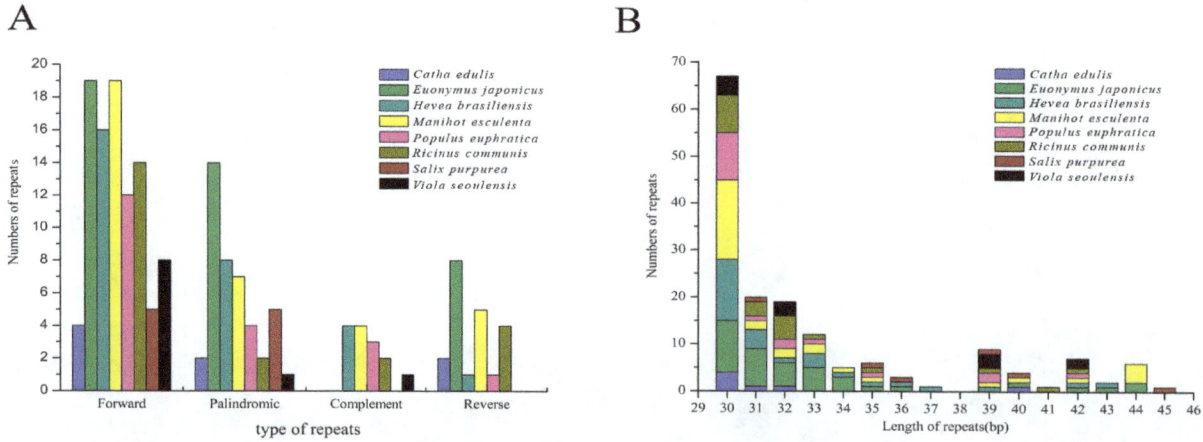

Figure 4. Analysis of repeat sequences in eight chloroplast genomes: (**A**) frequency of repeat types; and (**B**) frequency of the repeats by length ≥30 bp.

2.7. Chloroplast Genome Simple Sequence Repeats (SSRs)

Simple sequence repeats (SSRs) are sequences with motifs from 1 to 6 bp in length repeated multiple times (see methods for cutoff criteria), are found distributed throughout the cp genome, and are often used as markers for breeding studies, population genetics, and genetic linkage mapping [43,45].

A total of 278 SSRs were found in the *C. edulis* cp genome (Figure 5A and Table S4). These SSRs include 165 mononucleotide SSRs (59%), 43 dinucleotide SSRs (15%), 65 trinucleotide SSRs (23%), 3 tetranucleotide (0.01%), and 1 pentanucleotide SSR (0.003%) (Figure 5A and Table S4). Among the 165 SSRs, 98% of SSRs (161) are the AT type with copy number from 8 to 18 (Table S4). In these SSRs of the *C. edulis* cp genome, 89 SSRs were detected in protein-coding genes, 34 SSRs in introns, and 155 in intergenic regions (Figure 5B). In relation to the quadripartite, 195 SSRs were situated in the LSC, whereas 36 and 37 were identified in the SSC and IR, respectively (Figure 5C).

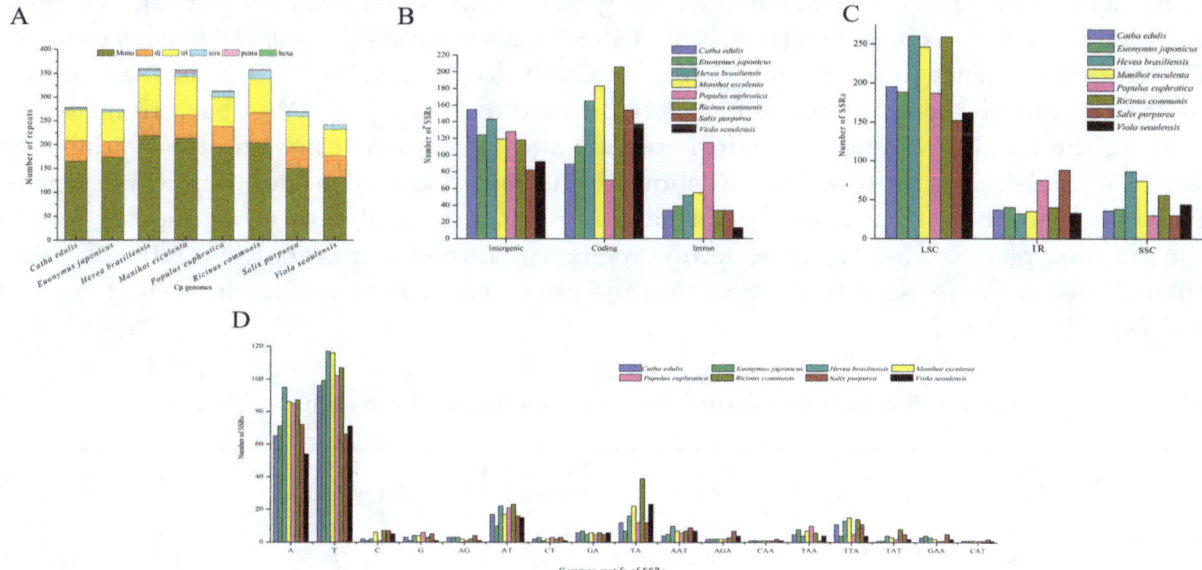

Figure 5. The distribution, type, and presence of simple sequence repeats (SSRs) in eight chloroplast genomes: (**A**) number of different SSR types detected in eight chloroplast genomes presence of SSRs at the LSC, SSC, and IR regions.; (**B**) frequency of SSRs in the protein-coding regions, intergenic spacers and intronic regions; (**C**) frequency of SSRs in the LSC, SSC, and IR regions; and (**D**) frequency of common motifs in the eight chloroplast genomes.

Among the eight species, *V. seoulensis* had the fewest SSRs (242) and *H. brasiliensis* had the most SSRs (360). *Salix purpurea* has the shortest cp genome (155,590 bp) with 270 SSRs and *R. communis* has the longest cp genome (163,161 bp) and 358 SSRs of those analyzed in this study suggesting that number of SSRs may affect genome length, but a strong correlation was not found in all species (Figure 5A). This result indicates that cp genome sizes were not obviously connected with the number of SSRs in these species. Additionally, an abundance of tetranucleotide SSRs were not found in the species studied and no pentanucleotide SSRs were found in *V. seoulensis* or hexanucleotide in *E. japonicus*, *R. communis* and *V. seoulensis* (Figure 5A). Among the eight species, most SSRs of *C. edulis* and *E. japonicus* were located in intergenic regions, most SSRs of *H. brasiliensis*, *M. esculenta*, *P. euphratica*, *R. communis*, and *V. seoulensis* in coding regions, and most SSRs of *S. purpurea* are in intronic regions (Figure 5B). Some SSRs were distributed in protein-coding regions such as *ycf1* and *rpoC2* (Table S4), which could also be employed as DNA markers for population level and genomic studies. Most SSRs in all eight-species were in the LSC region (Figure 5C). Common motifs in the eight-species studied generally consisted of polythymine (poly-T) or polyadenine (poly-A) (Figure 5D). The Euphorbiaceae species in this study all have more SSRs than the other species in this study as well as similar patterns of distribution in the genome. More work is needed to understand these patterns of SSR distribution in cp genomes. Lastly, the SSRs from this study should be valuable for phylogeographic studies and comparing phylogenetic relationships among Celastraceae species.

2.8. Highly Informative Coding Genes and Markers for Phylogenomic Analysis

Detecting highly informative and variable coding genes is important for DNA barcoding, marker development and phylogenomic analyses [46]. Coding genes such as *matK*, *rbcL* have been widely employed for barcoding applications [47,48] and phylogenetic reconstructions [49–51]. Based on compared complete cp genomes, additional informative markers were identified within the Celastraceae.

We aligned entire coding genes more than 200 bp in length to discover genes with the highest sequence identity index and the highest proportion of parsimony-informative sites, for the seven species in this study (Table 5, Table S5). In the coding regions, *matK* and *ycf1* have the largest proportion of parsimony information characters (16.83% and 16.80%, respectively). The *matK* gene is used as core DNA barcoding sequence under the suggestion of CBOL working group (CBOL is The Consortium for the Barcode of Life, an international initiative devoted to developing DNA barcoding as a global standard for the identification of biological species) and also in concert with other variable genes such as *ITS + psbA-trnH + matK* which was shown to have the highest species identification rate [52]. Given the high number of parsimony informative in *ycf1*, it may also serve as another core DNA barcode in future plant studies [14]. The coding regions identified in this analysis (Table 5) should be particularly informative for species identification and phylogenetic analyses due to the high percentage of variable sites.

Table 5. Ten highest informative sites of coding genes in eight species.

No.	Region	Length (bp) [1]	Aligned Length (bp) [2]	Conserved Sites	Parsimony Informative [3]	Parsimony Informative % [4]	CI. [5]	RI [6]	SI [7]
1	*matK*	1518	1575	1028	265	16.83	0.82	0.7	0.9
2	*ycf1*	5640	6327	3970	1063	16.80	0.82	0.6	0.8
3	*ccsA*	969	987	689	160	16.21	0.84	0.7	0.9
4	*accD*	1509	1401	242	227	16.20	0.83	0.7	0.8
5	*rps3*	648	663	467	107	16.14	0.82	0.7	0.9
6	*ndhF*	2232	2331	1606	368	15.79	0.81	0.6	0.8
7	*rps8*	405	411	294	64	15.57	0.8	0.7	0.9
8	*rpl22*	399	551	345	82	14.88	0.83	0.6	0.7
9	*petL*	96	96	70	14	14.58	0.9	0.8	0.9
10	*ndhD*	1503	1527	1116	207	13.56	0.82	0.7	0.9

[1] Length: refers to sequence length in *Catha edulis*; [2] Aligned length: refers to the alignment of seven other species considered in the comparative analysis (see Materials and Methods); [3] Number of parsimony informative sites; [4] Percentage of parsimony informative sites; [5] CI: Consistency Index; [6] RI: Retention Index; [7] SI: Sequence Identity.

2.9. Phylogenetic Analysis

Based on cp genomes, phylogenetic analyses have helped to resolve the relationships of many angiosperm lineages [53,54]. Previous phylogenetic work in Celastraceae was inferred based on nuclear (26S rDNA and ITS) together with morphological traits and chloroplast genes (*matK*, *trnL-F*) [10]. Our phylogenetic analyses included *C. edulis* and 28 species which were sampled based on relationships from NCBI database (Available online: http://www.ncbi.nlm.nih.gov/genomes/GenomesGroup.cgi? taxid=2759&opt=plastid) and the angiosperm tree of life (Available online: http://www.mobot.org/ mobot/research/apweb/) with *Glycine canescens*, *Glycine falcate*, *Trifolium aureum*, and *Trifolium boissieri* from Fabaceae as outgroup taxa. The phylogenetic tree indicated that *Catha* and *Euonymus* where most closely related based on 73 common protein-coding genes (Figure 6). Most branches of the phylogenetic tree had high bootstrap support with all three methods. This suggests that the full cp genome information could be very useful in resolving phylogenetic conflicts but phylogenetic analyses with many closely related species are needed to test the resolving power of chloroplast coding genes [55].

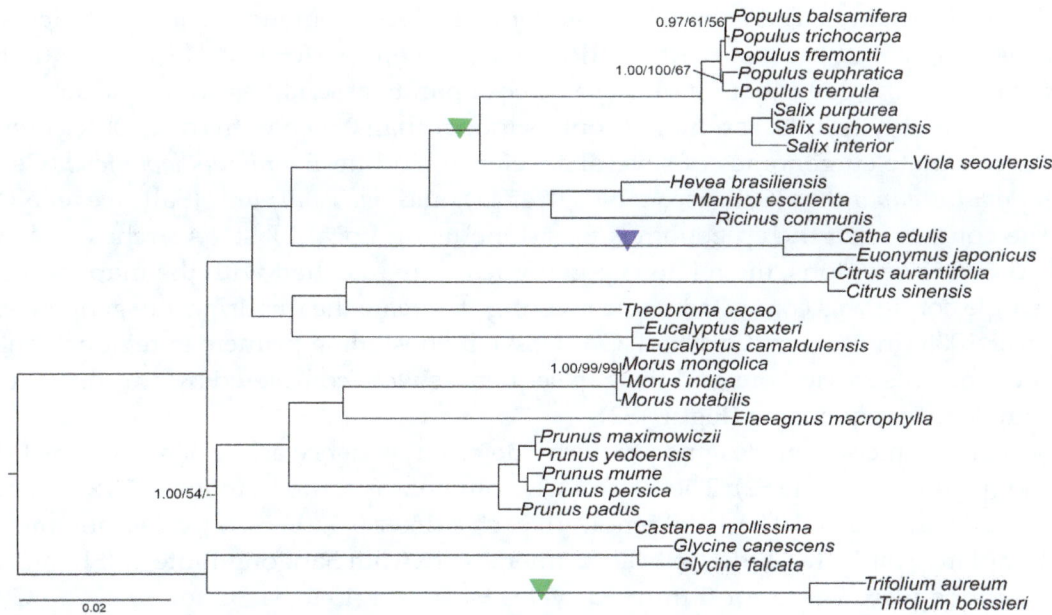

Figure 6. Phylogenetic tree based on 73 shared protein-coding genes was constructed for 33 species using three different methods, including Parsimony analysis, maximum likelihood (ML) and Bayesian inference (BI). All branches had bootstrap values or posterior probability of 100/1.00 except those labeled. The *rps16* gene losses are indicated with green triangles and the *rps16* intron loss is indicated with a purple triangle.

With a clearly resolved and strongly supported phylogeny, evolutionary patterns can be more clearly interpreted, such as gene or intron sequence loss/gain. Specifically, the intron loss of the *rps16* gene and loss of the whole *rps16* gene (Figure 6), were found in Celastraceae (*rps16* intron loss) and independently (*rps16* gene loss) in the genus *Trifolium* (Fabaceae), and the clade Salicaceae + Violaceae (Table S6). Gene and intron loss have been noted numerous times in land plant cp genomes [37]. From the phylogenetic tree, we were able to infer that the intron of *rps16* was lost in an ancestor to the Celastraceae independently from the two *rps16* gene loss events (Figure 6). Why only the *rps16* intron was lost in the Celastraceae and the entire gene in other closely related lineages is not known. Further study is needed to understand the underlying mechanisms of gene vs. intron loss in these related groups.

3. Materials and Methods

3.1. DNA Extraction and Sequencing

DNA for this project was obtained from aliquots of the extracts used in Tembrock et al., 2017. Total genomic DNA was used to build sequence libraries (Illumina Inc., San Diego, CA, USA), and was extracted from leaves using a *Catha* specific DNA extraction protocol described in Tembrock et al., 2017. At the Beijing Genomics Institute (BGI), an Illumina HiSeq 2000 sequencer was used to sequence paired-end (PE) sequencing libraries with an average 300 bp insert length. From this, over 10 million clean reads were passed through quality control with a 100 bp each read length. All other used species in this paper were listed in Table S1.

3.2. Chloroplast Genome Assembly and Sequence Analysis

The original Illumina reads were pre-processed, including the trimming and filtering of low-quality sequences with Trimmomatic v0.3 [56] in which the parameters used were as follows: minlen: 50; trailing: 3; leading: 3; and sliding window: 4:15. De novo assembly from *C. edulis* employed the default parameters (Available online: http://www.clcbio.com) in the CLC genomic workbench v7

(CLCbio, Hilden, Germany). Then, three independent de novo assemblies, which included single-end forward reads, single end reverse reads, and PE reads, were performed [18]. After that, a single assembly formed by the combination of these three separate assemblies was conducted. From the complete CLC assembly results, assembled contigs longer than 0.5 kb with over 100× coverage were compared to complete cp genomes of several species, including *Euonymus japonicus* (Celastraceae, KP189362), *Populus euphratica* (Salicaceae; NC_024747), and *Salix purpurea* (Salicaceae; NC_026722). Matching the contigs from the cp genomes was done using Local BlastN searches [57]. Using the conserved cp genome regions, the related cp genomes were matched with the mapped contigs [58] and then a single contig was connected to these contigs to create the quadripartite genome employing Contig Express 2003 (Invitrogen, Carlsbad, CA, USA). By designing primers in regions flanking gaps, PCR amplification was carried out and the gap sequences were completed by adding sequence data obtained from Sanger sequencing (Figure S2).

Additionally, primers were designed to verify de novo sequence assemblies, such as the junction regions of the cp genome (Table S2). The 40-μL PCR volume was setup as follows: 10× Taq buffer 4 μL, ddH$_2$O 33.3 μL, 10 mM dNTP 0.8 μL, 20 pmol/μL each primer 0.5 μL, 5 U/μL Taq polymerase 0.4 μL and DNA template 0.5 μL. Taq buffer, dNTP, primers were from Sangong Biotech (Shanghai, China). Cycling conditions were 94 °C for 5 min, 32 cycles 94 °C for 45 s, 54 °C for 45 s, 72 °C for 2 min and, a 10 min 72 °C final extension step. By combining the results of Sanger sequencing, the whole cp genome was used to map reference species to confirm the assembly with the uniformity of the iterative sequences.

Annotation of the transfer RNAs (tRNAs), protein-coding genes, and ribosomal RNAs (rRNAs) was first performed using DOGMA v1.2 (University of Texas at Austin, Austin, TX, USA) [59]. Then, the protein-coding gene positions in the draft annotation were verified and if necessary manually adjusted following alignment to the related species, *Euonymus japonicas* [58] to accurately determine the genes starting point, stop codons and exon borders. Finally, BLASTN searches and tRNAscan-SE v1.21 (University of California Santa Cruz, CA, USA) [60] were employed to verify both tRNA and rRNA genes.

A graphical cp genome map for *C. edulis* was completed using OGDraw (OrganellarGenomeDRAW) (V 1.2, Max Planck Institute of Molecular Plant Physiology, Am Mühlenberg, Germany) [61]. The annotated *C. edulis* cp genome reported and analyzed herein has been deposited in GenBank (KT861471).

3.3. Chloroplast Genomes Comparison

3.3.1. IR Expansion and Contraction

The changes in the size of the angiosperm cp genomes are mainly due to the contraction and expansion from the inverted repeat region, and the two single copy boundary areas. Four borders (J$_{LA}$, J$_{LB}$, J$_{SA}$, and J$_{SB}$) are present in the *C. edulis* cp genome and are situated in the middle of two IRs and two single copy regions [62]. The IR borders and neighboring genes of the two Celastrales species (*Catha edulis* and *Euonymus japonicus*) and six Malpighiales species cp genomes (*Hevea brasiliensis*, *Manihot esculea*, *Populus euphratica*, *Ricinus communis*, *Salix purpurea*, and *Viola seoulensis*) were compared in this study.

3.3.2. Repeat Analysis

Two methods were used to search repeats in *C. edulis* [63]. We identified simple sequence repeats (SSRs) using SSR Hunter v1.3 (Nanjing Agricultural University, Nanjing, China) [64] with cut-offs of eight copy number for mono-SSRs, four copy number for di-, three copy number for tri-, tetra-, penta- and hexanucleotide SSRs. To discover larger repeat regions, REPuter [65] was employed to find four possible repeats types: containing complement, forward, palindrome, and reverse repeats. Nested and low complexity repeats were not included in this study [66].

3.3.3. Dot-Plot Analysis

To identify the structural variations across all 14 genera, *Populus* (Salicaceae; Malpighiales), *Salix* (Salicaceae; Malpighiales), *Viola* (Violaceae; Malpighiales), *Hevea* (Euphorbiaceae; Malpighiales), *Manihot* (Euphorbiaceae; Malpighiales), *Ricinus* (Euphorbiaceae; Malpighiales), and *Euonymus* (Celastraceae; Celastrales), as well as outgroup genera *Prunus, Morus, Theobroma, Eucalyptus, Elaeagnus, Castanea*, and *Citrus*, we conducted the dot-plot analysis (based on a custom perl script) [13] between *C. edulis* and all 14 genera to visualize structural differences in two dimensional plots.

3.3.4. Verification of the *rps16* Intron Loss from Catha and Seven Other Celastraceae Genera

During annotation, the intron loss of *rps16* was found in the cp genome of *C. edulis*. To verify whether this intron loss happened throughout Celastraceae, two primers were designed (Forward-ACTTCGTTTGAGACGGTGTG, Reverse- AAAAACCCCGATTTCTTTGA) to amplify the entire *rps16* intron from *C. edulis* and seven other Celastraceae species (*Quetzalia stipitata, Mortonia diffusa, Microtropis triflora, Maytenus elliptica, Monimopetalum chinensis, Cassine aethiopica,* and *Parnassia glauca*). In *C. edulis*, the target *rps16* fragment without the intron is about 550 bp. Absence of the *rps16* intron was visualized on 0.8% agarose gels. The size of the fragment was determined by comparing it to a DNA size standard [67]. The *rps16* gene was sequenced using Sanger sequencing at the Beijing Genomics Institute (BGI).

3.3.5. Phylogenetic Analyses

The 73 common protein-coding genes of 26 species cp genomes, among them eight Rosales and four Fabales outgroup species, were aligned under the default parameters of Clustal X, with reading frames included by manual correction (Supplement data matrix) [68]. The phylogenetic tree based on these 73 common genes was inferred using three different methods. Implementation of Parsimony analysis, Bayesian inference (BI), and maximum likelihood (ML) were made in PAUP* 4.0b10 [69], MrBayes 3.1.2, and PHYML v 2.4.5 [70,71] respectively using the parameters from Wu et al. [18].

4. Conclusions

In this study, using next generation sequencing technology, we successfully completed the whole chloroplast genome for the economically important species *C. edulis*. In comparing the *C. edulis* cp genome with numerous closely related species, we found that it has a typical angiosperm cp genome structure and gene content. However, some unique features are reported here, such as the loss of the intron region from the *rps16* gene, and repeat structure and abundance. We also resolved the phylogenetic position of *C. edulis* with its relatives including the monophyly of Celastraceae. The whole cp genome of *C. edulis* provides a valuable genetic resource for further phylogenomic research, barcoding, and cp transformation in Celastraceae.

Acknowledgments: This research was supported by Zhejiang Provincial Natural Science Foundation of China under Grant No. LY17C160003. The sponsors had no role in data collection, study design, data analysis, or preparing the manuscript. We also thank the editor and the constructive comments of the four anonymous reviewers who helped us to improve this manuscript. We are grateful to Nels Johnson for his kinds help on manuscript editing and improvement.

Author Contributions: Conceived and designed the experiments: Zhiqiang Wu, Cuihua Gu; Performed the experiments: Zhiqiang Wu, Cuihua Gu; Analyzed the data: Zhiqiang Wu, Cuihua Gu, Luke R. Tembrock, Shaoyu Zheng; Contributed reagents/materials/analysis tools: Zhiqiang Wu, Cuihua Gu, Luke R. Tembrock; Wrote the paper: Zhiqiang Wu, Cuihua Gu, Luke R. Tembrock, Shaoyu Zheng.

References

1.   Al-Motarreb, A.; Baker, K.; Broadly, K.J. Khat: Pharmacological and medical aspects and its social use in Yemen. *Phytother. Res.* **2002**, *16*, 403–413. [CrossRef] [PubMed]

2.   Anderson, D.; Beckerleg, S.; Hailu, D.; Klein, A. *The Khat Controversy: Stimulating the Debate on Drugs*; Berg: Oxford, UK, 2007.

3.   Carrier, N.C.M. *The Social Life of a Stimulant*; Brill: Leiden, The Netherlands, 2007.

4.   Kennedy, J.G. The flower of paradise: The Institutional Use of the Drug Qat in North Yemen. *Q. Rev. Biol.* **1988**, *63*, 364–365.

5.   World Bank. *Yemen: Towards Qat Demand Reduction*; World Bank Document Report 39738-YE; World Bank: Washington, DC, USA, 2007.

6.   Gebissa, E. *Leaf of Allah: Khat & Agricultural Transformation in Harerge, Ethiopia*; James Currey Ltd.: Oxford, UK, 2004.

7.   Curto, M.A.; Tembrock, L.R.; Puppo, P.; Nogueira, M.; Simmons, M.P.; Meimberg, H. Evaluation of microsatellites of *Catha edulis* (qat; Celastraceae) identified using pyrosequencing. *Biochem. Syst. Ecol.* **2013**, *49*, 1–9. [CrossRef]

8.   Hagel, J.M.; Krezevski, K.; Sitrit, Y.; Marsolais, F.; Facchini, J.P.; Krizevski, R.; Lewinsohn, E. Expressed sequence tag analysis of khat (*Catha edulis*) provides a putative molecular biochemical basis for the biosynthesis of phenylpropylamino alkaloids. *Genet. Mol. Biol.* **2011**, *34*, 640–646. [CrossRef] [PubMed]

9.   Tembrock, L.R.; Broeckling, C.D.; Heuberger, A.L.; Simmons, M.P.; Stermitz, F.R.; Uvarov, J.M. Employing two-stage derivatisation and GC–MS to assay for cathine and related stimulant alkaloids across the Celastraceae. *Phytochem. Anal.* **2017**, *28*, 257–266. [CrossRef] [PubMed]

10.   Simmons, M.P.; Cappa, J.J.; Archer, R.H.; Ford, A.J.; Eichstedt, D.; Clevinger, C.C. Phylogeny of the Celastreae (Celastraceae) and the relationships of *Catha edulis* (qat) inferred from morphological characters and nuclear and plastid genes. *Mol. Phylogenet. Evol.* **2008**, *48*, 745–757. [CrossRef] [PubMed]

11.   Tembrock, L.R.; Simmons, M.P.; Richards, C.M.; Reeves, P.A.; Reilley, A.; Curto, M.A.; Al-Thobhani, M.; Varisco, D.M.; Simpson, S.; Ngugi, G.; et al. Phylogeography of the wild and cultivated stimulant plant qat (*Catha edulis*, Celastraceae) in areas of historical cultivation. *Am. J. Bot.* **2017**, *104*, 538–549. [CrossRef] [PubMed]

12.   Ravi, V.; Khurana, J.P.; Tyagi, A.K.; Khurana, P. An update on chloroplast genomes. *Plant Syst. Evol.* **2008**, *271*, 101–122. [CrossRef]

13.   Gu, C.H.; Tembrock, L.R.; Johnson, N.G.; Simmons, M.P.; Wu, Z.Q. The complete plastid genome of *Lagerstroemia fauriei* and loss of *rpl2* intron from *Lagerstroemia* (Lythraceae). *PLoS ONE* **2016**, *11*, e0150752. [CrossRef] [PubMed]

14.   Dong, W.; Xu, C.; Li, C.; Sun, J.; Zuo, Y.; Shi, S.; Cheng, T.; Guo, J.; Zhou, S. *ycf1*, the most promising plastid DNA barcode of land plants. *Sci. Rep.* **2015**, *5*, 8348. [CrossRef] [PubMed]

15.   Palmer, J.D. Comparative organization of chloroplast genomes. *Annu. Rev. Genet.* **1985**, *19*, 325–354. [CrossRef] [PubMed]

16.   Wicke, S.; Schneeweiss, G.M.; DePamphilis, C.W.; Müller, K.F.; Quandt, D. The evolution of the plastid chromosome in land plants: Gene content, gene order, gene function. *Plant Mol. Biol.* **2011**, *76*, 273–297. [CrossRef] [PubMed]

17.   Soltis, D.E.; Gitzendanner, M.; Stull, G.; Chester, M.; Chanderbali, A.; Jordon-Thaden, I.; Soltis, P.S.; Schnable, P.S.; Barbazuk, W.B. The potential of genomics in plant systematics. *Taxon* **2013**, *62*, 886–898. [CrossRef]

18.   Wu, Z.Q.; Tembrock, L.R.; Ge, S. Are Differences in Genomic Data Sets due to True Biological Variants or Errors in Genome Assembly: An Example from Two Chloroplast Genomes. *PLoS ONE* **2015**, *10*, e0118019. [CrossRef] [PubMed]

19.   CBOL. A DNA barcode for land plants. *Proc. Natl. Acad. Sci. USA* **2009**, *106*, 12794–12797.

20.   Day, A.; Goldschmidt-Clermont, M. The chloroplast transformation toolbox: Selectable markers and marker removal. *Plant Biotechnol. J.* **2011**, *9*, 540–553. [CrossRef] [PubMed]

21.   Shaw, J.; Lickey, E.B.; Beck, J.T.; Farmer, S.B.; Liu, W.; Miller, J.; Siripun, K.C.; Winder, C.T.; Schilling, E.E.; Small, R.L. The tortoise and the hare II: Relative utility of 21 noncoding chloroplast DNA sequences for phylogenetic analysis. *Am. J. Bot.* **2005**, *92*, 142–166. [CrossRef] [PubMed]

22.   Wu, Z.Q.; Ge, S. The phylogeny of the BEP clade in grasses revisited: Evidence from the whole-genome sequences of chloroplasts. *Mol. Phylogenet. Evol.* **2012**, *62*, 573–578. [CrossRef] [PubMed]

The Complete Chloroplast Genome of Catha edulis: A Comparative Analysis of Genome Features...

163

23. Cerutti, H.; Johnson, A.M.; Boynton, J.E.; Gillham, N.W. Inhibition of chloroplast DNA recombination and repair by dominant negative mutants of Escherichia coli RecA. *Mol. Cell. Biol.* **1995**, *15*, 3003–3011. [CrossRef] [PubMed]

24. Maliga, P. Plastid transformation in higher plants. *Annu. Rev. Plant Biol.* **2004**, *55*, 289–313. [CrossRef] [PubMed]

25. Maliga, P.; Staub, J.; Carrer, H.; Kanevski, I.; Svab, Z. *Homologous Recombination and Integration of Foreign DNA in Plastids of Higher Plants*; Paszkowski, J., Ed.; Kluwer Academic: Amsterdam, The Netherlands, 1994.

26. Svab, Z.; Maliga, P. High-frequency plastid transformation in tobacco by selection for a chimeric aadA gene. *Proc. Natl. Acad. Sci. USA* **1993**, *90*, 913–917. [CrossRef] [PubMed]

27. Yang, J.B.; Li, D.Z.; Li, H.T. Highly effective sequencing whole chloroplast genomes of angiosperms by nine novel universal primer pairs. *Mol. Ecol. Resour.* **2014**, *14*, 1024–1031. [CrossRef] [PubMed]

28. O'Brien, S.J.; Stanyon, R. Phylogenomics. Ancestral primate viewed. *Nature* **1999**, *402*, 365–366. [CrossRef] [PubMed]

29. Green, B.R. Chloroplast genomes of photosynthetic eukaryotes. *Plant J.* **2011**, *66*, 34–44. [CrossRef] [PubMed]

30. Su, H.; Hogenhout, S.A.; Al-sadi, A.M.; Kuo, C. Complete chloroplast genome sequence of Omani Lime (*Citrus aurantiifolia*) and comparative analysis within the Rosids. *PLoS ONE* **2014**, *9*, e113049. [CrossRef] [PubMed]

31. Redwan, R.M.; Saidin, A.; Kumar, S.V. Complete chloroplast genome sequence of MD-2 pineapple and its comparative analysis among nine other plants from the subclass Commelinidae. *BMC Plant Biol.* **2015**, *15*, 196. [CrossRef] [PubMed]

32. Tangphatsornruang, S.; Uthaipaisanwong, P.; Sangsrakru, D.; Chanprasert, J.; Yoocha, T.; Jomchai, N.; Tragoonrung, S. Characterization of the complete chloroplast genome of *Hevea brasiliensis* reveals genome rearrangement, RNA editing sites and phylogenetic relationships. *Gene* **2011**, *475*, 104–112. [CrossRef] [PubMed]

33. Raubeson, L.A.; Peery, R.; Chumley, T.W.; Dziubek, C.; Fourcade, H.M. Comparative chloroplast genomics: Analyses including new sequences from the angiosperms Nuphar advena and Ranunculus macranthus. *BMC Genom.* **2007**, *8*, 174. [CrossRef] [PubMed]

34. Kim, K.J.; Lee, H.L. Complete chloroplast genome sequences from Korean ginseng (*Panax ginseng* Nees) and comparative analysis of sequence evolution among 17 vascular plants. *DNA Res.* **2004**, *11*, 247–261. [CrossRef] [PubMed]

35. Ryzhova, N.N.; Kholda, O.A.; Kochieva, E.Z. Structure characteristics of the chloroplast *rps16* intron in Allium sativum and related Allium species. *Mol. Biol.* **2009**, *43*, 766–775. [CrossRef]

36. Schwarz, E.N.; Ruhlman, T.A.; Sabir, J.S.; Hajrah, N.H.; Alharbi, N.S.; Al-Malki, A.L.; Bailey, C.D.; Jansen, R.K. Plastid genome sequences of legumes reveal parallel inversions and multiple losses of *rps16* in papilionoids. *J. Syst. Evol.* **2015**, *53*, 458–468. [CrossRef]

37. Downie, S.R.; Olmstead, R.G.; Zurawski, G.; Soltis, D.E.; Soltis, S.; Watson, J.C.; Palmer, J.D. Six independent losses of the Chloroplast DNA *rpl2* intron in Dicotyledons: Molecular and Phylogenetic Implications. *Evolution* **1991**, *45*, 1245–1259. [CrossRef] [PubMed]

38. Tank, D.C.; Eastman, J.M.; Pennell, M.W.; Soltis, P.S.; Soltis, D.E.; Hinchliff, C.E.; Brown, J.W.; Sessa, E.B.; Harmon, L.J. Nested radiations and the pulse of angiosperm diversification: Increased diversification rates often follow whole genome duplications. *New Phytol.* **2015**, *207*, 454–467. [CrossRef] [PubMed]

39. CavalierSmith, T. Chloroplast evolution: Secondary symbiogenesis and multiple losses. *Curr. Biol.* **2002**, *12*, 62–64. [CrossRef]

40. Rubinsztein, D.C.; Amos, W.; Leggo, J.; Goodburn, S.; Jain, S.; Li, S.H.; Margolis, R.L.; Ross, C.A.; Ferguson-Smith, M.A. Microsatellite evolution—Evidence for directionality and variation in rate between species. *Nat. Genet.* **1995**, *10*, 337–343. [CrossRef] [PubMed]

41. Gemayel, R.; Cho, J.; Boeynaems, S.; Verstrepen, K.J. Beyond junk-variable tandem repeats as facilitators of rapid evolution of regulatory and coding sequences. *Genes* **2012**, *3*, 461–480. [CrossRef] [PubMed]

42. Voronova, A.; Belevich, V.; Jansons, A.; Rungis, D. Stress-induced transcriptional activation of retrotransposon-like sequences in the Scots pine (*Pinus sylvestris* L.) genome. *Tree Genet. Genomes* **2014**, *10*, 937–951. [CrossRef]

43. Timme, R.E.; Kuehl, J.V.; Boore, J.L.; Jansen, R.K. A comparative analysis of the Lactuca and Helianthus (Asteraceae) plastid genomes: Identification of divergent regions and categorization of shared repeats. *Am. J. Bot.* **2007**, *94*, 302–312. [CrossRef] [PubMed]

44. Nie, X.; Lv, S.; Zhang, Y.; Du, X.; Wang, L.; Biradar, S.S.; Tan, X.; Wan, F.; Weining, S. Complete chloroplast genome sequence of a major invasive species, crofton weed (*Ageratina adenophora*). *PLoS ONE* **2012**, *7*, e36869. [CrossRef] [PubMed]

45. Grassi, F.; Labra, M.; Scienza, A.; Imazio, S. Chloroplast SSR markers to assess DNA diversity in wild and cultivated grapevines. *Vitis* **2002**, *41*, 157–158.

46. Dong, W.; Liu, J.; Yu, J.; Wang, L.; Zhou, S. Highly variable chloroplast markers for evaluating plant phylogeny at low taxonomic levels and for DNA barcoding. *PLoS ONE* **2012**, *7*, e35071. [CrossRef] [PubMed]

47. Kress, W.J.; Erickson, D.L. A two-locus global DNA barcode for land plants: The coding *rbcL* gene complements the non-coding *trnH-psbA* spacer region. *PLoS ONE* **2007**, *2*, e508. [CrossRef] [PubMed]

48. Li, X.; Yang, Y.; Henry, R.J.; Rossetto, M.; Wang, Y.; Chen, S. Plant DNA barcoding: From gene to genome. *Biol. Rev.* **2014**, *90*, 157–166. [CrossRef] [PubMed]

49. Hilu, K.W.; Black, C.; Diouf, D.; Burleigh, J.G. Phylogenetic signal in *matK* vs. *trnK*: A case study in early diverging eudicots (angiosperms). *Mol. Phylogenet. Evol.* **2008**, *48*, 1120–1130. [CrossRef] [PubMed]

50. Kim, K.J.; Jansen, R.K. *ndhF* sequence evolution and the major clades in the sunflower family. *Proc. Natl. Acad. Sci. USA* **1995**, *92*, 10379–10383. [CrossRef] [PubMed]

51. Li, J. Phylogeny of *Catalpa* (Bignoniaceae) inferred from sequences of chloroplast *ndhF* and nuclear ribosomal DNA. *J. Syst. Evol.* **2008**, *46*, 341–348.

52. Yan, H.F.; Liu, Y.J.; Xie, X.F.; Zhang, C.Y.; Hu, C.M.; Hao, G.; Ge, X.J. DNA barcoding evaluation and its taxonomic implications in the species-rich genus *Primula* L. in China. *PLoS ONE* **2015**, *10*, e0122903. [CrossRef] [PubMed]

53. Jansen, R.K.; Cai, Z.; Raubeson, L.A.; Daniell, H.; Depamphilis, C.W.; Leebens-Mack, J.; Müller, K.F.; Guisinger-Bellian, M.; Haberle, R.C.; Hansen, A.K.; et al. Analysis of 81 genes from 64 plastid genomes resolves relationships in angiosperms and identifies genome-scale evolutionary patterns. *Proc. Natl. Acad. Sci. USA* **2007**, *104*, 19369–19374. [CrossRef] [PubMed]

54. Moore, M.J.; Bell, C.D.; Soltis, P.S.; Soltis, D.E. Using plastid genome-scale data to resolve enigmatic relationships among basal angiosperms. *Proc. Natl. Acad. Sci. USA* **2007**, *104*, 19363–19368. [CrossRef] [PubMed]

55. Gao, L.; Su, Y.J.; Wang, T. Plastid genome sequencing, comparative genomics, and phylogenomics: Current status and prospects. *J. Syst. Evol.* **2010**, *48*, 77–93. [CrossRef]

56. Bolger, A.M.; Lohse, M.; Usadel, B. Trimmomatic: A flexible trimmer for Illumina sequence data. *Bioinformatics* **2014**, *30*, 2114–2120. [CrossRef] [PubMed]

57. Camacho, C.; Coulouris, G.; Avagyan, V.; Ma, N.; Papadopoulos, J.; Bealer, K.; Madden, T.L. BLAST+: Architecture and applications. *BMC Bioinform.* **2009**, *10*, 421. [CrossRef] [PubMed]

58. Choi, K.S.; Park, S. The complete chloroplast genome sequence of *Euonymus japonicus* (Celastraceae). *Mitochondrial DNA* **2015**, *1736*, 1–2.

59. Wyman, S.K.; Jansen, R.K.; Boore, J.L. Automatic annotation of organellar genomes with DOGMA. *Bioinformatics* **2004**, *20*, 3252–3255. [CrossRef] [PubMed]

60. Schattner, P.; Brooks, A.N.; Lowe, T.M. The tRNAscan-SE, snoscan and snoGPS web servers for the detection of tRNAs and snoRNAs. *Nucleic Acids Res.* **2005**, *33*, 686–689. [CrossRef] [PubMed]

61. Lohse, M.; Drechsel, O.; Bock, R. OrganellarGenomeDRAW (OGDRAW): A tool for the easy generation of high-quality custom graphical maps of plastid and mitochondrial genomes. *Curr. Genet.* **2007**, *52*, 267–274. [CrossRef] [PubMed]

62. Wang, R.J.; Cheng, C.L.; Chang, C.C.; Wu, C.L.; Su, T.M.; Chaw, S.M. Dynamics and evolution of the inverted repeat-large single copy junctions in the chloroplast genomes of monocots. *BMC Evol. Biol.* **2008**, *8*, 36. [CrossRef] [PubMed]

63. Huang, H.; Shi, C.; Liu, Y.; Mao, S.Y.; Gao, L.Z. Thirteen Camellia chloroplast genome sequences determined by high-throughput sequencing: Genome structure and phylogenetic relationships. *BMC Evol. Biol.* **2016**, *14*, 151. [CrossRef] [PubMed]

64. Li, Q.; Wan, J.M. SSRHunter: Development of local searching software for SSR sites. *Yi Chuan* **2005**, *27*, 808–810. [PubMed]

65. Kurtz, S.; Choudhuri, J.V.; Ohlebusch, E.; Schleiermacher, C.; Stoye, J.; Giegerich, R. REPuter: The manifold applications of repeat analysis on a genomic scale. *Nucleic Acids Res.* **2001**, *29*, 4633–4642. [CrossRef] [PubMed]

66. Yang, Y.; Dang, Y.; Li, Q.; Lu, J.J.; Li, X.W.; Wang, Y.T. Complete Chloroplast genome sequence of poisonous and medicinal plant *Datura stramonium*: Organizations and implications for genetic engineering. *PLoS ONE* **2014**, *9*, e110656. [CrossRef] [PubMed]

67. Jansen, R.K.; Wojciechowski, M.F.; Sanniyasi, E.; Lee, S.B.; Daniell, H. Complete plastid genome sequence of the chickpea (*Cicer arietinum*) and the phylogenetic distribution of *rps12* and *clpP* intron losses among legumes (Leguminosae). *Mol. Phylogenet. Evol.* **2008**, *48*, 1204–1217. [CrossRef] [PubMed]

68. Simmons, M.P. Independence of alignment and tree search. *Mol. Phylogenet. Evol.* **2004**, *31*, 874–879. [CrossRef] [PubMed]

69. Swofford, D.L. Paup*: Phylogenetic Analysis Using Parsimony (and other methods). *Mccarthy* **1993**, 1–142.

70. Guindon, S.; Dufayard, J.F.; Lefort, V.; Anisimova, M. New alogrithms and methods to estimate maximum-likelihoods phylogenies: Assessing the performance of PhyML 30. *Syst. Biol.* **2010**, *59*, 307–321. [CrossRef] [PubMed]

71. Ronquist, F.; Teslenko, M.; Van Der Mark, P.; Ayres, D.L.; Darling, A.; Höhna, S.; Larget, B.; Liu, L.; Suchard, M.A.; Huelsenbeck, J.P. Mrbayes 3.2: Efficient bayesian phylogenetic inference and model choice across a large model space. *Syst. Biol.* **2012**, *61*, 539–542. [CrossRef] [PubMed]

11

Effects of TROL Presequence Mutagenesis on its Import and Dual Localization in Chloroplasts

Lea Vojta *, Andrea Čuletić † and Hrvoje Fulgosi

Laboratory for Plant Molecular Biology and Biotechnology, Division of Molecular Biology, Ruđer Bošković Institute, Bijenička cesta 54, 10000 Zagreb, Croatia; aculetic@bot.uni-kiel.de (A.Č.); fulgosi@irb.hr (H.F.)
* Correspondence: lvojta@irb.hr
† Current address: Biologie der Pflanzenzelle, Botanisches Institut, Christian-Albrechts-Universität zu Kiel, Am Botanischen Garten 1-9, 24118 Kiel, Germany.

Abstract: Thylakoid rhodanase-like protein (TROL) is involved in the final step of photosynthetic electron transport from ferredoxin to ferredoxin: NADP$^+$ oxidoreductase (FNR). TROL is located in two distinct chloroplast compartments—in the inner envelope of chloroplasts, in its precursor form; and in the thylakoid membranes, in its fully processed form. Its role in the inner envelope, as well as the determinants for its differential localization, have not been resolved yet. In this work we created six N-terminal amino acid substitutions surrounding the predicted processing site in the presequence of TROL in order to obtain a construct whose import is affected or localization limited to a single intrachloroplastic site. By using in vitro transcription and translation and subsequent protein import methods, we found that a single amino acid exchange in the presequence, Ala67 to Ile67 interferes with processing in the stroma and directs the whole pool of in vitro translated TROL to the inner envelope of chloroplasts. This result opens up the possibility of studying the role of TROL in the chloroplast inner envelope as well as possible consequence/s of its absence from the thylakoids.

Keywords: TROL; protein import; chloroplasts; dual localization; ATP; inner envelope membrane; thylakoids

1. Introduction

TROL (thylakoid rhodanase-like protein, At4g01050) is an integral membrane component of non-appressed thylakoid membranes, responsible for the anchoring of FNR (ferredoxin:NADP$^+$ oxidoreductase) [1]. This 66 kDa protein is firmly attached to the membrane by its two transmembrane helices, resisting extraction by high salt, urea, or high pH treatments [1]. The N-terminus of TROL consists of a presequence that directs the protein to chloroplasts. Two predicted transmembrane helices surround the centrally positioned inactive rhodanase-like domain, RHO, which is orientated towards the thylakoid lumen. The C-terminus of the protein resides in the cytosol and consists of a single hydrophobic FNR-binding region, ITEP (highly conserved module of TROL necessary for establishing high-affinity interaction with FNR), and a region upstream of the ITEP domain, designated PEPE (Pro-Val-Pro repeat-rich region), which is followed by a possible PVP hinge, proposed to introduce flexibility to the FNR-binding region. In our previous research we proposed that the TROL–FNR interaction is dynamic [2,3], in which binding and release of FNR from TROL can regulate the flow of photosynthetic electrons before the pseudo-cyclic electron transfer pathway becomes activated [4]. By studying Arabidopsis *trol* mutants we proposed that the TROL–FNR interaction is the bifurcating point between electron-conserving and electron-dissipating pathways [4].

In addition to being mainly located at the stroma thylakoids, in its fully processed form of 66 kDa, TROL can also be found embedded in the chloroplast inner envelope membrane (IM) in its non-processed form (70 kDa), which indicates a possible role in the electron transfer chain specific for this membrane [1]. This dual localization has been proven by the proteome analysis of chloroplast envelopes and thylakoid membranes [5] and was proposed to be dependent on the $NADP^+/NADPH$ ratio in the chloroplasts, as shown for the shuttling of the Tic62 protein [6]. Although there were attempts to investigate the structure and the function of TROL as the FNR anchor [1,7–9] its role in the IM remains undefined.

TROL is a nuclear-encoded protein, synthesized with a cleavable N-terminal presequence that targets the protein to and across the chloroplast envelope. During chloroplast protein import, the targeting sequences are sequentially decoded resulting in localization of the polypeptide to the appropriate organelle subcompartment [10]. Since TROL has been found both in the IM and the thylakoids [1,5], the determinants for this dual localization were the subject of investigation in this research.

Most of the IM proteins are directed to the general import pathway (Toc and Tic complex) through their cleavable transit peptides on their way to the chloroplasts. Some preproteins are released from the translocon at the level of the IM, as instructed by a hydrophobic stop-transfer signal in their sequence [11]. Others are first targeted to the stroma by their stroma-targeting presequence, and their processed mature form is subsequently re-exported into the IM, by so-called conservative sorting [11]. Proteins targeted to the thylakoid membrane require dual targeting signals that direct their import across the chloroplast envelope membranes and subsequent transport to the thylakoids. Stroma-targeting domains of preproteins are recognized and removed by the stromal processing peptidase (SPP) and lumen-targeting domain by a second processing protease [10,12]. A single bipartite transit sequence can carry the information for targeting to the thylakoids, with the stromal targeting domain located at the N-terminal region and the thylakoid lumen targeting domain at the C-terminal region of the presequence [10,13]. Preplastocyanin (prePC), and the subunits of the oxygen evolving complex (preOE16, preOE23, and preOE33) are configured as described. Integral membrane proteins, such as the precursor to the light-harvesting chlorophyll a/b binding protein (preLHCP) and the precursor to the 20-kDa subunit of the CP24 complex [13], contain information only for envelope transport within the presequence. The signals for targeting of these proteins to the thylakoids seem to reside within the primary structure of the mature polypeptide [14]. Since a single targeting domain is predicted for the transit sequence of TROL, this protein might belong to the latter group.

Targeting sequences are of various sizes, from 30 to 120 amino acids, and contain a high number of hydroxylated residues, but lacking acidic ones. The N-terminal 10–15 residues are devoid of Gly, Pro, and charged residues, a variable middle region is rich in Ser, Thr, Lys, and Arg, lacking acidic residues; and a C-terminal proteolytic processing site [15]. It contains a loosely conserved Ile/Val-x-Ala/Cys-Ala motif, recognized by SPP.

It has been postulated that chloroplast-localized proteins have various energy requirements for their import, according to their localization. ATP in the intermembrane space, usually less than 50 μM, drives the transport of the precursor across the outer envelope membrane [16,17], and import through the IM into the stroma progresses if the ATP concentration is higher than 100 μM [18,19]. This ATP is needed for the action of molecular chaperones in the stroma, which provide the driving force to complete import into the organelle [20]. Proteins localized in the same organellar compartment might have different energy needs for their import, depending on the import pathway they use [17]. Upon entering the stroma, the transit sequence is removed by SPP [20]. Further import into thylakoids requires additional translocators and energy demand. Four different mechanisms are known to target proteins from the stroma to the thylakoids [20–22].

The aim of this work was to introduce mutations in the presequence of the TROL protein, around the processing site, in an attempt to obtain a construct that would direct TROL to a single sub-chloroplast compartment, namely to a single membrane: either the inner envelope or the thylakoids. In addition, we wanted to characterize TROL import properties and requirements in more detail.

2. Results

After amino acid comparison between TROL from *A. thaliana* and other vascular plants, the N-terminal conserved region was determined around the SPP cleavage site. In this conserved region, AKSLTYEEALQQ (aa 67–78), we have chosen six potentially significant amino acids that could influence the import and localization of protein TROL in chloroplasts. Changes made to the presequece were as following: e1: 67Ala→67Ile, e2: 71Thr→71Asn, e3: 72Tyr→72Val, e4: 73Glu74Glu→73Gln74Gln, e5: 76Leu→76Thr, e6: 78Gln→78Val (Table 1). Hydrophobicity was checked for each amino acid substitution, according to Kyte and Doolittle [23] (Figure 1). For substitutions 73Glu74Glu→73Gln74Gln, the exchange of polarity has been compared according to Zimmerman et al. [24] (Figure 1). After the mutations were introduced into the gene *At4g01050*, constructs in the pZL1 vector were checked by DNA sequencing.

Table 1. Mutations introduced to the TROL presequence by the QuikChange Multi Site-Directed Mutagenesis and corresponding primer sequences.

Change	Mutation/Primer Name	Primer Sequence
e1	g199a_c200t	aagcagtgcaacagctcctattaaatccctgacgtacgag
e2	c212a_g213t	gctcctgctaaatccctgaattacgaggaagctctgcaac
e3	t214g_a215t	ctgctaaatccctgacggtcgaggaagctctgcaac
e4	g217c_g220c	gctaaatccctgacgtaccagcaagctctgcaacaatcta
e5	c226a_t227c	ccctgacgtacgaggaagctacgcaacaatctatgacca
e6	c232g_a233t	cgtacgaggaagctctgcaagtatctatgaccacttcttca
	TROLfor	caccatggaagctctgaaaaccgca
	TROLrev	gggctgcgatggcatcg

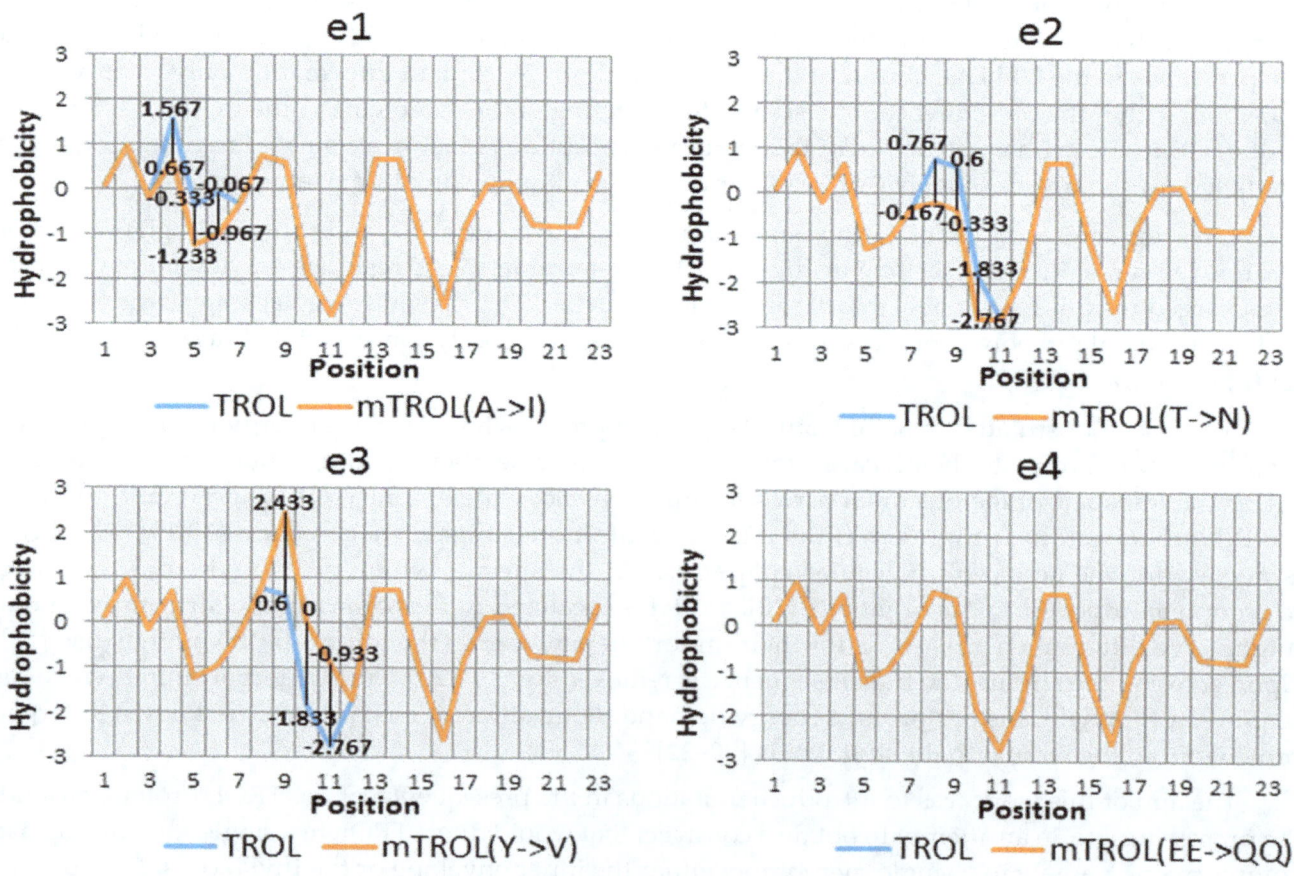

Figure 1. *Cont.*

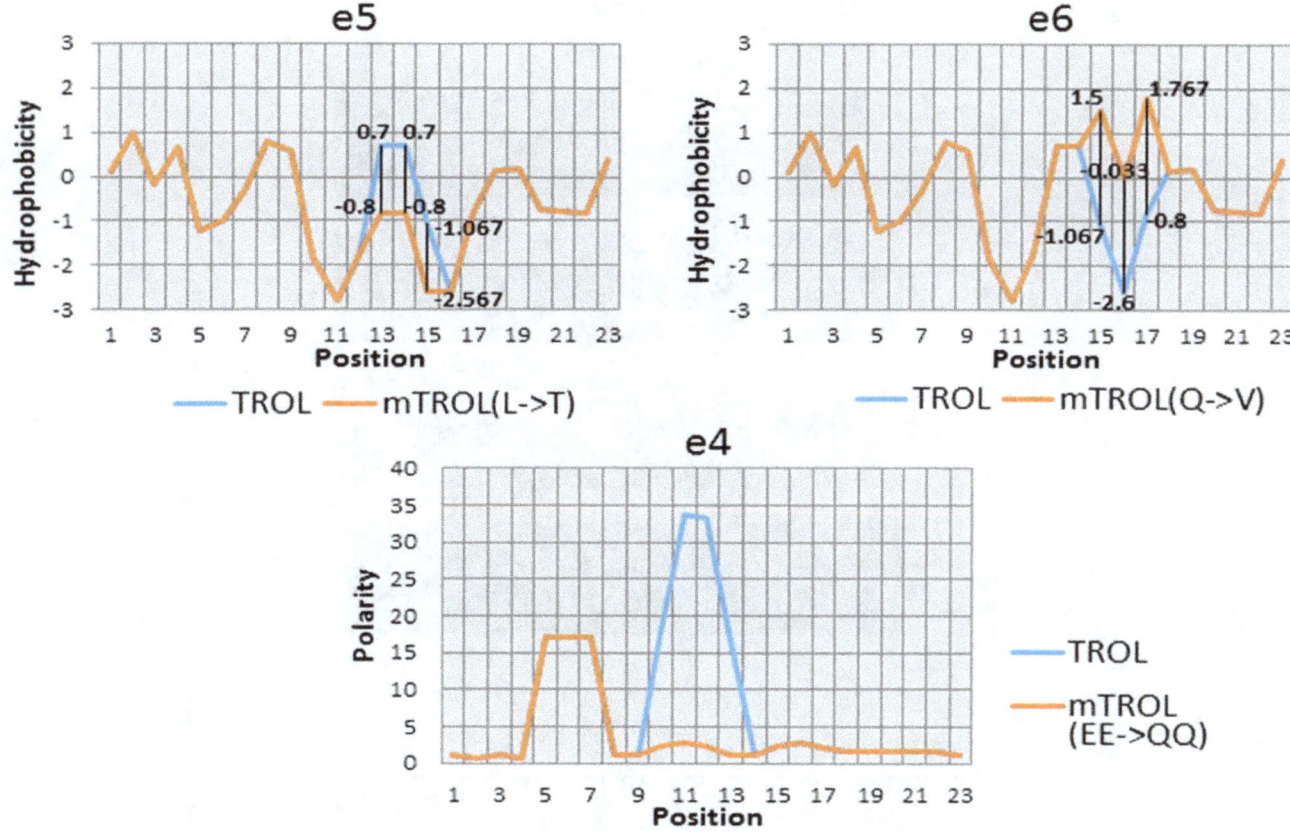

Figure 1. Kyte and Doolittle plots representing hydrophobicity change as a consequence of amino acid substitution/s in the TROL presequence [23]. Amino acids 67–78 of the partially conserved N-terminal part of the presequence around the predicted transit peptide cleavage site, AKSLTYEEALQQ, were substituted as follows: e1: 67Ala→67Ile, e2: 71Thr→71Asn, e3: 72Tyr→72Val, e4:73Glu74Glu→73Gln74Gln, e5: 76Leu→76Thr, e6:78Gln→78Val. For the e6 substitution, a polarity check according to Zimmerman was performed [24].

The wild type precursor of TROL and its e1–e6 presequence mutants were labeled with [35S]-methionine during in vitro translation and imported into isolated intact pea chloroplasts under various conditions. First, the standard import experiment was performed, in the absence or the presence of externally added 3 mM ATP. After a 20-min long import at 25 °C, chloroplasts were re-isolated on a Percoll cushion and treated with the protease thermolysin, to distinguish precursors loosely bound to the chloroplast envelope (early intermediates) from the firmly incorporated ones [17,18]. We observed that TROL requires ATP for successful processing and its import into the thylakoid membrane. After import with ATP, the majority of TROL was found in thylakoids, in its fully processed form, and just a small portion was located in the chloroplast envelope, in its precursor form, protected from thermolysin (Figure 2a, WT, lane 5). Presequence mutants e2, e3, e4, and e6 showed identical import behaviour as the wild type TROL (Figure 2a,b). In contrast, mutant e1 locates almost exclusively to the envelope membrane (Figure 2a, e1, lane 10, Figure 2b, lane 4). Only a very small portion was processed to the mature form and transported to the thylakoids. It seems that just a single amino acid substitution in the TROL presequence (67Ala→67Ile) results in almost exclusive envelope localization (Figure 2a,b). Presequence mutant e5, although having very similar import properties to the wild-type TROL, seems to have a slightly increased portion of envelope-bound form after import into chloroplasts (Figure 2a, e5, lanes 9 and 10, Figure 2b, lane 12).

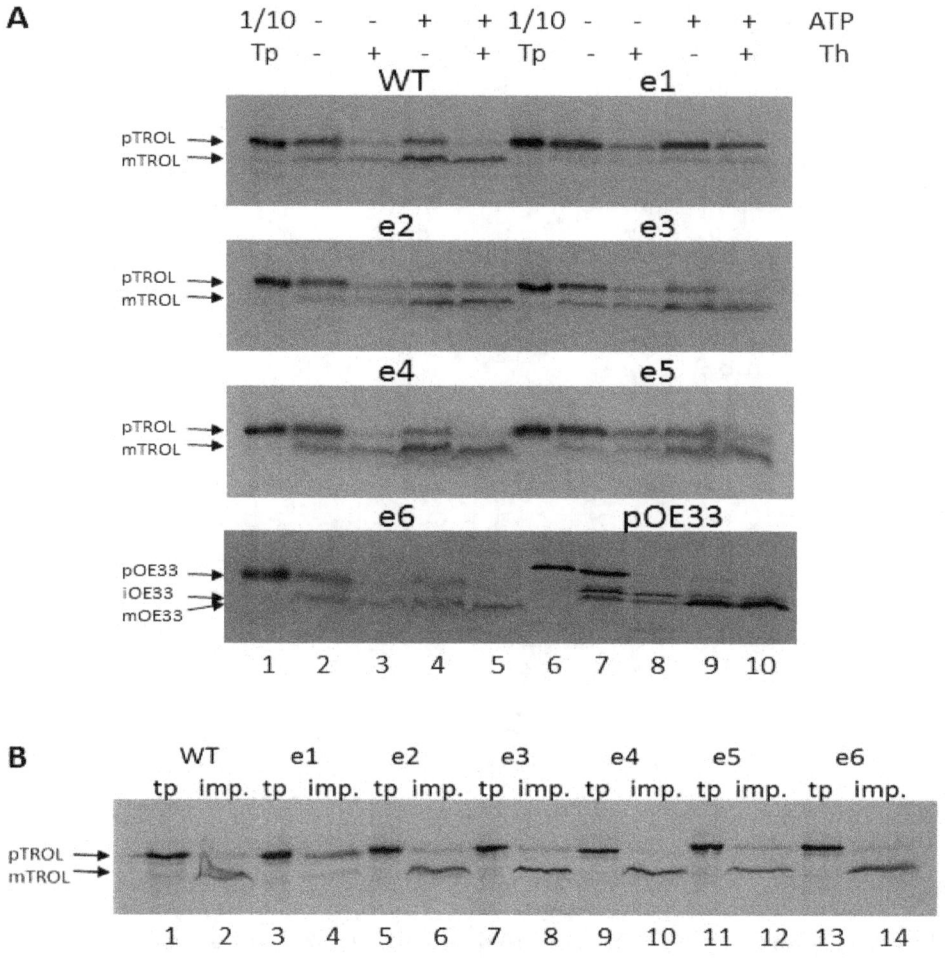

Figure 2. (**A**) Wt TROL and its presequence mutants e1–e6 are imported into pea chloroplasts. In vitro synthesized [^{35}S]-TROL and e1–e6, as well as the control protein pOE33 from thylakoid lumen were incubated with isolated intact chloroplasts at 25 °C for 20 min, in a standard import reaction containing 3 mM ATP (lanes 4, 5, 9, and 10) or without ATP (lanes 2, 3, 7, and 8). After import, samples were re-isolated on a Percoll cushion and treated with 0.5 µg thermolysin (Th) per µg chlorophyll (lanes 3, 5, 8, and 10). Untreated samples are shown in lanes 2, 4, 7, and 9. The results were analyzed by SDS/PAGE. Lanes 1 and 6 represent 10% of the translation product (Tp) used for the import reactions. The positions of pTROL, mTROL, pOE33, iOE33, and mOE33 are indicated by arrows. (**B**) Comparison of import of TROL and its presequence mutants e1–e6 into chloroplasts. In vitro synthesized [^{35}S]-TROL and e1–e6 were incubated with isolated intact chloroplasts at 25 °C for 20 min, in a standard import reaction containing 3 mM ATP. After import, samples were re-isolated on a Percoll cushion and treated with 0.5 µg thermolysin (Th) per µg chlorophyll (lanes 2, 4, 6, 8, 10, 12, and 14). The results were analyzed by SDS/PAGE. Lanes 1, 3, 5, 7, 9, 11, and 13 represent 10% of the respective translation product. The positions of pTROL and mTROL are indicated by arrows.

Further, we wanted to investigate the energy requirements for the import of TROL and its presequence mutants e1 and e5. After isolating chloroplasts from peas grown in the dark for at least 8 h, and incubating isolated intact chloroplasts in the dark and on ice to minimize internal ATP production, we added 0, 30, 300, or 3000 µM of external ATP to the import reaction. After 20 min. import at 25 °C and subsequent chloroplast re-isolation, we observed that even the smallest amount of ATP enabled the import of TROL into the thylakoids to some extent (Figure 3, lanes 3 and 5). Additional ATP resulted in a slight increase in the quantity of processed protein, while at the same time the portion of precursor form, located in the inner envelope, decreased (Figure 3, lanes 7 and 9). Mutant e5 seems to require higher amounts of external ATP to complete the import (Figure 3, e5). Also, there is more envelope form present in this mutant (Figure 3, lanes 5, 7, 9). Mutant e1 localizes almost exclusively to the chloroplast inner envelope, needing some external ATP (30–300 µM) for complete protection from thermolysin, as required for inner envelope incorporation (Figure 3, e1). As a control, oxygen evolving complex protein of 33 kDa (pOE33) was imported along with the TROL constructs. This is a well-characterized protein which is localized to the thylakoid lumen and forms a soluble translocation intermediate in the stroma. It uses the general import pathway into the chloroplasts and is therefore a suitable control for TROL import experiments. It seems that pOE33 requires more ATP to complete its import. With the lowest ATP concentration, stromal intermediate iOE33 is visible, and import seems to be complete after the addition of more than 300 µM ATP (Figure 3, pOE33). The higher energy need for pOE33 import results probably from its two processing events (in the stroma and in the thylakoid lumen) and its luminal localization.

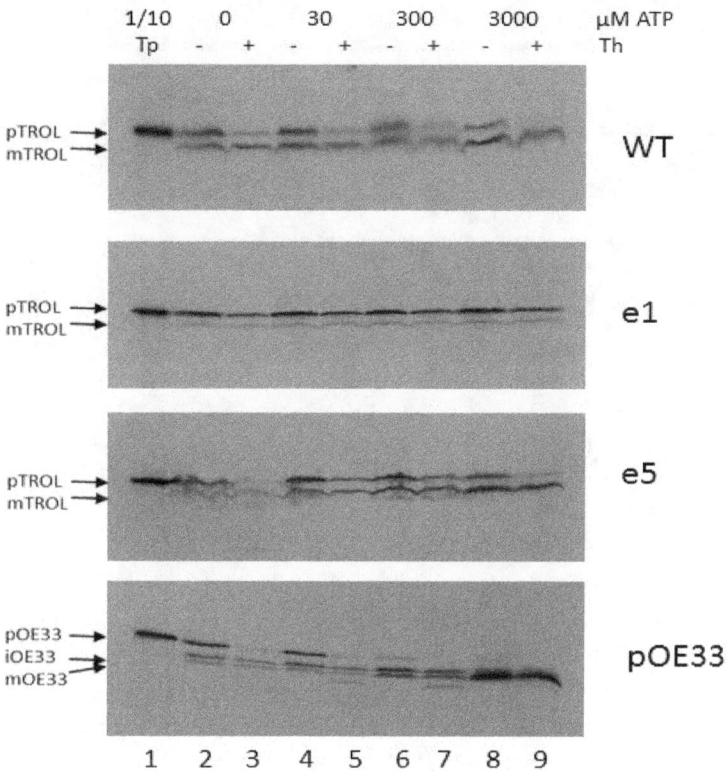

Figure 3. Energy requirement for import of wt TROL and its presequence mutants e1 and e5 into pea chloroplasts. Import into intact pea chloroplasts was performed under standard conditions, by incubating in vitro synthesized [^{35}S]-TROL, e1, e5, and control protein pOE33 from thylakoid lumen with chloroplasts corresponding to 20 µg chlorophyll at 25 °C. ATP-scale import into intact pea chloroplasts was performed using increasing concentrations of ATP from 0 to 3000 µM. After import, chloroplasts were re-isolated on a Percoll cushion and samples were treated with 0.5 µg thermolysin (Th) per µg chlorophyll (lanes 3, 5, 7 and 9). Untreated samples are shown in lanes 2, 4, 6, and 8. The results were analyzed by SDS/PAGE. The respective precursor, intermediate, and mature forms are indicated by arrow heads. Lane 1 represents 1/10 of the translation product (Tp) used for the import reaction.

We also wanted to explore how the import of TROL and e1 and e5 mutants proceeds on a temporal scale. Radioactively labeled precursors were added to the standard import reaction including 3 mM ATP and chloroplasts corresponding to 20 µg chlorophyll. Imports were performed for 0.5, 2, 5, 10, and 20 min at 25 °C. All samples were subsequently re-isolated on Percoll cushion, treated with thermolysin, lysed, and separated into membrane (P) and soluble fractions (S). After the analysis of radioactive signals on films, it was clearly visible that as time advances the amount of proteins in the pellet fractions increases for all tested precursors (Figure 4). Between 5 and 10 min of import are needed for around 50% of TROL import to be accomplished. For complete import, more than 10 min were required (Figure 4, lane 10). For the e1 mutant, the portion of imported protein that locates to the IM visibly increases with time, while the amount of mature, thylakoid-localized form starts to appear after 10 min. of import (Figure 4, lane 8). In this experiment, mutant e5 showed nearly identical properties to wt TROL.

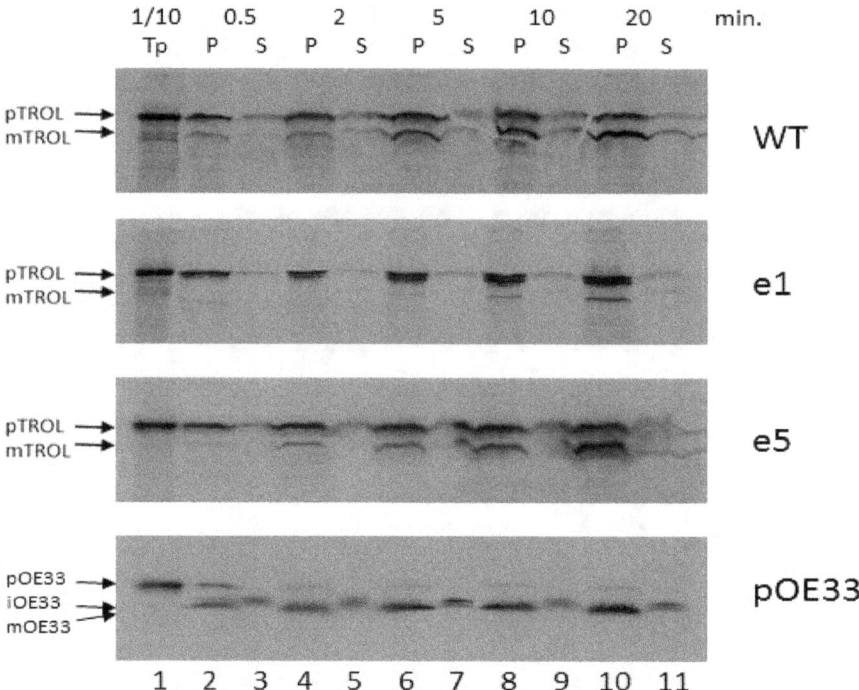

Figure 4. Time-scale import of wt TROL and e1 and e5 presequence mutants into intact pea chloroplasts. Radioactively labeled TROL, e1, e5, and control protein pOE33 were imported using increasing times in standard import reactions at 25 °C in the presence of 3 mM ATP. Import was performed for 0.5 (lanes 2 and 3), 2 (lanes 4 and 5), 5 (lanes 6 and 7), 10 (lanes 8 and 9), or 20 min (lanes 10 and 11). After import chloroplasts were separated into the pellet (P, lanes 2, 4, 6, 8, and 10) and soluble (S, lanes 3, 5, 7, 9, and 11) fractions. Lane 1 indicates 1/10 of the respective translation product (Tp) used for the import reaction. Precursor (p), intermediate (i), and mature (m) forms of TROL, e1, e5, and OE33 are indicated by arrows.

Finally, we investigated to what extent is TROL, imported both to the IM and the thylakoids, associated with the membranes. The strength and the nature of this association was tested by applying either 6 M Urea, 0.1 M Na_2CO_3 pH 11.5 for the separation of integral from peripheral membrane proteins, or 1 M NaCl, which decreases electrostatic interactions between proteins and charged lipids. 6 M Urea only partially extracted imported TROL from the membranes, mostly its envelope-located portion, while the mature forms from thylakoid membranes remained almost fully intact (Figure 5, lanes 2 and 3). Treatment of the envelopes with Na_2CO_3 at pH 11.5 converts membrane vesicles to sheets and disrupts protein–protein interactions, while protein–lipid interactions remain and the bilayer is otherwise intact. In this way we could determine if an integral membrane protein has achieved stable insertion into the bilayer. Carbonate treatment extracted only a very small portion of the tested proteins from the membrane, indicating that TROL and the tested mutants are strongly attached to the lipid bilayer (Figure 5, lanes 4 and 5). High salt did not extract TROL from the membranes (Figure 5, lanes 6 and 7), indicating strong ionic interactions between TROL and the membranes.

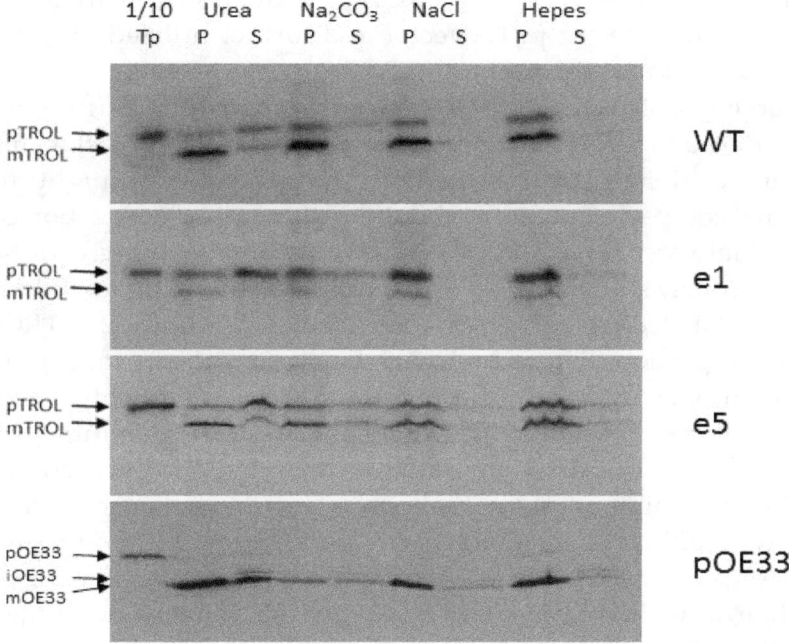

Figure 5. Extraction of wt TROL and its presequence mutants e1 and e5 from the membranes. In vitro synthesized [^{35}S]-TROL, e1, e5, and control protein pOE33 were incubated with isolated intact chloroplasts at 25 °C for 20 min, in a standard import reaction containing 3 mM ATP. Subsequent to import, chloroplasts were re-isolated, washed, and separated into membrane and soluble fractions. Isolated membranes, containing imported proteins, were treated with 6 M Urea in 10 mM HEPES/KOH pH 7.6 (lanes 2 and 3), 0.1 M Na_2CO_3 pH 11.5 (lanes 4 and 5), or 1 M NaCl (lanes 6 and 7). All incubations were performed for 20 min on RT. As a control, membranes were incubated solely in 10 mM HEPES/KOH pH 7.6 for 30 min on ice (lanes 8 and 9). Afterwards, samples were centrifuged at 265,000× *g* for 10 min at 4 °C, and both pellets (P, lanes 2, 4, 6, and 8) and the supernatants (S, lanes 3, 5, 7, and 9) were analyzed by SDS–PAGE and by exposure on X-ray films. Lane 1 represents 10% of the respective translation product (Tp) used for the import reactions. The positions of pTROL, mTROL, pOE33, iOE33, and mOE33 are indicated by arrows.

3. Discussion

In the N-terminal conserved region of TROL, around the SPP cleavage site, we have chosen six potentially significant amino acids for its import and localization in chloroplasts. Selection of the amino acids was largely based on their hydrophobic or hydrophilic character, which we tried to change by substitutions, in an attempt to interfere with stromal processing. The structure and influence of the neighboring amino acids have also been taken into consideration. We have chosen those amino acids whose neighboring hydrophobic/hydrophilic signals were as indiscernible as possible. In the e1 mutation, Ala67 was changed to Ile. Ala67 is a part of the AXS motif, predicted to represent a signal for cleavage of the presequence by the SPP. Ile is similar in nature to Ala, but contains three more methyl groups that make it more hydrophobic. In the e2 mutation, neutral and polar Thr71 was changed to hydrophobic Asn. The corresponding Kyte and Doolittle plot [23] resulted in a neutral to slightly hydrophilic character of the changed amino acid site. In e3, by Tyr72 to Val exchange, strong hydrophilicity was substituted by medium to strong hydrophobicity. In this example the structure has also been changed by removing the cyclic ring in Tyr, which could additionally influence the hydrophobicity. In e4, where Glu73 and Glu74 were changed to two Gln, hydrophobicity remained the same, but the loss of polarity (according to Zimmerman et al. [24]) could influence protein sorting. In e5 hydrophobic Leu76 was exchanged for neutral Thr, leading to a more hydrophilic character. In e6 Gln78 to Val exchange leads to a large hydrophilicity loss and change to hydrophobicity.

After successfully introduced mutations into TROL presequence, constructs e1–e6, as well as the wild type, were incorporated into the pZL1 vector and further utilized for investigation of import characteristics of this dually localized protein.

In organello import into isolated chloroplasts was performed to investigate chloroplast localization and integration of wt TROL and its presequence mutants into the chloroplast membranes. We expected that some of the amino acid substitutions made to the presequence might interfere with proper processing in the stromal compartment and/or result in alternative localization of TROL. The labeled precursor was imported into organelles and processed into a smaller mature protein of around 66 kDa (Figure 2, mTROL), previously shown to co-purify entirely with the thylakoid fraction [1]. A weaker signal of the size of the labeled precursor was also detected (Figure 2, pTROL). This signal was protected from protease digestion (Figure 2a, lane 5, Figure 2b, lane 2), indicating a portion of TROL located in the IM of chloroplasts, in its non-processed form of 70 kDa [1,5]. In contrast to the wt, the TROL e1 comprised almost entirely of the IM portion, indicating an influence of the 67Ala/67Ile presequence mutation on TROL localization. In this mutant the SPP recognition motif AKS has been changed to IXS, leading to inhibition of stromal processing and the increment of the IM portion of TROL. Compared to wt TROL, it is obvious that for the e1 mutant not only import into thylakoids is impaired, but the IM portion of the protein is highly increased. Since this portion is protected from thermolysin action, as well as from extraction with high salt and carbonate concentrations, we conclude that envelope-TROL is firmly incorporated into the membrane.

Experiments using increasing external ATP concentrations indicate that TROL requires more than 300 mM ATP for completion of its import into thylakoids, while only 30 mM is necessary for IM incorporation, as visible after thermolysin treatment (Figure 3). This result implies the stop-transfer mechanism of TROL import and its lateral insertion to the IM [11,22]. Time-scale experiments in the presence of 3 mM ATP indicate that the incorporation of TROL into the IM happens very early, and the smaller thylakoid form starts to appear after 2–5 min of import, reaching a maximum between 10–20 min (Figure 4). We observed that after addition of 3 mM ATP in wt, but also in e1, to some extent, there is much less of the IM form of TROL compared to the 300 mM experiment. The distribution of TROL between these two membranes might be influenced by the energy distribution between thylakoids and envelope compartments. TROL could be switching the intensity of its action between the IM (role unknown) and the thylakoids (regulation of photosynthetic electron transfer), influenced/directed by the energetic state of those compartments.

Jurić et al. [1] have shown that imported TROL could not be extracted from membranes by high salt, urea, or high pH treatments, indicating that At4g01050 is an integral thylakoid membrane protein. The same extraction procedures were used to investigate the membrane incorporation character of the e1–e6 TROL mutants (Figure 5). Only 6 M urea extracted a portion of protein, mainly the IM incorporated one, while TROL in the thylakoids remained fully protected. Mutation e1 causes visible changes in TROL localization, directing most of the protein to the IM of chloroplasts. Once there, the incorporated TROL resists the extraction procedures in the same way as the wild type protein, in which only urea solubilized a portion of the protein from the membranes. After import, the e5 mutation exhibits slightly more IM portion of TROL than the wt, but not significant enough to use it in further experiments.

TROL is not an isolated case in terms of dual localization in chloroplasts. Seventeen other proteins have been identified in both chloroplastic envelope and thylakoid membranes [22]. All of them are predicted to carry an N-terminal signal for chloroplast import. These proteins belong to one of the following groups according to their function: protein transport, tetrapyrrole biosynthesis, membrane dynamics, and transport of nucleotides and inorganic phosphate [22]. One of these proteins is Tic62, a redox sensor in the inner envelope and the thylakoids, proposed to anchor FNR, just like TROL [7].

The nature and the mechanism of dual localization of TROL remains unresolved. The inner envelope of chloroplasts and thylakoid membrane share a similar lipid composition, but perform very different functions. The way in which TROL incorporates into these membranes is at the moment just a speculation. Up to now, no pathway that would catalyse dual targeting has been found. All known nuclear-encoded thylakoid proteins and some IEM proteins are targeted to the respective membranes via a stromal intermediate. TROL, as a dually localized protein, could also use a similar pathway. This has been confirmed by the import properties of the e1 mutant. However, absence of stromal processing for IM located TROL favours the possibility of lateral insertion to the IM. Stromal processing seems to be the key moment for further sorting of TROL to thylakoids. Tha4 and Hcf106 components of the tat pathway are integral membrane proteins that insert into thylakoid membrane by unassisted insertion [25]. The same pathway use PSII subunits W, X, and Y [26], PSI subunit K [27], and the CFoII subunit of the ATP synthase [28], and these have been shown to insert into the membrane by the unassisted pathway. As already mentioned, some IM proteins insert laterally into the membrane by a stop transfer mechanism upon entering the chloroplast. Some of the dual-localized integral proteins, like TROL, may be similarly held at the envelope first, then sorted to the thylakoids. Low ATP demands for insertion of TROL into the IM points to this conclusion.

Future prospects for the e1 mutant would be the production of transgenic *A. thaliana* plants containing TROL located only/mostly to the IM of chloroplasts. In this way we could study the effect of the absence of TROL from thylakoids on photosynthesis and subsequent electron transfer/dissipation events, as well as the so far unknown role of this protein in the inner envelope membrane.

4. Materials and Methods

4.1. TROL Presequence Substitutions

Using the QuikChange Multi Site-Directed Mutagenesis method, various mutations were introduced into the TROL presequence. Amino acids 67–78 of the presequence, AKSLTYEEALQQ, represent a partially conserved N-terminal part of the sequence, around the predicted transit peptide cleavage site. In this sequence we have chosen six amino acids potentially significant for TROL processing that might influence its import into and localization inside the chloroplasts. Changes made to the presequence were as following:

e1: 67Ala→67Ile, e2: 71Thr→71Asn, e3: 72Tyr→72Val, e4:73Glu74Glu→73Gln74Gln, e5: 76Leu→76Thr, e6:78Gln→78Val. Hydrophobicity was checked for each amino acid substitution, according to Kyte and Doolittle [23] (Figure 1). For substitutions 73Glu74Glu→73Gln74Gln, exchange polarity has been compared according to Zimmerman et al. [24] (Figure 1).

Subsequent to the introduction of mutations into the TROL presequence by PCR, constructs were transformed into competent bacteria, multiplied, purified, and checked by restriction enzymes and DNA sequencing.

4.2. In Vitro Transcription and Translation

The coding region for TROL from *Arabidopsis thaliana* was cloned into the vector pZL1 under the control of the T7 promoter and pOE33 from *Pisum sativum* was cloned into the vector pGEM4Z under the control of the SP6 promoter. Transcription and translation were carried out using the TNT® Quick Coupled Transcription/Translation System (Promega, Madison, WI, USA) in the presence of [^{35}S]-methionine (185 MBq, PerkinElmer, Boston, MA, USA) for radioactive labelling. After translation, the reaction mixture was centrifuged at $50,000 \times g$ for 20 min at 4 °C and the post-ribosomal supernatant was used for import experiments.

4.3. Chloroplast Isolation and Protein Import

Chloroplasts were isolated from leaves of 8 days old pea seedlings (*P. sativum* var. Letin, Agricultural Institute Osijek, Osijek, Croatia) and purified through Percoll density gradients as described [11,17]. A standard import reaction contained chloroplasts equivalent to 20 µg chlorophyll in 100 µL import buffer (330 mM sorbitol, 50 mM HEPES/KOH pH 7.6, 3 mM $MgSO_4$, 10 mM Met, 10 mM Cys, 20 mM K-gluconate, 10 mM $NaHCO_3$, 2% BSA (*w/v*)), up to 3 mM ATP and maximal 10%

(v/v) [^{35}S]-labeled translation products. Import reactions were initiated by the addition of translation product and carried out for 20 min at 25 °C, unless indicated otherwise. Reactions were terminated by separation of chloroplasts from the reaction mixture by centrifugation through a 40 % (v/v) Percoll cushion. Chloroplasts were washed once in 330 mM sorbitol, 50 mM HEPES/KOH pH 7.6, and 0.5 mM CaCl$_2$, lysed in 10 mM HEPES/KOH pH 7.6 for 30 min on ice and separated into membrane and soluble fractions by centrifugation at 265,000× g for 10 min at 4 °C. Import products were separated by SDS–PAGE and radiolabeled proteins analysed by exposure on X-ray films.

For the purpose of investigating the energy requirement, prior to import, ATP was depleted from chloroplasts and the translation product. For chloroplast isolation, plants were taken from the dark and isolated intact chloroplasts were further incubated in the dark on ice for 30 min. to diminish internal ATP production. For the import experiment, 0, 30, 30, and 3000 μM ATP was used and chloroplasts corresponding to 20 μg chlorophyll, in a 20 min import reaction.

Some experiments included chloroplast protease posttreatment, by using thermolysin after import. Thermolysin in concentration of 0.5 μg per μg chlorophyll was applied for 20 min on ice. The reaction was stopped by adding 5 mM EDTA. Chloroplasts were pelleted and resuspended in Laemmli buffer [29].

4.4. Membrane Extraction of Imported Proteins

Subsequent to import of 20 min at 25 °C, chloroplasts were re-isolated, washed, and separated to the membrane and soluble fractions. Isolated membranes, containing imported proteins, were treated with 6 M Urea in 10 mM HEPES/KOH pH 7.6, 0.1 M Na$_2$CO$_3$ pH 11.5, or 1 M NaCl. All incubations were performed for 20 min on RT. As a control, membranes were incubated solely in 10 mM HEPES/KOH pH 7.6 for 30 min on ice. Afterwards, samples were centrifuged at 265,000× g for 10 min. at 4 °C, and both pellets and the supernatants were analysed by SDS–PAGE and by the exposure on X-ray films.

Acknowledgments: This work has been funded by a Grant IP-2014-09-1173 from the Croatian Science Foundation to Hrvoje Fulgosi. We thank Dr. Mary Sopta for the language editing and critical reading of the manuscript.

Author Contributions: Lea Vojta designed the import experiments, grew experimental plants, isolated chloroplasts, performed in vitro transcription and translation and all import experiments. Lea Vojta also analyzed results and wrote the manuscript. Andrea Čuletić evaluated the character of substituted amino acids and performed molecular cloning. Hrvoje Fulgosi conceived this research and designed presequence amino acid substitutions and analyzed and discussed the results.

Abbreviations

TROL	thylakoid rhodanase-like protein
FNR	ferredoxin:NADP$^+$ oxidoreductase
RHO	rhodanase-like domain
ITEP	highly conserved module of TROL necessary for establishing high-affinity interaction with FNR
PEPE	Pro-Val-Pro repeat-rich region
IM	inner envelope membrane
SPP	stromal processing peptidase

References

1.   Jurić, S.; Hazler-Pilepić, K.; Tomašić, A.; Lepeduš, H.; Jeličić, B.; Puthiyaveetil, S.; Bionda, T.; Vojta, L.; Allen, J.F.; Schleiff, E.; et al. Tethering of ferredoxin:NADP$^+$ oxidoreductase to thylakoid membranes is mediated by novel chloroplast protein TROL. *Plant J.* **2009**, *60*, 783–794. [CrossRef] [PubMed]

2.   Vojta, L.; Fulgosi, H. Energy conductance from thylakoid complexes to stromal reducing equivalents. In *Advances in Photosynthesis—Fundamental Aspects*; Najafpour, M.M., Ed.; InTech: Rijeka, Croatia, 2012; pp. 175–190, ISBN 978-953-307-928-8.

3. Vojta, L.; Horvat, L.; Fulgosi, H. Balancing chloroplast redox status—Regulation of FNR binding and release. *Period. Biol.* **2012**, *114*, 25–31.

4. Vojta, L.; Carić, D.; Cesar, V.; Antunović Dunić, J.; Lepeduš, H.; Kveder, M.; Fulgosi, H. TROL-FNR interaction reveals alternative pathways of electron partitioning in photosynthesis. *Sci. Rep.* **2015**, *5*, 10085. [CrossRef] [PubMed]

5. Peltier, J.B.; Ytterberg, A.J.; Sun, Q.; van Wijk, K.J. New functions of the thylakoid membrane proteome of *Arabidopsis thaliana* revealed by a simple, fast, and versatile fractionation strategy. *J. Biol. Chem.* **2004**, *279*, 49367–49383. [CrossRef] [PubMed]

6. Stengel, A.; Benz, P.; Balsera, M.; Soll, J.; Bölter, B. TIC62 redox-regulated translocon composition and dynamics. *J. Biol. Chem.* **2008**, *283*, 6656–6667. [CrossRef] [PubMed]

7. Benz, J.P.; Lintala, M.; Soll, J.; Mulo, P.; Bölter, B. A new concept for ferredoxin-NADPH oxidoreductase binding to plant thylakoids. *Trends Plant Sci.* **2010**, *15*, 608–613. [CrossRef] [PubMed]

8. Twachtmann, M.; Altmann, B.; Muraki, N.; Voss, I.; Okutani, S.; Kurisu, G.; Hase, T.; Hanke, G.T. N-terminal structure of maize ferredoxin: NADP$^+$ reductase determines recruitment into different thylakoid membrane complexes. *Plant Cell* **2012**, *24*, 2979–2991. [CrossRef] [PubMed]

9. Vojta, L.; Fulgosi, H. Data supporting the absence of FNR dynamic photosynthetic membrane recruitment in *trol* mutants. *Data Brief* **2016**, *7*, 393–396. [CrossRef] [PubMed]

10. Cline, K.; Henry, R. Import and routing of nucleus-encoded chloroplast proteins. *Annu. Rev. Cell Dev. Biol.* **1996**, *12*, 1–26. [CrossRef] [PubMed]

11. Vojta, L.; Soll, J.; Bölter, B. Requirements for a conservative protein translocation pathway in chloroplasts. *FEBS Lett.* **2007**, *581*, 2621–2624. [CrossRef] [PubMed]

12. Teixeira, P.F.; Glaser, E. Processing peptidases in mitochondria and chloroplasts. *BBA Mol. Cell Res.* **2013**, *1833*, 360–370. [CrossRef] [PubMed]

13. Schnell, D.J. Protein targeting to the thylakoid membrane. *Annu. Rev. Plant Physiol. Plant Mol. Biol.* **1998**, *49*, 97–126. [CrossRef] [PubMed]

14. Lamppa, G.K. The chlorophyll a/b-binding protein inserts into the thylakoids independent of its cognate transit peptide. *J. Biol. Chem.* **1988**, *263*, 14996–14999. [PubMed]

15. Von Heijne, G.; Steppuhn, J.; Herrmann, R.G. Domain structure of mitochondrial and chloroplast targeting peptides. *Eur. J. Biochem.* **1989**, *180*, 535–545. [CrossRef] [PubMed]

16. Scott, S.V.; Theg, S.M. A new chloroplast protein import intermediate reveals distinct translocation machineries in the two envelope membranes: Energetics and mechanistic implications. *J. Cell Biol.* **1996**, *132*, 63–75. [CrossRef] [PubMed]

17. Vojta, L.; Soll, J.; Bölter, B. Protein transport in chloroplasts—Targeting to the intermembrane space. *FEBS J.* **2007**, *274*, 5043–5054. [CrossRef] [PubMed]

18. Theg, S.M.; Scott, S.V. Protein import into chloroplasts. *Trends Cell Biol.* **1993**, *3*, 186–190. [CrossRef]

19. Schnell, D.J.; Blobel, G. Identification of intermediates in the pathway of protein import into chloroplasts and their localization to envelope contact sites. *J. Cell Biol.* **1993**, *120*, 103–115. [CrossRef] [PubMed]

20. Soll, J.; Robinson, C.; Heins, L. The import and sorting of protein into chloroplasts. In *Protein Targeting, Transport and Translocation*; Dalbey, R., von Heijne, G., Eds.; Elsevier: Amsterdam, The Netherlands, 2002; pp. 240–267, ISBN 978-0-12-200731-6.

21. Jarvis, P.; Robinson, C. Mechanisms of protein import and routing in chloroplasts. *Curr. Biol.* **2004**, *14*, R1064–R1077. [CrossRef] [PubMed]

22. Klasek, L.; Inoue, K. Dual protein localization to the envelope and thylakoid membranes within the chloroplast. *Int. Rev. Cell Mol. Biol.* **2016**, *323*, 231–263. [CrossRef] [PubMed]

23. Kyte, J.; Doolittle, R.F. A simple method for displaying the hydropathic character of a protein. *J. Mol. Biol.* **1982**, *157*, 105–132. [CrossRef]

24. Zimmerman, J.M.; Eliezer, N.; Simha, R. The characterization of amino acid sequences in proteins by statistical methods. *J. Theor. Biol.* **1968**, *21*, 170–201. [CrossRef]

25. Fincher, V.; Dabney-Smith, C.; Cline, K. Functional assembly of thylakoid deltapH-dependent/Tat protein transport pathway components in vitro. *Eur. J. Biochem.* **2003**, *270*, 4930–4941. [CrossRef] [PubMed]

26. Woolhead, C.A.; Thompson, S.J.; Moore, M.; Tissier, C.; Mant, A.; Rodger, A.; Henry, R.; Robinson, C. Distinct Albino3-dependent and -independent pathways for thylakoid membrane protein insertion. *J. Biol. Chem.* **2001**, *276*, 40841–40846. [CrossRef] [PubMed]

27. Mant, A.; Woolhead, C.A.; Moore, M.; Henry, R.; Robinson, C. Insertion of PsaK into the thylakoid membrane in a "Horseshoe" conformation occurs in the absence of signal recognition particle, nucleoside triphosphates, or functional albino3. *J. Biol. Chem.* **2001**, *276*, 36200–36206. [CrossRef] [PubMed]

28. Michl, D.; Robinson, C.; Shackleton, J.B.; Herrmann, R.G.; Klösgen, R.B. Targeting of proteins to the thylakoids by bipartite presequences: CFoII is imported by a novel, third pathway. *EMBO J.* **1994**, *13*, 1310–1317. [PubMed]

29. Laemmli, U.K. Cleavage of structural proteins during the assembly of the head of bacteriophage T4. *Nature* **1970**, *227*, 680–685. [CrossRef] [PubMed]

12

Comparative Analysis of the Chloroplast Genomes of the Chinese Endemic Genus *Urophysa* and their Contribution to Chloroplast Phylogeny and Adaptive Evolution

Deng-Feng Xie, Yan Yu, Yi-Qi Deng, Juan Li, Hai-Ying Liu, Song-Dong Zhou and Xing-Jin He *

Key Laboratory of Bio-Resources and Eco-Environment of Ministry of Education, College of Life Sciences, Sichuan University, Chengdu 610065, Sichuan, China; df_xie2017@163.com (D.-F.X.); yyu@scu.edu.cn (Y.Y.); yiqiden@gmail.com (Y.-Q.D.); lijuanxxn@163.com (J.L.); lhy921180@163.com (H.-Y.L.); songdongzhou@aliyun.com (S.-D.Z.)
* Correspondence: xjhe@scu.edu.cn

Abstract: *Urophysa* is a Chinese endemic genus comprising two species, *Urophysa rockii* and *Urophysa henryi*. In this study, we sequenced the complete chloroplast (cp) genomes of these two species and of their relative *Semiquilegia adoxoides*. Illumina sequencing technology was used to compare sequences, elucidate the intra- and interspecies variations, and infer the phylogeny relationship with other Ranunculaceae family species. A typical quadripartite structure was detected, with a genome size from 158,473 to 158,512 bp, consisting of a pair of inverted repeats separated by a small single-copy region and a large single-copy region. We analyzed the nucleotide diversity and repeated sequences components and conducted a positive selection analysis by the codon-based substitution on single-copy coding sequence (CDS). Seven regions were found to possess relatively high nucleotide diversity, and numerous variable repeats and simple sequence repeats (SSR) markers were detected. Six single-copy genes (*atpA*, *rpl20*, *psaA*, *atpB*, *ndhI*, and *rbcL*) resulted to have high posterior probabilities of codon sites in the positive selection analysis, which means that the six genes may be under a great selection pressure. The visualization results of the six genes showed that the amino acid properties across each column of all species are variable in different genera. All these regions with high nucleotide diversity, abundant repeats, and under positive selection will provide potential plastid markers for further taxonomic, phylogenetic, and population genetics studies in *Urophysa* and its relatives. Phylogenetic analyses based on the 79 single-copy genes, the whole complete genome sequences, and all CDS sequences showed same topologies with high support, and *U. rockii* was closely clustered with *U. henryi* within the *Urophysa* genus, with *S. adoxoides* as their closest relative. Therefore, the complete cp genomes in *Urophysa* species provide interesting insights and valuable information that can be used to identify related species and reconstruct their phylogeny.

Keywords: *Urophysa*; *Semiaquilegia adoxoides*; cp genome; repeat analysis; SSRs; positive selection analysis; phylogeny

1. Introduction

The genus *Urophysa* (Ranunculaceae) is a Chinese endemic genus with only two species, *Urophysa rockii* Ulbr. and *Urophysa henryi* (Oliv.) Ulbr. *U. rockii* is an extremely rare species with fewer than 2000 individuals living in Jiangyou, a Sichuan province of China, and *U. henryi* is distributed in Guizhou, south Chongqing, north Hunan, and west Hubei [1]. The two species' natural populations are restricted to small and isolated areas separated by high mountains and deep valleys and grow in steep and karstic cliffs with dramatically shrinking and fragmenting natural distributions [2]. In addition, the plants are collected for Chinese traditional medicine for the treatment of contusions and bruises, which contributed to the decline of their populations [3]. Previous studies on the genus *Urophysa* are scarce and mainly focused on the endangered *U. rockii*, its growing environment and conservation strategies [4], its biological and ecological characteristics, and its reproductive biology [5,6]. A recent study suggested that the uplift of the Yungui Plateau played an important role in the species divergence of *Urophysa* [2]. However, the chloroplast DNA (cpDNA) phylogeny showed inconsistency with the nuclear ribosomal DNA (nrDNA). Hence, to gain a better insight into the relationship of these two species and understand their genome structure so as to facilitate their speciation process and the conservation of *U. rockii*, we assembled and characterized the complete chloroplast genome sequence of *U. rockii* and *U. henryi* using the Illumina paired-end sequencing reads.

The angiosperm cp genome is one of the three DNA genomes (the other two are nuclear and mitochondrial genome), is uniparentally inherited, and has a high conserved circular DNA arrangement [7]. It is widely considered an informative and valuable resource for investigating evolutionary biology because of its relatively stable genome structure, gene content, and gene order [8–13]. The cp genome of plants always ranges from 115 to 210 kb and has a quadripartite structure that is typically composed of two copies of inverted repeat (IR) regions, which are separated by a large single-copy (LSC) region and a small single-copy (SSC) region [14–16]. Because of its compact size, less recombination, and maternal inheritance, the cp genome has been used to generate genetic markers for phylogenetic analysis [17,18], molecular identification [19], and divergence dating [20]. Especially, the low evolutionary rate of the cp genome in taxa that are not very young makes it an ideal system for assessing plant phylogeny [21].

In the present study, we report the complete chloroplast genome sequences of these two *Urophysa* species and their relative *Semiaquilegia adoxoides* for the first time. Combining previously reported cp genome sequences, we performed phylogenetic analyses according to the whole cp genome and shared single-copy genes. Our findings will contribute to our understanding of the evolutionary history of the genus *Urophysa*. Additionally, highly variable regions and genes that were detected to be under positive selection could be employed to develop potential markers for phylogenetic analyses or candidates for DNA barcoding in future studies.

2. Results and Discussion

2.1. Complete Chloroplast Genomes of Three Species

The complete chloroplast genome of *U. rockii*, *U. henryi*, and *S. adoxoides* showed a single circular molecule with a typical quadripartite structure (Figure 1). The sizes of the *U. rockii*, *U. henry*, and *S. adoxoides* cp genomes were found to be 158,512 bp, 158,303, and 158,340 bp, respectively, which are in the range of most angiosperm plastid genomes [22]. The cp genome consists of a pair of IRs (IRa and IRb, with length 26,473–26,584 bp), separated by a LSC (87,031–87,202 bp) region and one SSC (18,192–18,220 bp) region (Table 1). The GC content of each species was very similar in the whole cp genome and the same region (LSC, SSC, and IR), but in the IR regions it was clearly higher than in the other regions, possibly because of the high GC content of the rRNA (55.8%) that was located in the IR regions (Table 2). These results are similar to a previously reported high GC percentage in IR regions [23–25].

The genomes contain 87 coding genes, 36 transfer RNA genes (tRNA), and 8 ribosomal RNA genes (rRNA) (Table 3). Most of the genes occur as a single copy in LSC or SSC regions, while 18 genes are duplicated in the IR regions, including seven protein-coding genes (*ndhB, rpl2, rpl23, rps7, rps12, rps19, ycf2*), seven tRNA species (*trnA-UGC, trnI-CAU, trnI-GAU, trnL-CAA, trnN-GUU, trnR-ACG, and trnV-GAC*) and four rRNA species (*rrn4.5, rrn5, rrn16, and rrn23*). The gene *ycf1* straddles the SSC and IRs, while *rps12* locates its first exon in the LSC region and two other exons in the IRs. The LSC region comprises 63 protein-coding genes and 21 tRNA genes, whereas the SSC and IR regions include 12 and 7 protein-coding genes, with one and seven tRNA, respectively. The protein-coding genes present in the *U. rockii* cp genome include 9 genes encoding large ribosomal proteins (*rpl2*, *rpl14*, *rpl16*, *rpl20*, *rpl22*, *rpl23*, *rpl32*, *rpl33*, *rpl36*) and 12 genes encoding small ribosomal proteins (*rps2*, *rps3*, *rps4*, *rps7*, *rps8*, *rps11*, *rps12*, *rps14*, *rps15*, *rps16*, *rps18*, *rps19*). There are 5 genes encoding phytosystem I subunits (*psaA*, *psaB*, *psaC*, *psaI*, *psaJ*), along with 15 genes related to photosystem II subunits (*psbA*, *psbB*, *psbC*, *psbD*, *psbE*, *psbF*, *psbH*, *psbI*, *psbJ*, *psbK*, *psbL*, *psbM*, *psbN*, *psbT*, *psbZ*) (Table 3). Six genes (*atpA*, *atpB*, *atpE*, *atpF*, *atpH*, *atpI*) encode ATP synthase and electron transport chain components (Table 3). A similar pattern of protein-coding genes is also present in *U. henryi* and *S. adoxoides*. There are eight intron-containing genes, six of which contain one intron; only the genes *clpP* and *ycf3* have two introns (Table S1). All these eight genes possess at least two exons, and *ycf3* has three exons. The *rps16* gene has the longest intron (866 bp), and *rpoC1* has the longest exon (1613 bp).

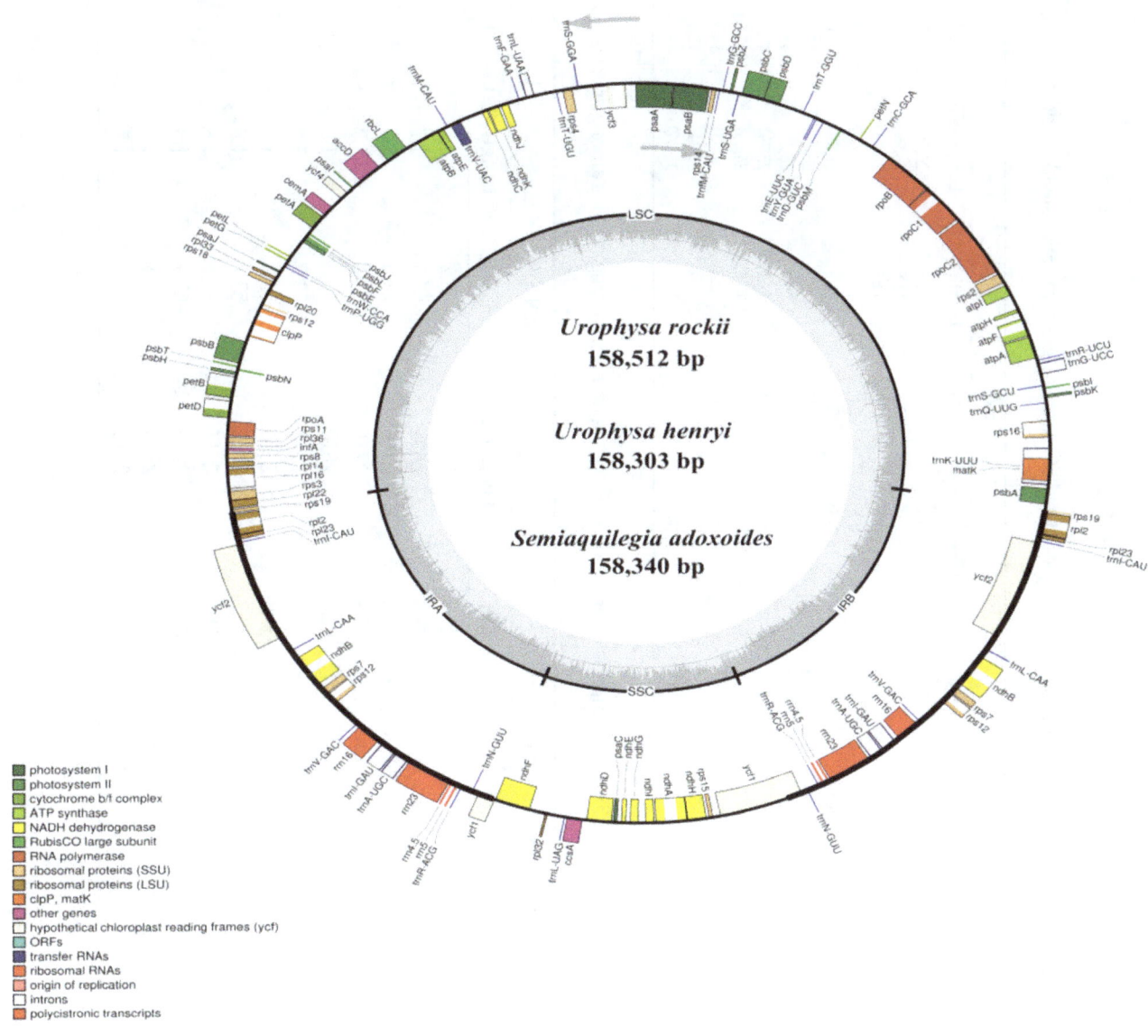

Figure 1. Gene maps of the *Urophysa rockii*, *Urophysa henryi* and *Semiquilegia adoxoides* chloroplast (cp) genomes. Genes shown inside the circle are transcribed clockwise, and those outside are transcribed counterclockwise. Genes belonging to different functional groups are color-coded. The darker gray color in the inner circle corresponds to the GC content, and the lighter gray color corresponds to the AT content. SSU: small subunit; LSU: large subunit; ORF: open reading frame.

Table 1. Summary of complete chloroplast genomes. LSC, large single-copy; SSC, small single-copy; IR, inverted repeat

Species	LSC			SSC			IR			Total	
	Length (bp)	GC%	Length (%)	Length (bp)	GC%	Length (%)	Length (bp)	GC%	Length (%)	Length (bp)	GC%
U. rockii	87,128	37.2	55.0	18,216	32.5	11.5	26,584	43.7	16.8	158,512	38.8
U. henryi	87,031	37.2	55.0	18,260	32.6	11.5	26,506	43.6	16.7	158,303	38.8
S. adoxoides	87,202	37.2	55.1	18,192	32.5	11.5	26,473	43.7	16.7	158,340	38.9
Tsuga chinensis	88,522	36.3	55.3	18,405	32.0	11.5	26,632	43.1	16.6	160,191	38.1
Aconitum austrokoreense	86,362	36.2	55.4	16,948	32.7	10.9	26,291	43.0	16.9	155,892	38.1
A. kusnezoffii	86,335	36.2	55.4	16,945	32.7	10.9	26,291	43.0	16.9	155,862	38.1
A. volubile	86,348	36.2	55.4	16,944	32.6	10.9	26,290	43.0	16.9	155,872	38.1
Ranunculus macranthus	84,637	36.0	54.6	18,909	31.0	12.2	25,791	43.5	16.6	155,129	37.9
R. occidentalis	83,532	35.9	54.1	21,269	31.6	13.8	24,831	43.6	16.1	154,474	37.8
R. austro-oreganus	83,582	35.9	54.1	21,249	31.6	13.8	24,831	43.6	16.1	154,493	37.8
Clematis terniflora	79,328	36.3	49.7	18,110	31.4	11.4	31,045	42.0	19.5	159,528	38.0
Coptis chinensis	84,567	36.4	54.4	17,376	32.1	11.2	26,762	43.0	17.2	155,484	38.2

Table 2. Comparison of the sizes of coding and non-coding regions among species.

Species	Protein-Coding			tRNA			rRNA		
	Length (bp)	GC%	Length (%)	Length (bp)	GC%	Length (%)	Length (bp)	GC%	Length (%)
U. rockii	78,867	39.2	49.8	2687	53.2	1.7	8602	55.8	5.4
U. henryi	78,769	39.2	49.8	2695	53.3	1.7	8602	55.8	5.4
S. adoxoides	78,498	39.3	49.6	2706	53.6	1.7	8602	55.8	5.4
T. chinensis	78,903	38.4	49.3	2716	53.1	1.7	9050	55.4	5.6
A. austrokoreense	79,575	38.3	51.0	2810	53.0	1.8	9050	55.4	5.8
A. kusnezoffii	78,294	38.4	50.2	2813	52.9	1.8	9046	55.3	5.8
A. volubile	79,560	38.3	51.0	2810	53.0	1.8	9050	55.5	5.8
R. macranthus	78,615	38.2	50.7	2738	53.1	1.8	7559	55.2	4.9
R. occidentalis	69,294	38.6	44.9	2717	53.1	1.8	9050	55.4	5.9
R. austro-oreganus	74,355	38.1	48.1	2796	52.9	1.8	9050	55.4	5.9
C. terniflora	81,819	38.3	51.3	2718	53.4	1.7	9050	55.4	5.7
C. chinensis	71,637	39.0	46.1	2716	53.2	1.7	9050	55.5	5.8

Table 3. List of genes encoded in two *Urophysa* species and *S. adoxoides*.

Category for Genes	Group of Genes	Name of Genes
Self-replication	transfer RNAs	*trnA-UGC* *, *trnC-GCA*, *trnD-GUC*, *trnE-UUC*, *trnF-GAA*, *trnfM-CAU*, *trnG-GCC*, *trnG-UCC*, *trnI-CAU* *, *trnI-GAU* *, *trnK-UUU*, *trnL-CAA* *, *trnL-UAA*, *trnL-UAG*, *trnM-CAU*, *trnN-GUU* *, *trnP-UGG*, *trnQ-UUG*, *trnR-ACG* *, *trnR-UCU*, *trnS-GCU*, *trnS-GGA*, *trnS-UGA*, *trnT-GGU*, *trnT-UGU*, *trnV-GAC* *, *trnV-UAC*, *trnW-CCA*, *trnY-GUA*
	ribosomal RNAs	*rrn4.5* *, *rrna5* *, *rrn16* *, *rrn23* *
	RNA polymerase	*rpoA*, *rpoB*, *rpoC1*, *rpoC2*
	Small subunit of ribosomal proteins (SSU)	*rps2*, *rps3*, *rps4*, *rps7* *, *rps8*, *rps11*, *rps12* *, *rps14*, *rps15*, *rps16*, *rps18*, *rps19* *
	Large subunit of ribosomal proteins (LSU)	*rpl2* *, *rpl14*, *rpl16*, *rpl20*, *rpl22*, *rpl23* *, *rpl32*, *rpl33*, *rpl36*
Genes for photosynthesis	Subunits of NADH-dehydrogenase	*ndhA*, *ndhB* *, *ndhC*, *ndhD*, *ndhE*, *ndhF*, *ndhG*, *ndhH*, *ndhI*, *ndhJ*, *ndhK*
	Subunits of photosystem I	*psaA*, *psaB*, *psaC*, *psaI*, *psaJ*
	Subunits of photosystem II	*psbA*, *psbB*, *psbC*, *psbD*, *psbE*, *psbF*, *psbH*, *psbI*, *psbJ*, *psbK*, *psbL*, *psbM*, *psbN*, *psbT*, *psbZ*
	Subunits of cytochrome b/f complex	*petA*, *petB*, *petD*, *petG*, *petL*, *petN*
	Subunits of ATP synthase	*atpA*, *atpB*, *atpE*, *atpF*, *atpH*, *atpI*
	Large subunit of rubisco	*rbcL*
Other genes	Tanslational initiation factor	*infA*
	Protease	*clpP*
	Maturase	*matK*
	Subunit of Acetyl-CoA-carboxylase	*accD*
	Envelope membrane protein	*cemA*
	C-type cytochrome synthesis gene	*ccsA*
Genes of unknown function	hypothetical chloroplast reading frames (ycf)	*ycf1* *, *ycf2* *, *ycf3*, *ycf4*

* Gene with two copies.

2.2. Repeat Analysis

Chloroplast repeats are potentially useful genetic resources to investigate population genetics and biogeography of allied taxa [26]. Analyses of various cp genomes revealed that repeat sequences are essential to induce indels and substitutions [27]. Repeat analysis of the *U. rockii* cp genome revealed 22 palindromic repeats, 23 forward repeats, 5 reverse, and 1 complement repeats. Among them, 16 palindromic, 18 forward, and 5 reverse repeats are 20–40 bp in length. Six palindromic and five forward repeats are 41–60 in length (Figure 2). Similarly, 23 and 25 palindromic repeats, 21 and 22 forward repeats, 5 and 2 reverse repeats, and 1 complement repeats were detected, and the detailed repeats length distributions are shown in Figure 2. The number and length of the repeats indicate that *U. rockii* is more similar to *U. henryi* than to *S. aquilegia*. Previous studies suggested that the slipped-strand mispairing and improper recombination of repeat sequences can result in sequence variation and genome rearrangement [28–30]. These repeats are informative sources for developing genetic markers for phylogenetic and population studies [31].

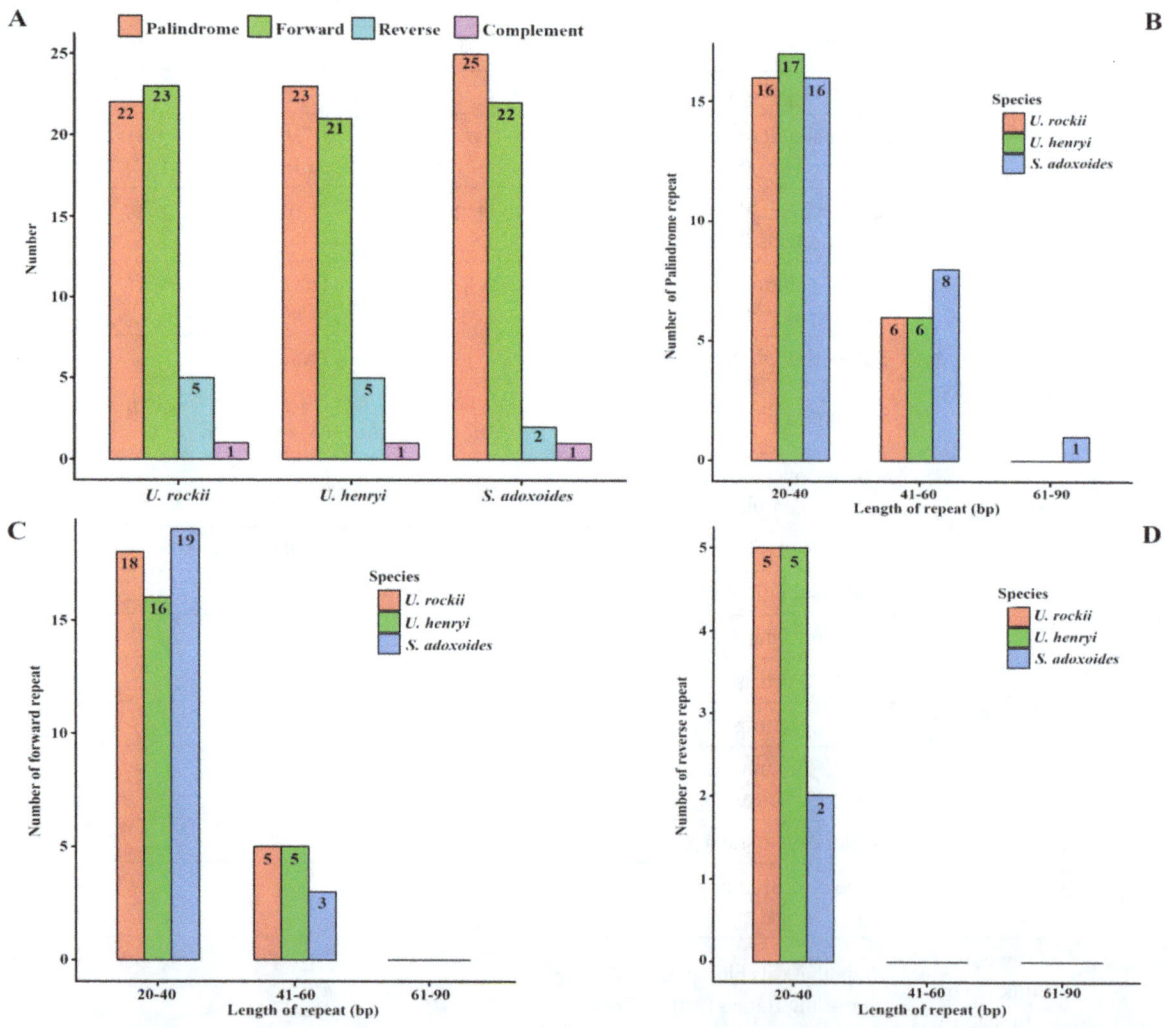

Figure 2. Analysis of repeated sequences in *U. rockii*, *U. henryi*, and *S. adoxoides* chloroplast genomes. (**A**) Total of four repeat types; (**B**) Frequency of the palindromic repeat by length; (**C**) Frequency of the forward repeat by length; (**D**) Frequency of the reverse repeat by length.

Simple sequence repeats (SSRs) in the cp genome can be highly variable at the intra-specific level and are therefore often used as genetic markers in population genetic and evolutionary studies [12,32–34]. Because of a high polymorphism rate at the species level, SSRs have been recognized as one of the main sources of molecular markers and have been extensively researched in phylogenetic and biogeographic studies of populations [35–37]. In this study, we analyzed the SSRs in the cp genomes. Five categories of perfect SSRs (mono-, di-, tri-, tetra-, and penta-nucleotide repeats) were detected in the cp genome of these three species, with an overall length ranging from 10 to 26 bp (Figure 3, Table S2). Certain parameters were set, because SSRs of 10 bp or longer are prone to slipped-strand mispairing, which is believed to be the main mutational mechanism for polymorphism [38–40].

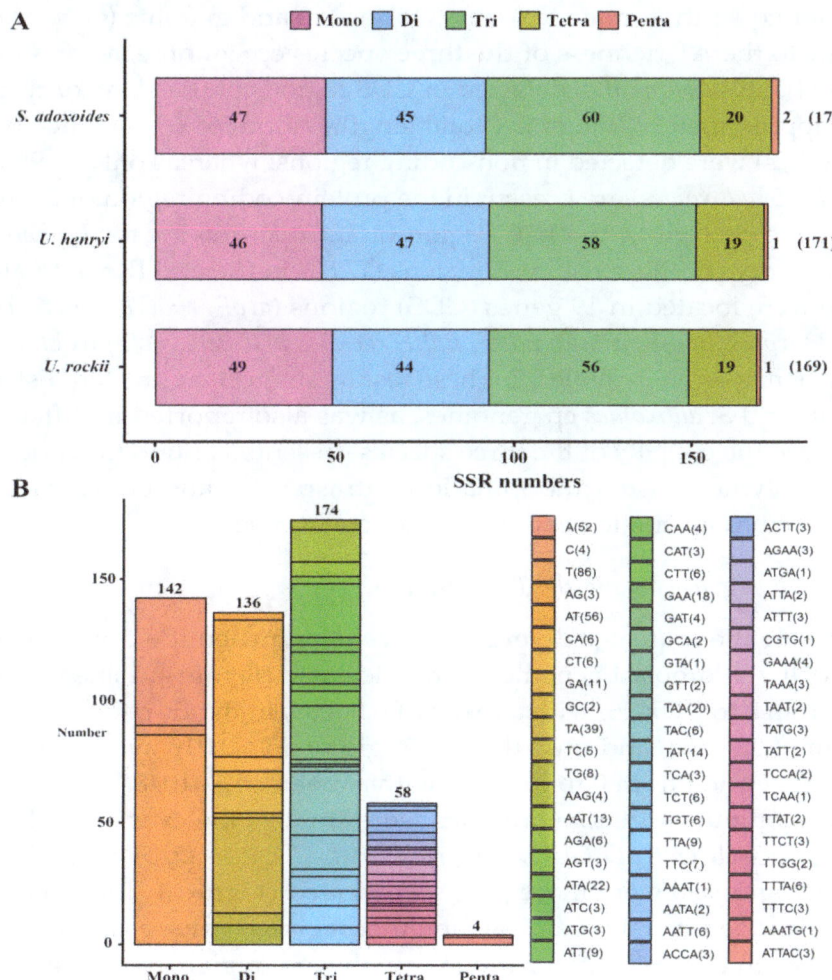

Figure 3. Analysis of simple sequence repeats (SSRs) in chloroplast genomes of the three species. (**A**) Number of different SSR types detected in each species; (**B**) type and frequency of each identified SSR.

A total of 169 microsatellites were detected in the *U. rockii* cp genome on the basis of the SSR analysis. Similarly, 171 and 174 SSRs were detected in *U. henryi* and *S. adoxoides*, respectively (Figure 3A). The most abundant were tri-nucleotide repeats, which accounted for about 33.85% of the total SSRs, and whose number varies from 56 in *U. rockii* to 60 in *S. adoxoides*, followed by mono-nucleotide repeats (27.63%), di-nucleotide repeats (26.46%), and tetra-nucleotides repeats (11.28%). Penta-nucleotide repeats were the least abundant (0.78%; Figure 3, Table S2). Most previous studies revealed that the richness of SSR types varies between species. In *Quercus* species, mono-nucleotide repeats are the most abundant, accounting for about 80% of the total SSRs [34]. In the cp genome of *Forthysia*, the number of di-nucleotide repeat is the highest [41]. Tri-nucleotide SSRs are most abundant in *Nicotiana species*, accounting for approximately 43.03% [42]. These results suggest that different repeats may contribute to the genetic variations differently among species. Thus, the SSR information will be important for understanding the genetic diversity status of *Urophysa* and its relatives.

In *U. rockii*, more than 96.2% mono-nucleotides are composed of A/T, and a majority of di-nucleotides (84.9%) is composed of A/T (Figure 3B, Table S2), which is consistent with *U. henryi* (97.8% mono-nucleotides and 83.0% di-nucleotides) and *S. aquilegia* (97.9% mono-nucleotides and 85.6% di-nucleotides). Our findings are comparable to previously reported observations that SSRs found in the chloroplast genome are generally composed of poly-thymine (polyT) or poly-adenine (polyA) repeats and infrequently contain tandem cytosine (C) and guanine (G) repeats [43]. Therefore, these SSRs contribute to the AT richness of the three species cp genome, as previously reported for different species [43,44]. SSRs were also detected in CDS regions of the *U. rockii* cp genome. The CDS regions account for approximately 49% of the total length. About 68.6% of SSRs (68.4% for *U. henryi* and 67.2% for *S. adoxoides*) were detected in non-coding regions, whereas only 28.9% of SSRs (29.2% for *U. henryi* and 30.5% for *S. adoxoides*) are present in the protein-coding region of *U. rockii*. Furthermore, about 62.1% of SSRs are present in the LSC region of *U. rockii* (66.1% for *U. henryi* and 68.9% for *S. adoxoides*), and a minority of SSRs exist in IR regions (17.8% in IRa and IRb in total). It was observed that 49 SSRs (28.9%) were located in 19 genes (CDS) regions (*atpF, rpoC1, rpoC2, rps14, rps15, rps19, psaB, psaA, rbcL, rpl33, rpl22, ndhB, ndhD, ndhF, ndhH, ccsA, ycf1, ycf2, ycf3*) in *U. rockii*. The detailed SSR location information is listed in Table S2. These results suggest an uneven distribution of SSRs in the *U. rockii, U. henryi,* and *S. adoxoides* cp genomes, as was also reported in different angiosperm cp genomes [44]. Moreover, the cp SSRs of the three species presented abundant variation and are useful for detecting genetic polymorphisms at population, intraspecific, and cultivar levels, as well as for comparing more distant phylogenetic relationships among species.

2.3. Genomes Sequence Divergence among the Three Species

In order to calculate the sequence divergence level, the nucleotide diversity values in the LSC, SSC, and IR regions of the chloroplast genomes were calculated (Figure 4, Table S3). In the LSC regions, these values varied from 0 to 0.05496, with a mean of 0.00705, in the IR regions they varied from 0 to 0.01265, with a mean of 0.00363, and only the SSC region had >0.010 average sequence nucleotide diversity, and its values varied from 0 to 0.02369, with a mean of 0.01048. All these results indicated that the differences among these genome regions were small. However, some highly variable loci, including *trnK-UUU, trnG-UCC, trnD-GUC, atpF, rps4, trnL-UAA, accD, cemA, rpl36, rpl22, rps19, ndhF, trnL-UAG, ccsA, ndhA,* and *ycf3* were more precisely located (Figure 4, Table S3). All these regions displayed higher nucleotide diversity values than other regions (value > 0.015). Twelve of these loci were found to be located in the LSC region, and four in the SSC region, but the nucleotide diversity in the IR regions appeared small, less than 0.015. Among these loci, *atpF, accD, ndhF, rpl22, ccsA,* and *ycf3* have been detected as highly variable regions in different plants [19,23,45,46]. On the basis of these results, we believe that *accD, rps4, ccsA, rpl36,* and *ndhF*, which have comparatively high sequence deviation, are good sources for interspecies phylogenetic analysis, as shown in previous studies [42,44].

Expansion and contraction at the borders of IR regions is the main reason for size variations in the cp genome and plays a vital role in its evolution [39,47,48]. The IR/LSC and IR/SSC junction regions were compared to identify IR expansion or contraction. The *rps19, ndhF, ycf1,* and *psbA* genes were located in the junctions of the LSC/IRa, IRa/SSC, SSC/IRb, and IRb/LSC regions, respectively (Figure 5). Despite the similar length of these three species IR regions, from 26,473 to 26,584 bp, some IR expansion and contraction were observed. The *rps19* gene traverses the LSC and IRb regions (LR line), with 104 bp located in the IR region. The RS line (the junction line between IRb and SSC) is located between *ycf1* and *ndhF*, and the variation in distances between the RS line and *ndhF* ranges from 33 to 36 bp across the three species. The SR line (the junction line between SSC and IRa) intersects the ycf1 gene, the SSC and IRa regions are the same in *U. rockii* and *U. henryi* (4259 bp in SSC and 1081 bp in IRb), while different in *S. adoxoides* (4229 bp in SSC and 1084 bp in IRb) (Figure 5). The distance between the *psbA* and RL line varies from 386 to 403 bp.

Compared to species of other genera, the IRb/SSC and SSC/IRa regions of *Urophysa* showed an expansion in *ycf1*, but a contraction in *rps19* (Figure 5). The expansion and contraction detected in the IR regions may act as a primary mechanism in creating the length variation of the cp genomes in *U. rockii*, *U. henryi*, and *S. adoxoides*, as previous studies suggested [32,34,42,49].

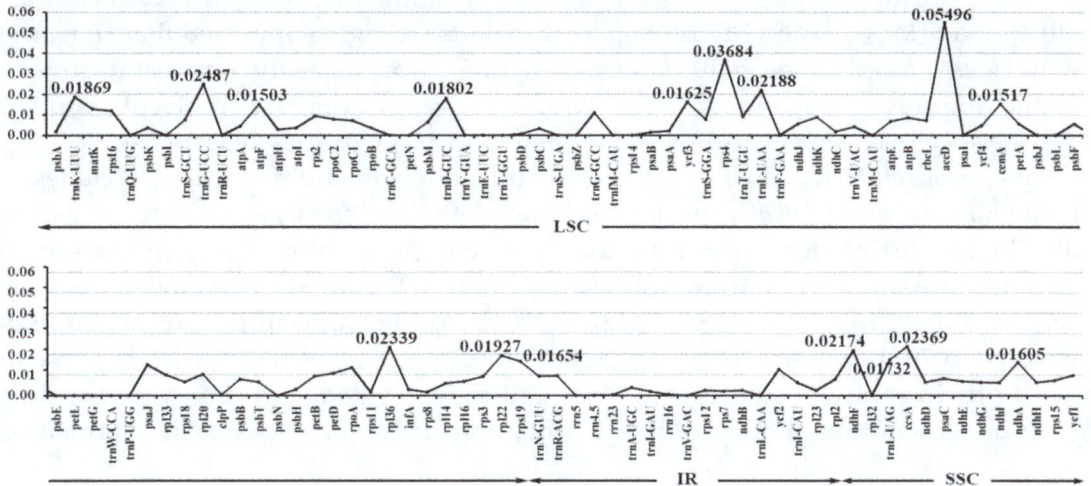

Figure 4. The nucleotide diversity of the whole chloroplast genomes of the three species. LSC: large single-copy region; IRs: inverted repeats region; SSC: small single-copy region.

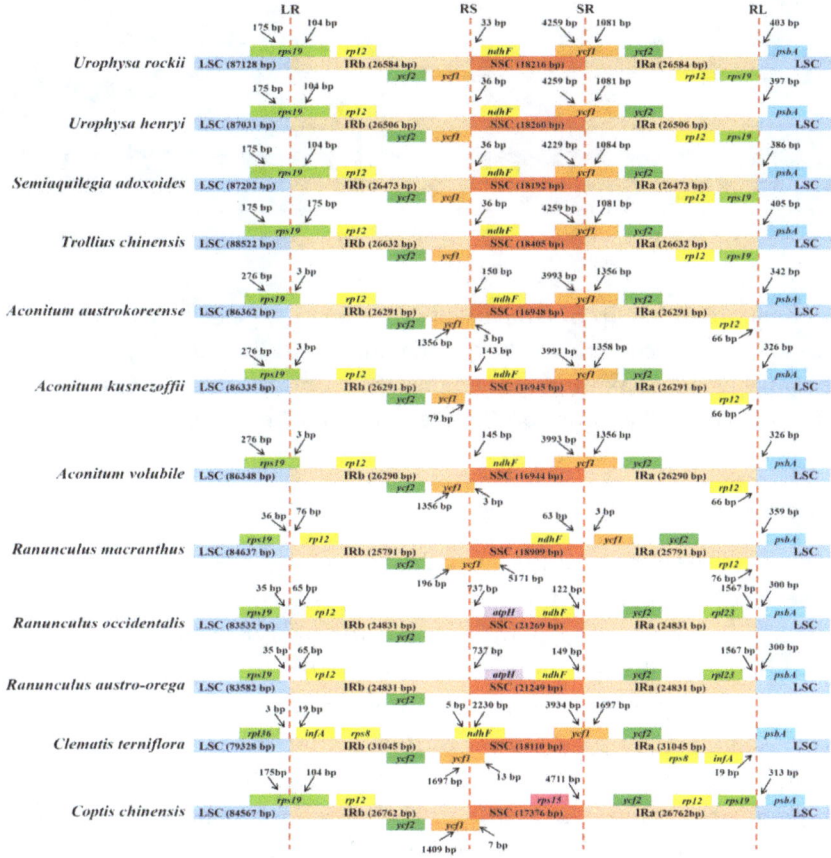

Figure 5. Comparison of the borders of the LSC, SSC, and IR regions of the chloroplast genomes of the three species. LR: junction line between LSC and IRb; RS: junction line between IRb and SSC; SR: junction line between SSC and IRa; RL: junction line between IRa and LSC.

2.4. Phylogenetic Analysis

To study the phylogenetic position of *U. rockii* and *U. henryi* within the Ranunculaceae family, we used 79 single-copy genes shared by the cp genomes of 12 Ranunculaceae members, representing seven genera (Figure 6). For Bayesian inference (BI) and maximum parsimony (MP), the posterior probabilities and bootstrap values were very high for each lineage, with all values ≥98%. Both the maximum likelihood (ML), BI, and MP phylogenetic results strongly supported that *U. rockii* is closely clustered with *U. henryi* within the genus *Urophysa*, with *S. adoxoides* as their closest relative with 100% bootstrap value (Figure 6), which is consistent with the results of previous molecular studies [50–52]. Furthermore, the species in each genus formed a single clade. The first clade is formed by species of the genera *Urophysa*, *Semiaquilegia*, and *Trollius*, the second clade was divided into two clades: one clade includes the *Ranunculus* and *Clematis* species, and the other clade consists of just the *Aconitum* species. Additionally, the topological structures from the whole complete chloroplast genome sequences and the CDS sequences are similar to that from single-copy genes (Figure S1), and all lineages possess high bootstrap values. These results suggest that there is no conflict among the entire genome data set, CDS sequences, and 79 shared single-copy genes of these cp genomes. Furthermore, these results are in accord with previous phylogeny research [53]. All these phylogenetic analyses are substantially increasing our understanding of the evolutionary relationship among species in Ranunculaceae.

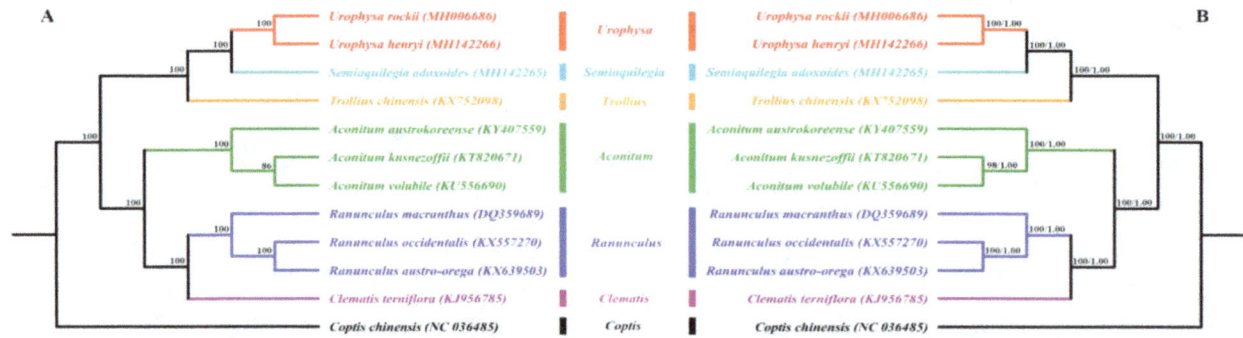

Figure 6. Phylogenetic relationship of *Urophysa* with related species based on 79 single-copy genes shared by all cp genomes. Tree constructed by (**A**) maximum likelihood (ML) with the bootstrap values of ML above the branches; (**B**) maximum parsimony (MP) and Bayesian inference (BI) with bootstrap values of MP and posterior probabilities of BI above the branches, respectively.

2.5. Positive Selected Analysis

Of 57 single-copy CDS genes initially considered for the positive selection analysis (Table S4), 47 were eventually selected (Table 4). No significant positive selection was detected for all genes (*p*-value > 0.05), but six genes that possess high posterior probabilities for codon sites were found in the Bayesian Empirical Bayes (BEB) test (*atpA*, *rpl20*, *psaA*, *atpB*, *ndhI*, and *rbcL*) (Figure 7, Figure S2 and Table 4). Previous studies suggested that codon sites with a high posterior probability should be regarded as positively selected sites [54], which means that these six genes may be under positive selection pressure [55]. After Jalview visualization, the results of the amino acid properties across each column of all species revealed that many amino acids vary between different genera, such as the 88th amino acid (G in *U. rockii* and *U. henryi*, R in other species) of the *rpl20* gene (Figure 7A) and other amino acids (marked with red blocks in Figure 7A). In the *ndhI* gene, two amino acids (the A in 168th and the P in 174th) were specific for *U. rockii* and *U. henryi*, and three amino acids (the 9th, 148th, and 165th, marked with red blocks in Figure 7B) were only possessed by *U. rockii*, *U. henryi*, and *S. adoxoides*. The amino acid properties of the other four genes (*atpA*, *atpB*, *rbcL*, and *psaA*) are shown in Figure S2. As we know, most amino acids may be under strong structural and functional constraints and not free to change [55]. We detected six genes with high posterior probability in codon site and many different amino acids among species, which may play an important role in *Urophysa* species evolution and environment adaptation. Populations of *U. rockii* and *U. henryi* are distributed only in karst regions of southern China, and the karst environments are characterized by low soil water content, insufficient light, and poor nutrient availability, which might have exerted strong selective forces on plant evolution [56].

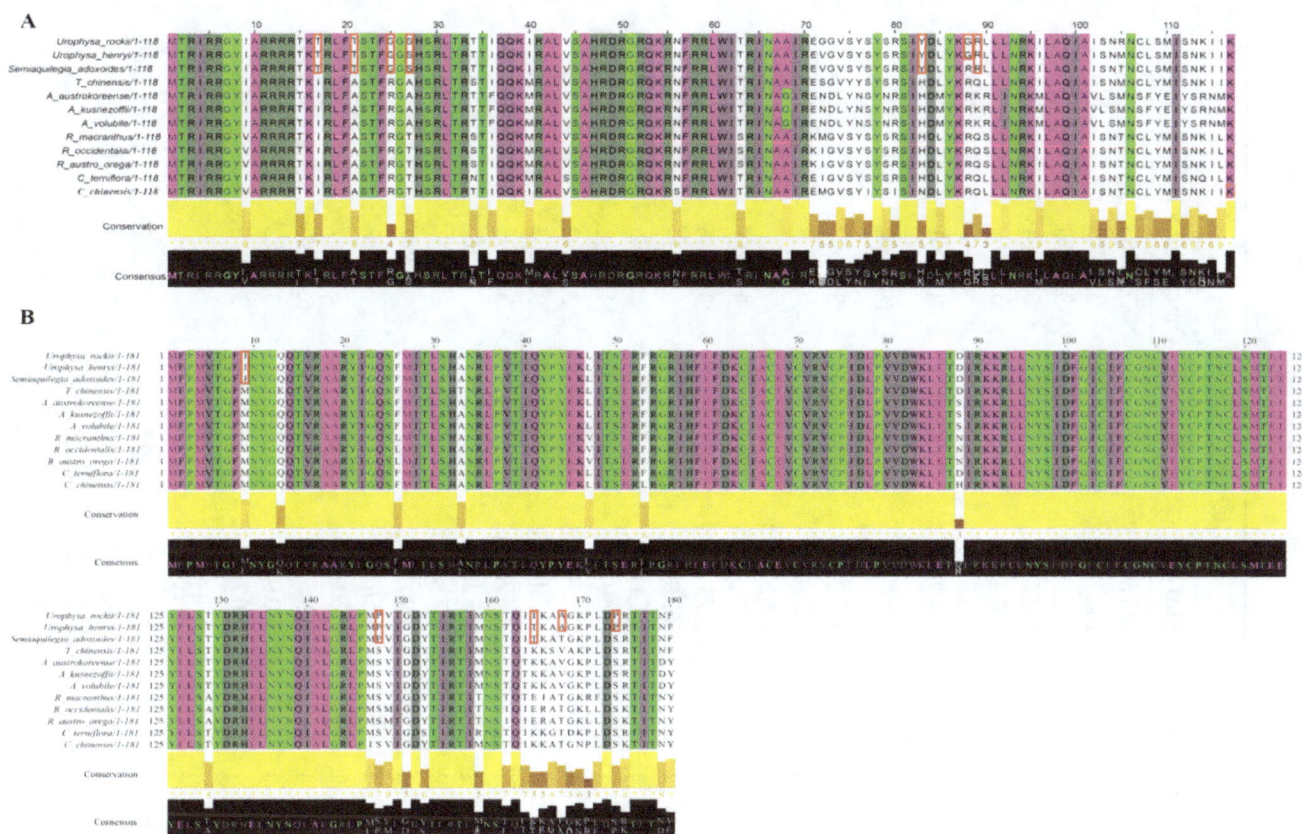

Figure 7. Two of the amino acids sequences that showed positive selection in the branch-site model test. (**A**) Amino acids sequences of the *rpl20* gene; (**B**) amino acids sequences of the *ndhI* gene. The red blocks represent the different amino acids.

Table 4. The potential positive selection test based on the branch-site model.

Gene Name	Null Hypothesis			Alternative Hypothesis			Significance Test		
	lnL	df	Omega ($\omega = 1$)	lnL	df	Omega ($\omega > 1$)	BEB	NEB	p-Value
psbI	−188.6475	26	1	−188.6475	27	3.40383	NA	NA	1
psbL	−164.11693	26	1	−164.1169	27	3.40719	NA	NA	1
rps14	−621.64162	26	1	−621.6416	27	3.40833	NA	NA	1
psaI	−214.67663	26	1	−214.6766	27	3.38764	NA	NA	1
atpH	−434.45059	26	1	−434.4506	27	3.35869	NA	NA	1
psaJ	−318.52192	26	1	−318.5219	27	3.4089	NA	NA	1
atpE	−868.20243	26	1	−868.2024	27	3.40891	NA	NA	1
atpA	−3297.629	26	1	−3297.41	27	69.43581	220, E, 0.794	NA	5.04×10^{-1}
petN	−126.25816	26	1	−126.2582	27	3.40693	NA	NA	1
rps11	−920.92455	26	1	−920.9246	27	1	NA	NA	1
psbT	−216.52331	26	1	−216.5233	27	1	NA	NA	1
ndhG	−1238.1161	26	1	−1238.116	27	3.33667	NA	NA	9.99×10^{-1}
ycf4	−1275.4093	26	1	−1275.409	27	3.40886	NA	NA	1
rps18	−567.98294	26	1	−567.9829	27	3.39414	NA	NA	1
petB	−1274.0507	26	1	−1274.051	27	3.403	NA	NA	1
rpl20	−1000.285	26	1	−999.941	27	112.30316	88, R, 0.683	NA	4.07×10^{-1}
psbN	−223.7602	26	1	−223.7602	27	3.40292	NA	NA	1
psbF	−198.46733	26	1	−198.4673	27	3.38407	NA	NA	1
petG	−206.74878	26	1	−206.7488	27	3.42095	NA	NA	1
psbK	−375.13705	26	1	−375.1371	27	3.4063	NA	NA	1
rpl36	−267.8099	26	1	−267.8099	27	1	NA	NA	1
rps2	−1620.734	26	1	−1620.734	27	3.40891	NA	NA	1
psbM	−179.71897	26	1	−179.719	27	3.4064	NA	NA	1
rpoB	−6830.0894	26	1	−6830.089	27	3.40847	NA	NA	9.99×10^{-1}
psaA	−4245.754	26	1	−4245.49	27	63.47379	28, R, 0.778	NA	4.66×10^{-1}
psbH	−540.92362	26	1	−540.9236	27	3.40123	NA	NA	1
ndhE	−616.75534	26	1	−616.7553	27	3.40218	NA	NA	1
atpB	−3133.747	26	1	−3133.75	27	1	115, N, 0.828	NA	1
ndhI	−1307.986	26	1	−1307.68	27	575.22179	174, S, 0.696	NA	4.35×10^{-1}
cemA	−1787.561	26	1	−1787.561	27	3.40891	NA	NA	1
ndhJ	−1001.4075	26	1	−1001.407	27	1	NA	NA	1
psbJ	−209.10513	26	1	−209.1051	27	3.38566	NA	NA	1

Table 4. *Cont.*

Gene Name	Null Hypothesis			Alternative Hypothesis			Significance Test		
	InL	df	Omega ($\omega = 1$)	InL	df	Omega ($\omega > 1$)	BEB	NEB	*p*-Value
petA	−1331.3789	26	1	−1331.379	27	3.4089	NA	NA	1
psbC	−2760.6743	26	1	−2760.674	27	1	NA	NA	1
ndhH	−2643.2896	26	1	−2643.29	27	1	NA	NA	9.98×10^{-1}
rbcL	−2937.477	26	1	−2937.41	27	5.22178	440, E, 0.736	NA	7.20×10^{-1}
clpP	−1301.1173	26	1	−1301.117	27	3.40876	NA	NA	1
ndhC	−731.03212	26	1	−731.0321	27	3.33544	NA	NA	1
ycf3	−935.76375	26	1	−935.7638	27	3.40891	NA	NA	1
psbD	−1922.7755	26	1	−1922.775	27	3.38592	NA	NA	1
psbA	−1960.3785	26	1	−1960.379	27	3.39639	NA	NA	1
petL	−172.24809	26	1	−172.2481	27	3.40087	NA	NA	1
rpl33	−413.59385	26	1	−413.5939	27	3.4089	NA	NA	1
psbE	−435.90511	26	1	−435.9051	27	3.40785	NA	NA	1
psaC	−498.98549	26	1	−498.9855	27	3.408	NA	NA	1
atpI	−1445.5558	26	1	−1445.556	27	3.39588	NA	NA	1
psaB	−4069.2947	26	1	−4069.295	27	3.41513	NA	NA	1

Bold types are positively selected sites. BEB: Bayesian Empirical Bayes; NEB: Naïve Empirical Bayes; Amino acid: (E: Glu; R: Arg; N: Asn; S: Ser).

However, five of the abovementioned six genes are involved in photosynthesis (*atpA, psaA, atpB, ndhI,* and *rbcL*) (Table 3). The gene *rpl20* is involved in translation, which is an important part of protein synthesis [57]. The genes *atpA* and *atpB* participate in ATP synthesis, which is the main source of energy for the functioning of living cells and all multicellular organisms [58]. Additionally, *rbcL* is the gene for the Rubisco large subunit protein, which is an important component of photosynthetic electron transport [59,60]. Most previous research has revealed that positive selection of the *rbcL* gene in land plants may be a common phenomenon [61]. All these genes might play important roles when founder effects occur in populations; both changes in selection pressures and genetic drift result in the rapid shift of these genes to a new, coadapted combination. Therefore, all these genes under positive selection give an indication of why *U. rockii* and *U. henryi* could adapt to the harsh environment of karst (characterized by low soil water content, periodic water deficiency, and poor nutrient availability). Moreover, the results of the gene effectiveness test (*rbcL* and *rpl20*) (Figure S3) suggested that these genes can distinguish the species of *Urophysa* and its relatives and can be used for future phylogenetic analyses. The six genes will not only provide insights into chloroplast genome evolution of species of *Urophysa*, but also offer valuable genetic markers for population phylogenomic studies of *Urophysa* and its close lineages.

3. Materials and Methods

3.1. Plant Materials and DNA Extraction

Fresh leaves of *U. rockii, U. henryi,* and *S. aquilegia* were collected from Jiangyou (Sichuan, China; coordinates: 31°59′ N, 104°51′ E), Yichang (Hubei, China; coordinates: 30°42′ N, 111°17′ E), and Nanchuan (Chongqing, China; coordinates: 30°04′ N, 90°33′ E), respectively. The fresh leaves from each site were immediately dried with silica gel for further DNA extraction. The total genomic DNA was extracted from leaf tissues with a modified Cetyl Trimethyl Ammonium (CTAB) method [62].

3.2. Chloroplast Genome Sequencing and Assembling

All cp genomes were sequenced using an Illumina Hiseq 2500 platform by Biomarker Technologies, Inc. (Beijing, China) In order to eliminate the interference from mitochondrial or nuclear DNAs, all the cp genome reads were extracted by mapping all raw reads to the reference cp genome of *Trollius chinensis* (KX752098) with Burrows Wheeler Alignment (BWA) [63]. High-quality reads were obtained using the CLC Genomics Workbench v7.5 (CLC Bio, Aarhus, Denmark) with the default parameters set. A few gaps in the assembled cp genomes were corrected by Sanger sequencing. The primers were designed using Lasergene 7.1 (DNASTAR, Madison, WI, USA). Primer synthesis and the sequencing of the polymerase chain reaction products were conducted by Sangon Biotech (Shanghai, China). The primers and amplifications are shown in Supplementary Table S5.

3.3. Genome Annotation and Analysis

The complete cp genomes were annotated using the online program DOGMA [64]. The annotation results were checked manually, and the codon positions were adjusted by comparing to a previously homologous gene from various chloroplast genomes present in the database using Geneious R11 (Biomatters, Ltd., Auckland, New Zealand). Furthermore, the OGDRAW1 program [65] was used to draw the circular plastid genome maps. GC content and codon usage were analyzed by the MEGA 6 software [66]. The complete cp genomes of *U. rockii, U. henryi,* and *S. adoxoides* are deposited in the GenBank under the accession numbers MH006686, MH142266, and MH142265, respectively.

3.4. Repeat Sequence Characterization and SSRs

Perl script MISA [67] was used to search for microsatellites (mono-, di-, tri-, tetra-, penta-, and hexa-nucleotides) loci in the cp genomes. The minimum numbers (thresholds) of the SSRs were 10, 5, 4, 3, 3, and 3 for mono-, di-, tri-, tetra-, penta-, and hexa-nucleotides, respectively. All the repeats were manually verified, and redundant results were removed. REPuter was employed to identify repeat sequences, including palindromic, forward, reverse, and complement, within the cp genome [68]. The following conditions for repeat identification were used: (1) Hamming distance of 3; (2) 90% or greater sequence identity; (3) a minimum repeat size of 30 bp.

3.5. Phylogenetic Analysis

Phylogenetic analysis was conducted using the single-copy genes of the three taxa, together with nine species downloaded from the NCBI GenBank (Tables S6 and S7). The sequences were aligned using MAFFT v5 [69] in GENEIOUS R11 (Biomatters, Ltd.) with the default parameters set and were manually adjusted in MEGA 6.0 [66]. Maximum parsimony (MP) analyses were conducted using PAUP [70]. All characters were equally weighted, gaps were treated as missing, and character states were treated as unordered. Heuristic search was performed with MULPARS option, tree bisection-reconnection (TBR) branch swapping, and random stepwise addition with 1000 replications. The maximum likelihood (ML) analyses were performed using RAxML 8.0 [71]. For ML analyses, the best-fit model, general time reversible (GTR) + G was used with 1000 bootstrap replicates. Bayesian inference (BI) was performed with Mrbayes v3.2 [72]. The Markov chain Monte Carlo (MCMC) analysis was run for 1×10^8 generations. The trees were sampled at every 1000 generations with the first 20% discarded as burn-in. The remaining trees were used to build a 50% majority-rule consensus tree. The stationarity was considered to be reached when the average standard deviation of split frequencies remained below 0.001. Additionally, in order to test the utility of different cp regions, phylogenetic analyses were performed for the complete chloroplast genome sequences and the CDS sequences, respectively.

3.6. Chloroplast Genome Nucleotide Diversity and Positive Selected Analysis

The cp genome sequences were aligned using MAFFT v5 [69] and adjusted manually. Furthermore, a sliding window analysis was conducted for nucleotide diversity in LSC, SSC, and IR regions of the cp genomes using the DnaSP version 5.1 [73]. In addition, to identify the genes under positive selection in *U. rockii* and *U. henryi*, endemic to special karst environment, an optimized branch-site model [74] combined with Bayesian Empirical Bayes (BEB) methods [55] were used by comparison with their relatives. We firstly extracted all CDS sequences from *U. rockii*, *U. henryi*, *S. adoxoides*, and nine closely related species downloaded from GenBank (Table S6). The single-copy CDS sequences between these twelve species were obtained (see the Table S4). Each single-copy CDS sequence of these twelve species was aligned according to their amino acid sequence alignment generated by MUSCLE [75], and the "number of gaps" in the alignments was further checked. Then, the alignments of the corresponding DNA codon sequences were further trimmed by TRIMAL [76], and the bona fide alignments were used to support the subsequent positive selection analysis. The optimized branch-site model in the CODEML program implemented in the PAML 4 package [77] was used to assess potential positive selection affecting individual codons along a specifically designated lineage, which was set as *U. rockii* and *U. henryi*. Selective pressure is measured by the ratio (ω) of the nonsynonymous substitution rate (dN) to the synonymous substitutions rate (dS). A ratio $\omega > 1$ indicates positive selection, $\omega = 1$ implies neutral selection, and $\omega < 1$ suggests negative selection [78]. Log-likelihood values were calculated in an alternative branch-site model (Model = 2; NSsites = 2; and Fix = 0) that allowed ω to vary among different codons along particular lineages and a neutral branch-site model (Model = 2; NSsites = 2; Fix = 1; Fix ω = 1) that confined the codon sites under neutral selection ($\omega = 1$) on the basis of the likelihood ratio tests (LRT). The right-tailed chi-square test was performed to calculate the p values based on the difference in log-likelihood values between the alternative model and the neutral model with one degree of freedom to assess the model fit. Then, the p values were further adjusted according to multiple statistical tests [79]. A gene with an adjusted p value smaller than 0.05 and with positively selected sites was considered a positively selected gene (PSG). Moreover, in order to identify specific amino acid sites that are potentially under positive selection, a BEB method was implemented to calculate the posterior probabilities for sites classes. Codon sites with a high posterior probability were regarded as positively selected sites [54]. Jalview [80] was used to view the amino acid sequences of positively selected genes. In the end, in order to test the effectiveness of genes under positive selection, we randomly chose two genes to conduct the phylogenetic analyses.

Author Contributions: D.-F.X., Y.Y., S.-D.Z., and X.-J.H. conceived and designed the experiment; D.-F.X., J.L., and S.-D.Z. collected the materials; D.-F.X., Y.-Q.D., Y.Y., and H.-Y.L. participated in data analysis and manuscript drafting; D.-F.X., Y.-Q.D., X.-J.H., and S.-D.Z. revised the manuscript; all authors read and approved the final manuscript.

Acknowledgments: We acknowledge Fang-Yu Jin, Hao Li, Fu-Min Xie, and Xin Yang for their help in materials collection.

References

1.   Fu, D.Z.; Orbelia, R.R. *Flora of China*; Science Press: Beijing, China, 2001; Volume 6, pp. 277–278.

2.   Xie, D.F.; Li, M.J.; Tan, J.B.; Price, M.; Xiao, Q.Y.; Zhou, S.D.; He, X.J. Phylogeography and genetic effects of habitat fragmentation on endemic *Urophysa* (Ranunculaceae) in Yungui Plateau and adjacent regions. *PLoS ONE* **2017**, *12*, e0186378. [CrossRef] [PubMed]

3.   Du, B.G.; Zhu, D.Y.; Yang, Y.J.; Shen, J.; Yang, F.L.; Su, Z.Y. Living situation and protection strategies of endangered *Urophysa rockii*. *Jiangsu J. Agri. Sci.* **2010**, *1*, 324–325.

4.   Wang, J.X.; He, X.J.; Xu, W.; Meng, W.K.; Su, Z.Y. Preliminary study on *Urophysa rockii*. II. Biological characteristics, ecological characteristics and community analysis. *J. Sichuan For. Sci. Technol.* **2011**, *32*, 28–39.

5.   Zhang, Y.X.; Hu, H.Y.; He, X.J. Genetic diversity of *Urophysa rockii* Ulbrich, an endangered and rare species, detected by ISSR. *Acta Bot. Boreal.-Occident. Sin.* **2013**, *33*, 1098–1105.

6.   Zhang, Y.X.; Hu, H.Y.; Yang, L.J.; Wang, C.B.; He, X.J. Seed dispersal and germination of an endangered and rare species *Urophysa rockii* (Ranunculaceae). *Acta Bot. Boreal.-Occident. Sin.* **2013**, *35*, 303–309.

7.   Park, M.; Park, H.; Lee, H.; Lee, B.H.; Lee, J. The complete plastome sequence of an antarctic bryophyte *Sanionia uncinata* (hedw.) loeske. *Int. J. Mol. Sci.* **2018**, *19*, 709. [CrossRef] [PubMed]

8.   Dong, W.P.; Liu, H.; Xu, C.; Zuo, Y.J.; Chen, Z.J.; Zhou, S.L. A chloroplast genomic strategy for designing taxon specific DNA mini-barcodes: A case study on ginsengs. *BMC Genet.* **2014**, *15*, 138. [CrossRef] [PubMed]

9.   Curci, P.L.; de Paola, D.; Danzi, D.; Vendramin, G.G.; Sonnante, G. Complete chloroplast genome of the multifunctional crop Globe artichoke and comparison with other Asteraceae. *PLoS ONE* **2015**, *10*, e0120589. [CrossRef] [PubMed]

10.   Downie, S.R.; Jansen, R.K. A comparative analysis of whole plastid genomes from the Apiales: Expansion and contraction of the inverted repeat, mitochondrial to plastid transfer of DNA, and identification of highly divergent noncoding regions. *Syst. Bot.* **2015**, *40*, 336–351. [CrossRef]

11.   Nadachowska-Brzyska, K.; Li, C.; Smeds, L.; Zhang, G.J.; Ellegren, H. Temporal dynamics of avian populations during pleistocene revealed by whole-genome sequences. *Curr. Biol.* **2015**, *25*, 1375–1380. [CrossRef] [PubMed]

12.   Suo, Z.L.; Li, W.Y.; Jin, X.B.; Zhang, H.J. A new nuclear DNA marker revealing both microsatellite variations and single nucleotide polymorphic loci: A case study on classification of cultivars in *Lagerstroemia indica* L. *J. Microb. Biochem. Technol.* **2016**, *8*, 266–271. [CrossRef]

13.   Saina, J.K.; Li, Z.Z.; Gichira, A.W.; Liao, Y.Y. The complete chloroplast genome sequence of tree of heaven (*Ailanthus altissima* (mill.) (Sapindales: Simaroubaceae), an important pantropical tree. *Int. J. Mol. Sci.* **2018**, *19*, 929. [CrossRef] [PubMed]

14.   Yurina, N.P.; Odintsova, M.S. Comparative structural organization of plant chloroplast and mitochondrial genomes. *Genetika* **1998**, *34*, 5–22.

15.   Jansen, R.K.; Raubeson, L.A.; Boore, J.L.; DePamphilis, C.W.; Chumley, T.W.; Haberle, R.C.; Wyman, S.K.; Alverson, A.; Peery, R.; Herman, S.J.; et al. Methods for obtaining and analyzing whole chloroplast genome sequences. *Method Enzymol.* **2005**, *395*, 348–384.

16.   Jansen, R.K.; Ruhlman, T.A. Plastid Genomes of Seed Plants. In *Genomics of Chloroplasts and Mitochondria*; Bock, R., Knoop, V., Eds.; Springer: Dordrecht, The Netherlands, 2012; pp. 103–126.

17.   Choi, K.S.; Chung, M.G.; Park, S. The complete chloroplast genome sequences of three *Veroniceae* species (Plantaginaceae): Comparative analysis and highly divergent regions. *Front. Plant Sci.* **2016**, *7*, 355. [CrossRef] [PubMed]

18. Dong, W.L.; Wang, R.N.; Zhang, N.Y.; Fan, W.B.; Fang, M.F.; Li, Z.H. Molecular evolution of chloroplast genomes of orchid species: Insights into phylogenetic relationship and adaptive evolution. *Int. J. Mol. Sci.* **2018**, *19*, 716. [CrossRef] [PubMed]

19. Dong, W.; Liu, J.; Yu, J.; Wang, L.; Zhou, S. Highly variable chloroplast markers for evaluating plant phylogeny at low taxonomic levels and for DNA barcoding. *PLoS ONE* **2012**, *7*, e35071. [CrossRef] [PubMed]

20. Krak, K.; Vít, P.; Belyayev, A.; Douda, J.; Hreusová, L.; Mandák, B. Allopolyploid origin of *Chenopodium album* s. str. (Chenopodiaceae): A molecular and cytogenetic insight. *PLoS ONE* **2016**, *11*, e0161063. [CrossRef] [PubMed]

21. Smith, D.R. Mutation rates in plastid genomes: They are lower than you might think. *Genome Biol. Evol.* **2015**, *7*, 1227–1234. [CrossRef] [PubMed]

22. Jansen, R.K.; Cai, Z.; Raubeson, L.A.; Daniell, H.; Depamphilis, C.W.; Leebensmack, J.; Müller, K.F.; Guisinger-Bellian, M.; Haberle, R.C.; Chumley, T.W.; et al. Analysis of 81 genes from 64 plastid genomes resolves relationships in angiosperms and identifies genome-scale evolutionary patterns. *Proc. Natl. Acad. Sci. USA* **2007**, *104*, 19369–19374. [CrossRef] [PubMed]

23. Qian, J.; Song, J.; Gao, H.; Zhu, Y.; Xu, J.; Pang, X. The complete chloroplast genome sequence of the medicinal plant *Salvia miltiorrhiza*. *PLoS ONE* **2013**, *8*, e57607. [CrossRef] [PubMed]

24. Asaf, S.; Waqas, M.; Khan, A.L.; Khan, M.A.; Kang, S.M.; Imran, Q.M.; Shahzad, R.; Bilal, S.; Yun, B.W.; Lee, I.J.; et al. The complete chloroplast genome of wild rice (*Oryza minuta*) and its comparison to related species. *Front. Plant Sci.* **2017**, *8*, 304. [CrossRef] [PubMed]

25. Gu, C.; Tembrock, L.R.; Zheng, S.; Wu, Z. The complete chloroplast genome of *Catha edulis*: A comparative analysis of genome features with related species. *Int. J. Mol. Sci.* **2018**, *19*, 525. [CrossRef] [PubMed]

26. Huang, J.; Chen, R.; Li, X. Comparative analysis of the complete chloroplast genome of four known *Ziziphus* species. *Genes* **2017**, *8*, 340. [CrossRef] [PubMed]

27. Yi, X.; Gao, L.; Wang, B.; Su, Y.J.; Wang, T. The complete chloroplast genome sequence of *Cephalotaxus oliveri* (Cephalotaxaceae): Evolutionary comparison of *Cephalotaxus* chloroplast DNAs and insights into the loss of inverted repeat copies in gymnosperms. *Genome Biol. Evol.* **2013**, *5*, 688–698. [CrossRef] [PubMed]

28. Cavalier-Smith, T. Chloroplast evolution: Secondary symbiogenesis and multiple losses. *Curr. Biol.* **2002**, *12*, 62–64. [CrossRef]

29. Asano, T.; Tsudzuki, T.; Takahashi, S.; Shimada, H.; Kadowaki, K. Complete nucleotide sequence of the sugarcane (*Saccharum officinarum*) chloroplast genome: A comparative analysis of four monocot chloroplast genomes. *DNA Res.* **2004**, *11*, 93–99. [CrossRef] [PubMed]

30. Timme, R.E.; Kuehl, J.V.; Boore, J.L.; Jansen, R.K. A comparative analysis of the *Lactuca* and *Helianthus* (Asteraceae) plastid genomes: Identification of divergent regions and categorization of shared repeats. *Am. J. Bot.* **2007**, *94*, 302–312. [CrossRef] [PubMed]

31. Nie, X.J.; Lv, S.Z.; Zhang, Y.X.; Du, X.H.; Wang, L.; Biradar, S.S.; Tan, X.F.; Wan, F.H.; Weining, S. Complete chloroplast genome sequence of a major invasive species, crofton weed (*Ageratina adenophora*). *PLoS ONE* **2012**, *7*, e36869. [CrossRef] [PubMed]

32. Dong, W.P.; Xu, C.; Li, D.L.; Jin, X.B.; Lu, Q.; Suo, Z.L. Comparative analysis of the complete chloroplast genome sequences in psammophytic *Haloxylon* species (Amaranthaceae). *Peer J.* **2016**, *4*, e2699. [CrossRef] [PubMed]

33. Kaur, S.; Panesar, P.S.; Bera, M.B.; Kaur, V. Simple sequence repeat markers in genetic divergence and marker-assisted selection of rice cultivars: A review. *Crit. Rev. Food Sci. Nutr.* **2015**, *55*, 41–49. [CrossRef] [PubMed]

34. Yang, Y.; Zhou, T.; Duan, D.; Yang, J.; Feng, L.; Zhao, G. Comparative analysis of the complete chloroplast genomes of five *Quercus* species. *Front. Plant Sci.* **2016**, *7*, 959. [CrossRef] [PubMed]

35. Powell, W.; Morgante, M.; McDevitt, R.; Vendramin, G.G.; Rafalski, J.A. Polymorphic simple sequence repeat regions in chloroplast genomes-applications to the population genetics of pines. *Proc. Natl. Acad. Sci. USA* **1995**, *92*, 7759–7763. [CrossRef] [PubMed]

36. Provan, J.; Corbett, G.; McNicol, J.W.; Powell, W. Chloroplast DNA variability in wild and cultivated rice (*Oryza* spp.) revealed by polymorphic chloroplast simple sequence repeats. *Genome* **1997**, *40*, 104–110. [CrossRef] [PubMed]

37. Pauwels, M.; Vekemans, X.; Gode, C.; Frerot, H.; Castric, V.; Saumitou-Laprade, P. Nuclear and chloroplast

DNA phylogeography reveals vicariance among European populations of the model species for the study of metal tolerance, *Arabidopsis halleri* (Brassicaceae). *New Phytol.* **2012**, *193*, 916–928. [CrossRef] [PubMed]

38. Rose, O.; Falush, D. A threshold size for microsatellite expansion. *Mol. Biol. Evol.* **1998**, *15*, 613–615. [CrossRef] [PubMed]

39. Raubeson, L.A.; Peery, R.; Chumley, T.W.; Dziubek, C.; Fourcade, H.M.; Boore, J.L.; Jansen, R.K. Comparative chloroplast genomics: Analyses including new sequences from the angiosperms *Nuphar advena* and *Ranunculus macranthus*. *BMC Genom.* **2007**, *8*, 174. [CrossRef] [PubMed]

40. Huotari, T.; Korpelainen, H. Complete chloroplast genome sequence of *Elodea Canadensis* and comparative analyses with other monocot plastid genomes. *Gene* **2012**, *508*, 96–105. [CrossRef] [PubMed]

41. Wang, W.B.; Yu, H.; Wang, J.H.; Lei, W.J.; Gao, J.H.; Qiu, X.P.; Wang, J.S. The complete chloroplast genome sequences of the medicinal plant *Forsythia suspensa* (Oleaceae). *Int. J. Mol. Sci.* **2017**, *18*, 2288. [CrossRef] [PubMed]

42. Asaf, S.; Khan, A.L.; Khan, A.R.; Waqas, M.; Kang, S.M.; Khan, M.A.; Lee, S.M.; Lee, I.J. Complete chloroplast genome of *Nicotiana otophora* and its comparison with related species. *Front. Plant Sci.* **2016**, *7*, 447. [CrossRef] [PubMed]

43. Kuang, D.Y.; Wu, H.; Wang, Y.L.; Gao, L.M.; Zhang, S.Z.; Lu, L. Complete chloroplast genome sequence of *Magnolia kwangsiensis* (Magnoliaceae): Implication for DNA barcoding and population genetics. *Genome* **2011**, *54*, 663–673. [CrossRef] [PubMed]

44. Chen, J.; Hao, Z.; Xu, H.; Yang, L.; Liu, G.; Sheng, Y. The complete chloroplast genome sequence of the relict woody plant *Metasequoia glyptostroboides* Hu et Cheng. *Front. Plant Sci.* **2015**, *6*, 447. [CrossRef] [PubMed]

45. Kim, K.J.; Lee, H.L. Complete chloroplast genome sequences from Korean ginseng (*Panax schinseng* Nees) and comparative analysis of sequence evolution among 17 vascular plants. *DNA Res.* **2004**, *11*, 247–261. [CrossRef] [PubMed]

46. Hu, Y.; Woeste, K.E.; Zhao, P. Completion of the chloroplast genomes of five Chinese *Juglans* and their contribution to chloroplast phylogeny. *Front. Plant Sci.* **2017**, *7*, 1955. [CrossRef] [PubMed]

47. Wang, R.J.; Cheng, C.L.; Chang, C.C.; Wu, C.L.; Su, T.M.; Chaw, S.M. Dynamics and evolution of the inverted repeat-large single copy junctions in the chloroplast genomes of monocots. *BMC Evol. Biol.* **2008**, *8*, 36. [CrossRef] [PubMed]

48. Yang, M.; Zhang, X.; Liu, G.; Yin, Y.; Chen, K.; Yun, Q. The complete chloroplast genome sequence of date palm (*Phoenix dactylifera* L.). *PLoS ONE* **2010**, *5*, e12762. [CrossRef] [PubMed]

49. Li, Z.Z.; Saina, J.K.; Gichira, A.W.; Kyalo, C.M.; Wang, Q.F.; Chen, J.M. Comparative genomics of the balsaminaceae sister genera *Hydrocera triflora* and *Impatiens pinfanensis*. *Int. J. Mol. Sci.* **2018**, *19*, 319. [CrossRef] [PubMed]

50. Li, C.Y. *Classification and Systematics of the Aquilegiinae Tamura*; The Chinese Academy of Science: Beijing, China, 2006.

51. Bastida, J.M.; Alcántara, J.M.; Rey, P.J.; Vargas, P.; Herrera, C.M. Extended phylogeny of *Aquilegia*: The biogeographical and ecological patterns of two simultaneous but contrasting radiations. *Plant Syst. Evol.* **2010**, *284*, 171–185. [CrossRef]

52. Fior, S.; Li, M.; Oxelman, B.; Viola, R.; Hodges, S.A.; Ometto, L.; Varotto, C. Spatiotemporal reconstruction of the *Aquilegia* rapid radiation through next-generation sequencing of rapidly evolving cpDNA regions. *New Phytol.* **2013**, *198*, 579–592. [CrossRef] [PubMed]

53. Wei, W.; Lu, A.M.; Yi, R.; Endress, M.E.; Chen, Z.D. Phytogeny and classification of Ranunculales: Evidence from four molecular loci and morphological data. *Perspect. Plant Ecol. Evol. Syst.* **2009**, *11*, 81–110.

54. Lan, Y.; Sun, J.; Tian, R.M.; Bartlett, D.H.; Li, R.S.; Wong, Y.H.; Zhang, W.P.; Qiu, J.W.; Xu, T.; He, L.S.; et al. Molecular adaptation in the world's deepest-living animal: Insights from transcriptome sequencing of the hadal amphipod *Hirondellea gigas*. *Mol. Ecol.* **2017**, *26*, 3732–3743. [CrossRef] [PubMed]

55. Yang, Z.; Wong, W.S.; Nielsen, R. Bayes empirical Bayes inference of amino acid sites under positive selection. *Mol. Biol. Evol.* **2005**, *22*, 1107–1118. [CrossRef] [PubMed]

56. Ai, B.; Gao, Y.; Zhang, X.; Tao, J.; Kang, M.; Huang, H. Comparative transcriptome resources of eleven *Primulina* species, a group of 'stone plants' from a biodiversity hot spot. *Mol. Ecol. Resour.* **2015**, *15*, 619–632. [CrossRef] [PubMed]

57. Muto, A.; Ushida, C. Transcription and translation. *Methods Cell Biol.* **1995**, *48*, 483.

58. Romanovsky, Y.M.; Tikhonov, A.N. Molecular energy transducers of the living cell. Proton ATP synthase: A rotating molecular motor. *Physics-Uspekhi* **2010**, *53*, 931–956. [CrossRef]

59. Allahverdiyeva, Y.; Mamedov, F.; Mäenpää, P.; Vass, I.; Aro, E.M. Modulation of photosynthetic electron transport in the absence of terminal electron acceptors: Characterization of the rbcL deletion mutant of tobacco. *Biochim. Biophys. Acta Bioenerg.* **2005**, *1709*, 69–83. [CrossRef] [PubMed]

60. Piot, A.; Hackel, J.; Christin, P.A.; Besnard, G. One-third of the plastid genes evolved under positive selection in PACMAD grasses. *Planta* **2018**, *247*, 255–266. [CrossRef] [PubMed]

61. Kapralov, M.V.; Filatov, D.A. Widespread positive selection in the photosynthetic Rubisco enzyme. *BMC Evol. Biol.* **2007**, *7*, 73–82. [CrossRef] [PubMed]

62. Doyle, J.J.; Doyle, J.L. A rapid DNA isolation procedure for small quantities of fresh leaf tissue. *Phytochem Bull.* **1987**, *19*, 11–15.

63. Li, H.; Durbin, R. Fast and accurate short read alignment with Burrows-Wheeler Transform. *Bioinformatics* **2009**, *25*, 1754–1760. [CrossRef] [PubMed]

64. Wyman, S.K.; Jansen, R.K.; Boore, J.L. Automatic annotation of organellar genomes with DOGMA. *Bioinformatics* **2004**, *20*, 3252–3255. [CrossRef] [PubMed]

65. Lohse, M.; Drechsel, O.; Kahlau, S.; Bock, R. Organellar genome draw—A suite of tools for generating physical maps of plastid and mitochondrial genomes and visualizing expression data sets. *Nucleic Acids Res.* **2013**, *41*, 575. [CrossRef] [PubMed]

66. Kumar, S.; Nei, M.; Dudley, J.; Tamura, K. MEGA: A biologist centric software for evolutionary analysis of DNA and protein sequences. *Brief. Bioinform.* **2008**, *9*, 299–306. [CrossRef] [PubMed]

67. Thiel, T.; Michalek, W.; Varshney, R.; Graner, A. Exploiting EST databases for the development and characterization of gene derived SSR-markers in barley (*Hordeum vulgare* L.). *Theor. Appl. Genet.* **2003**, *106*, 411–422. [CrossRef] [PubMed]

68. Kurtz, S.; Choudhuri, J.V.; Ohlebusch, E.; Schleiermacher, C.; Stoye, J.; Giegerich, R. REPuter: The manifold applications of repeat analysis on a genomic scale. *Nucleic Acids Res.* **2001**, *29*, 4633–4642. [CrossRef] [PubMed]

69. Katoh, K.; Standley, D.M. MAFFT multiple sequence alignment software version 7: Improvements in performance and usability. *Mol. Biol. Evol.* **2013**, *30*, 772–780. [CrossRef] [PubMed]

70. Swofford, D.L. *PAUP*. Phylogenetic Analysis Using Parsimony (*and Other Methods)*; Version 4b10; Sinauer: Sunderland, MA, USA, 2003.

71. Stamatakis, A. RAxML-VI-HPC: Maximum likelihood-based phylogenetic analyses with thousands of taxa and mixed models. *Bioinformatics* **2006**, *22*, 2688–2690. [CrossRef] [PubMed]

72. Ronquist, F.; Teslenko, M.; van der Mark, P.; Ayres, D.L.; Darling, A.; Hohna, S.; Larget, B.; Liu, L.; Suchard, M.A.; Huelsenbeck, J. MrBayes 3.2: Efficient Bayesian phylogenetic inference and model choice across a large model space. *Syst. Biol.* **2012**, *61*, 539–542. [CrossRef] [PubMed]

73. Librado, P.; Rozas, J. DnaSP v5: A software for comprehensive analysis of DNA polymorphism data. *Bioinformatics* **2009**, *25*, 1451–1452. [CrossRef] [PubMed]

74. Yang, Z.; dos Reis, M. Statistical properties of the branch-site test of positive selection. *Mol. Biol. Evol.* **2011**, *28*, 1217–1228. [CrossRef] [PubMed]

75. Edgar, R.C. MUSCLE: Multiple sequence alignment with high accuracy and high throughput. *Nucleic Acids Res.* **2004**, *32*, 1792–1797. [CrossRef] [PubMed]

76. Capella-Gutierrez, S.; Silla-Martínez, J.M.; Gabaldon, T. TrimAl: A tool for automated alignment trimming in large-scale phylogenetic analyses. *Bioinformatics* **2009**, *25*, 1972–1973. [CrossRef] [PubMed]

77. Yang, Z. PAML 4: Phylogenetic analysis by maximum likelihood. *Mol. Biol. Evol.* **2007**, *24*, 1586–1591. [CrossRef] [PubMed]

78. Yang, Z.; Nielsen, R. Codon-substitution models for detecting molecular adaptation at individual sites along specific lineages. *Mol. Biol. Evol.* **2002**, *19*, 908–917. [CrossRef] [PubMed]

79. Benjamini, Y.; Hochberg, Y. Controlling the false discovery rate: A practical and powerful approach to multiple testing. *J. R. Stat. Soc. B* **1995**, *57*, 289–300.

80. Clamp, M.; Cuff, J.; Searle, S.M.; Barton, G.J. The Jalview java alignment editor. *Bioinformatics* **2004**, *20*, 426–427. [CrossRef] [PubMed]

13

Chloroplast Protein Turnover: The Influence of Extraplastidic Processes and Including Autophagy

Masanori Izumi [1,2,3,*] **and Sakuya Nakamura** [2]

[1] Frontier Research Institute for Interdisciplinary Sciences, Tohoku University, Sendai 980-8578, Japan

[2] Department of Environmental Life Sciences, Graduate School of Life Sciences, Tohoku University, Sendai 980-8577, Japan

[3] Precursory Research for Embryonic Science and Technology (PRESTO), Japan Science and Technology Agency, Kawaguchi 332-0012, Japan

* Correspondence: m-izumi@ige.tohoku.ac.jp

Abstract: Most assimilated nutrients in the leaves of land plants are stored in chloroplasts as photosynthetic proteins, where they mediate CO_2 assimilation during growth. During senescence or under suboptimal conditions, chloroplast proteins are degraded, and the amino acids released during this process are used to produce young tissues, seeds, or respiratory energy. Protein degradation machineries contribute to the quality control of chloroplasts by removing damaged proteins caused by excess energy from sunlight. Whereas previous studies revealed that chloroplasts contain several types of intraplastidic proteases that likely derived from an endosymbiosed prokaryotic ancestor of chloroplasts, recent reports have demonstrated that multiple extraplastidic pathways also contribute to chloroplast protein turnover in response to specific cues. One such pathway is autophagy, an evolutionarily conserved process that leads to the vacuolar or lysosomal degradation of cytoplasmic components in eukaryotic cells. Here, we describe and contrast the extraplastidic pathways that degrade chloroplasts. This review shows that diverse pathways participate in chloroplast turnover during sugar starvation, senescence, and oxidative stress. Elucidating the mechanisms that regulate these pathways will help decipher the relationship among the diverse pathways mediating chloroplast protein turnover.

Keywords: autophagy; chlorophagy; chloroplasts; Rubisco-containing bodies; photooxidative damage; plants; senescence; sugar starvation; ubiquitin proteasome system; vacuole

1. Introduction

Chloroplasts are a type of plastid in plants and algae. In land plants, chloroplasts are present in green tissues, such as leaves, that are required for photosynthetic energy production. Within chloroplasts, thylakoid membranes contain pigments and proteins that form the light harvesting complex and electron transport chain, and the stroma contains soluble proteins that mediate the assimilation of carbon dioxide (CO_2) via the Calvin cycle. In mature plant leaves, assimilated nutrients are largely stored in chloroplasts as photosynthetic proteins. For instance, the nitrogen in chloroplasts accounts for around 75% of the total leaf nitrogen in C3 species [1]. The CO_2-fixing enzyme ribulose-1,5-bisphosphate carboxylase/oxygenase (Rubisco) in the stroma is especially abundant, accounting for 10–30% of the total leaf nitrogen, and constituting around half of the total soluble proteins in leaves [2,3]. Chloroplast proteins are degraded, and the amino acids and other molecules released during this process are reutilized in growth. Leaf senescence is a well-established developmental process during which chloroplast proteins are degraded en masse; the released amino acids are remobilized to generate juvenile tissues and produce seeds [4,5].

A portion of the photoassimilate is accumulated in chloroplasts as starch during the day, and is degraded at night to produce sucrose as the major source for respiratory energy production within mitochondria [6]. Since the availability of solar energy fluctuates under the ever-changing environment, plants occasionally need alternatives to sugars for producing the energy required for continuous growth. Stress conditions can also interfere with photosynthetic energy production; for instance, stomatal closure due to drought stress inhibits CO_2 intake and thereby reduces photosynthetic activity in leaves [7,8]. Plants must metabolically produce alternative energy sources to survive under photosynthesis-limited conditions. Amino acids derived from chloroplast protein degradation via catabolic pathways can serve as alternative respiratory substrates [9].

Chloroplast protein degradation is also vital for maintaining chloroplast function, as chloroplast proteins constantly accumulate damage caused by sunlight during photosynthesis. Photoinhibition occurs when the photosynthetic apparatus is damaged by excess energy from strong visible light (with wavelengths of between 400 and 700 nm) [10–13]. Although chloroplasts cannot use ultraviolet-B (UVB; with wavelengths of between 280 and 315 nm) for photosynthesis, various macromolecules, such as proteins, lipids, and nucleotides, directly absorb UVB, which may result in cumulative damage [14]. To maintain photosynthetic activity and avoid the overproduction of reactive oxygen species (ROS) in response to sunlight irradiation, damaged components within chloroplasts must be removed.

Turnover of chloroplastic components is required for efficient nutrient recycling during plant senescence, respiratory energy production under photoassimilate-starved conditions, and quality control in individual chloroplasts under photooxidative damage. Chloroplasts contain various types of intraplastidic proteases, which are thought to have been derived from an endosymbiosed prokaryotic ancestor of chloroplasts [15,16]. Recent studies of chloroplast protein turnover have further demonstrated the contribution of extraplastidic protein degradation systems to nutrient or energy recycling and the removal of damaged proteins. In this review, we describe the extraplastidic pathways that facilitate the degradation of chloroplastic components, and compare the physiological roles of these pathways and the environmental and developmental stimuli that activate them.

2. Autophagic Degradation of Rubisco-Containing Bodies

Autophagy is an evolutionarily conserved process in eukaryotes whereby the cell sequesters a portion of cytoplasm, including organelles, for subsequent transport into lytic organelles [17–19]. During autophagy, a nascent double membrane-bound vesicle called an autophagosome encloses a portion of the cytoplasm. The outer membrane of autophagosomes then fuses with the vacuolar or lysosomal membrane to release the inner-membrane structures, referred to as autophagic bodies, into the vacuolar or lysosomal lumen for digestion. The basic mechanism of autophagosome formation was described in the budding yeast *Saccharomyces cerevisiae* through the identification of autophagy (*ATG*) genes [20].

The *ATG* genes required for the initiation or elongation of autophagosomal membranes are referred as core *ATGs* (*ATG1–10, 12–14, 16, 18*), and these genes are also required for all types of autophagy [17]. Many core ATGs function in two conjugation cascades that are required for ATG8 lipidation and autophagosomal membrane elongation. ATG7 and ATG10 conjugate ATG12 to ATG5, and the resulting ATG12-ATG5 conjugate then interacts with ATG16 to form the ATG12-ATG5-ATG16 complex. ATG8 is processed by the protease ATG4. The resulting mature ATG8 is activated by ATG7, transferred to ATG3, and is eventually conjugated with phosphatidylethanolamine with the aid of the ATG12-ATG5-ATG16 complex. Orthologues of the yeast core *ATGs* are conserved in plant species [21–23], and studies of autophagy-deficient *atg* mutants of *Arabidopsis thaliana* show that they have similar functions [24–32].

During leaf senescence, the amount of chloroplast stromal proteins, including Rubisco, decreases prior to the reduction in the number of chloroplasts [33–35]. Therefore, stromal proteins appear to be degraded either inside or outside the chloroplast without the breakdown of the entire chloroplast. An immuno-electron microscopy (EM) analysis of Rubisco degradation in senescing wheat (*Triticum aestivum*) leaves revealed the presence of cytosol-localized small vesicles that contained Rubisco, but not thylakoid proteins such as light-harvesting chlorophyll a/b protein of Photosystem II (LHC II), α, β-subunits of coupling factor 1 in ATPase, or cytochrome f [36]. These vesicles, which are around 1 µm in diameter and are frequently surrounded by autophagosome-like double membranes, were originally referred to as Rubisco-containing bodies (RCBs). The development of live-cell imaging techniques using fluorescent protein markers allowed for the visualization of RCBs in vivo in Arabidopsis and rice (*Oryza sativa*) leaves expressing stroma-targeted green fluorescent protein (GFP) or GFP-labeled Rubisco [37,38]. This technique further demonstrated that RCBs are not produced in the mutant *atg5* or *atg7* lines, and that RCBs labeled with stroma-targeted red fluorescent proteins (RFPs) are co-localized with an autophagosomal marker, GFP-ATG8. These observations revealed that RCBs are a type of autophagic body that delivers a portion of the stromal proteins into the vacuole. Thus, the RCB pathway was established as an autophagic process that mobilizes stromal proteins to the vacuole (Figure 1a).

Endosomal sorting complex required for transport (ESCRT) proteins are part of an evolutionarily conserved system that is responsible for the remodeling of endosomal membranes in eukaryotes [39]. A recent study in Arabidopsis indicated that the ESCRT-III paralogs charged multivesicular body protein 1A (CHMP1A) and CHMP1B are required for the delivery of RCBs to the vacuole [40]. In *chmp1a chmp1b* double mutant plants, RCBs were produced but accumulated in the cytoplasm; therefore, CHMP1 proteins are required for the vacuolar sorting of chloroplast-derived RCBs or the fusion of autophagosomes enclosing RCBs. How a portion of stroma is separated as RCBs, and how RCBs are then recruited for autophagic transport remain unclear.

The RCB pathway is particularly active in sugar-starved, excised Arabidopsis leaves in darkness or the presence of photosynthesis inhibitors [41]. Starch is the major carbohydrate form for energy storage. The starchless mutants, *phosphoglucomutase* (*pgm*) and *ADP-glucose pyrophosphorylase1* (*adg1*), which lack starch, exhibited enhanced production of RCBs [41,42]. Moreover, starchless and *atg* double mutants exhibited reduced growth and enhanced cell death during developmental senescence compared to the respective single mutants [42]. These results indicate that the RCB pathway plays a role in the response to sugar starvation. Recent studies found that in the sugar-starved leaves of Arabidopsis plants maintained in complete darkness for several days, autophagy deficiency compromises the release of free amino acids, especially free branched chain amino acids (BCAAs) like isoleucine, leucine, and valine [43,44]. Arabidopsis mutants with defects in the enzymes involved in BCAA catabolism have reduced tolerance to sugar starvation due to prolonged complete darkness [9,45–49]; thus, BCAAs are a particularly important energy source for mitochondrial respiration as alternatives to sugars. The RCB pathway might supply free amino acids, especially BCAAs, derived from vacuolar degradation of stromal proteins as an alternative energy source during periods of impaired photosynthesis (Figure 1a).

Photosynthetic energy production can be perturbed by various types of suboptimal conditions, including shading, flooding, or drought. The importance of core autophagy machinery during submergence-induced hypoxia or draught stress was reported in Arabidopsis plants [50,51]. The RCB pathway might alleviate the energy limitation that is caused by some types of abiotic stresses.

RCB production is also activated during accelerated leaf senescence induced in leaves that were individually covered to impair photosynthesis [52]. This activation of senescence corresponds to chloroplast shrinkage. In addition to direct observations of RCBs labeled with stroma-localized fluorescent proteins, the activity of the RCB pathway can be monitored by biochemical detection of free GFP or RFP derived from vacuolar degradation of Rubisco-GFP or -RFP fusion proteins, which are mobilized to the vacuole via RCBs [53]. This technique indicated that autophagy contributes substantially to the degradation of Rubisco in individually darkened leaves and in those shaded by the leaves of neighboring Arabidopsis plants [53]. Such biochemical methods of monitoring the RCB pathway have also been established in rice plants, and have shown that Rubisco is degraded via RCBs in individually darkened rice leaves [38]. In autophagy-deficient *atg* mutant rice plants, *osatg7*, Rubisco degradation was attenuated in senescing leaves, which is consistent with the partly compromised nitrogen remobilization from lower leaves to newly developing upper leaves [54]. These findings further indicate that the RCB pathway mediates nitrogen remobilization from older leaves that cannot acquire sufficient light due to shading of developing leaves by upper tissues.

Analyses of Arabidopsis and maize (*Zea mays*) plants harboring the *atg* mutation indicated that autophagy contributes to nitrogen remobilization from vegetative tissues to reproductive tissues, including seeds [55–57]. However, such a role for autophagy in rice plants was not evaluated, because autophagy-deficient rice plants exhibit male sterility due to impaired pollen maturation [58].

3. Chlorophagy: Degradation of Entire Chloroplasts

Whereas the amount of stromal proteins decreases during the earlier stages of leaf senescence in wheat or barley (*Hordeum vulgare*) plants, the number of chloroplasts per cell decreases during the later stages [33–35]. In individually darkened leaves of wild-type Arabidopsis plants, RCB production and subsequent shrinkage of chloroplasts occur during the earlier stages of senescence, and the chloroplast population decreases during the later stages of senescence [52]. This decrease in chloroplast number is suppressed in *atg4* mutants. Some isolated vacuoles from the darkened leaves of wild-type plants contained chloroplasts that exhibited chlorophyll autofluorescence signals. These findings suggest that shrunken chloroplasts, which are produced through the active separation of their components in the RCB pathway, become the targets of autophagic transport as entire organelles, a process known as chlorophagy [59] (Figure 1b).

In yeast and mammals, autophagy is also recognized as a major quality control system for organelles through the selective removal of dysfunctional organelles [19]. In Arabidopsis *atg* plants, oxidized peroxisomes containing aggregated catalase accumulate in the cytoplasm of senescing leaves [60–62]. During germination, enzymes in peroxisomes catalyze β-oxidation and the glyoxylate cycle, thereby allowing lipids stored in seeds to be used as energy before photosynthetic machinery within chloroplasts are developed. As photosynthetic growth is established several days after germination, peroxisomes are remodeled to carry out the glycolate pathway, which is required for photorespiration. This functional conversion of peroxisomes was partly compromised in Arabidopsis *atg* plants in which peroxisome aggregates accumulate in mesophyll cells containing mature chloroplasts [63,64]. Thus, plant peroxisomes are likely targets of a process of selective autophagy known as pexophagy during senescence or seedling development. Autophagic degradation of the endoplasmic reticulum (ER) during ER stress due to tunicamycin treatment was also observed in Arabidopsis roots [65,66]. Selective degradation of ER by autophagy termed ER-phagy may function in plants.

A recent study investigated the involvement of autophagy in the turnover of chloroplasts under photooxidative stress conditions and demonstrated that chlorophagy is induced in Arabidopsis leaves damaged by UVB exposure [67]. A subset of the chloroplasts in the cytoplasm of UVB-damaged *atg5* and *atg7* plants exhibited irregular shapes and disorganized thylakoid structures. Chlorophagy was also induced by chloroplast damage caused by exposure to strong visible light or natural sunlight. Therefore, chlorophagy may remove entire photo-damaged chloroplasts by transporting them into the vacuole [67,68] (Figure 1c).

The chloroplast-targeted RCB pathway and chlorophagy differ in individually darkened leaves and in leaves subjected to UVB damage [67,69]. During sugar starvation in individually darkened leaves, RCBs were observed after 1 d of treatment, whereas chlorophagy was rarely observed during 3 days of dark treatment. By contrast, in leaves subjected to UVB-mediated oxidative stress, chlorophagy was actively induced 2 days after treatment without prior RCB production. These observations suggest that the induction of these two types of autophagy is individually controlled by distinct upstream mechanisms in response to environmental or developmental conditions (Figure 1).

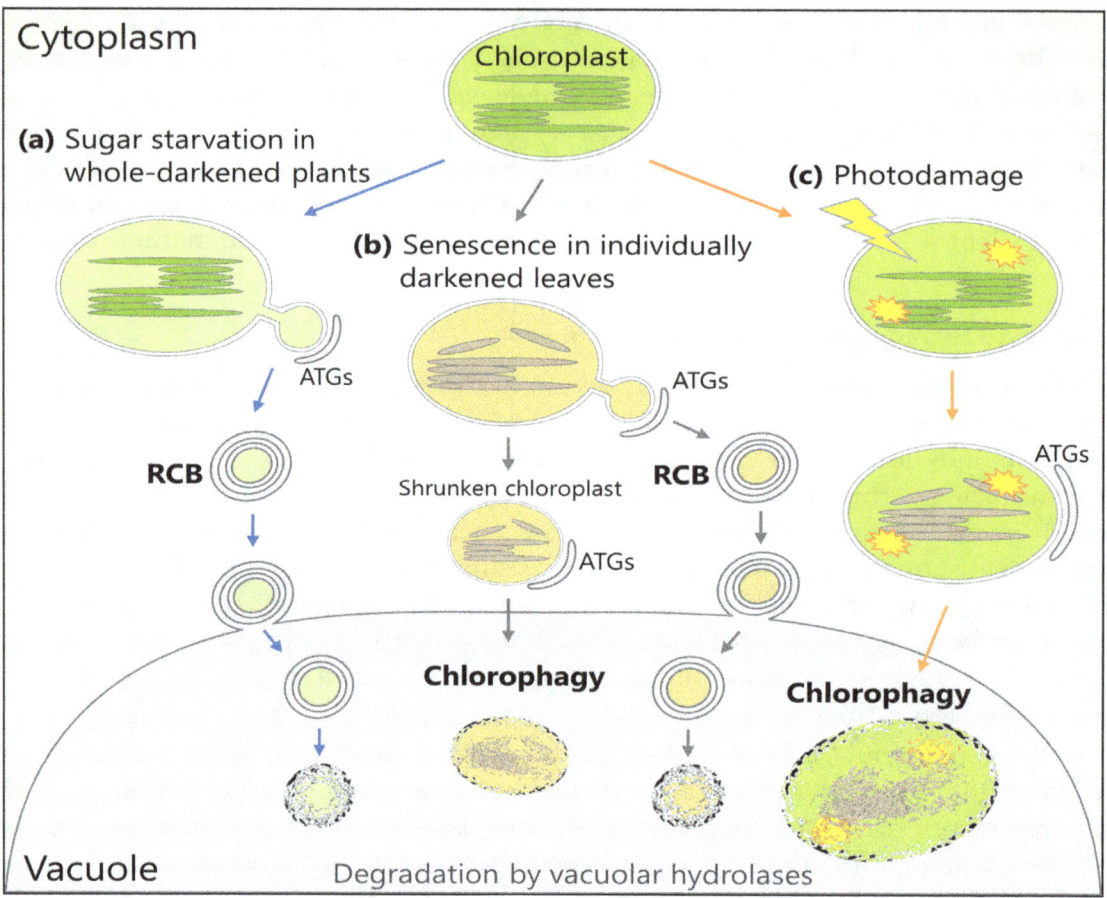

Figure 1. Schematic model for the Rubisco-containing body (RCB) pathway and chlorophagy forms of chloroplast-related autophagy. (**a**) When photosynthetic energy production of whole plants is impaired due to complete darkness, a portion of the chloroplast stroma is transported to the central vacuole via RCBs, which are a type of autophagic compartment that specifically contains stromal proteins. The RCB pathway can facilitate the recycling of amino acids as an energy source. (**b**) When senescence is accelerated in individually darkened leaves, the active production of RCBs leads to chloroplast shrinkage, thereby allowing the transport of entire chloroplasts to the vacuole via chlorophagy. (**c**) Photodamage from exposure to ultraviolet-B (UV-B), strong visible light, or natural sunlight causes chloroplasts to collapse. The collapsed chloroplasts are then transported to the vacuole without prior activation of RCBs. This process is suggested to serve as a quality control mechanism that removes damaged chloroplasts.

4. ATI Body-Mediated Chloroplast Degradation

ATG8 is a core ATG protein that builds up the autophagosomal membrane by conjugating with phosphatidylethanolamine [70]. In yeast, several types of organelle-targeted autophagy are controlled by ATG proteins containing an ATG8-interacting motif (AIM) [71]. ATG32 triggers the removal of dysfunctional or excess mitochondria by interacting with autophagosomal membrane-anchored ATG8 on the mitochondrial outer envelope [72,73]. ATG39 and ATG40 were also identified as ATG8-interacting proteins that control nucleus- or ER-targeted selective autophagy, respectively [74].

ATG8-interacting protein 1 (ATI1) and ATI2 were identified in a yeast two-hybrid screen for candidates that interact with the Arabidopsis ATG8 isoform, ATG8f [75]. These proteins were found to associate with plastids in addition to the ER as small vesicles of approximately 1 μm in diameter, which are referred to as ATI bodies [76]. A screen of potential ATI1-interacting proteins and microscopy observations of fluorescent marker proteins indicated that plastid-associated ATI bodies transport some thylakoid, stroma, and envelope proteins into the vacuole, especially under dark-induced energy limitation [76]. These delivery cargos differ from those of the RCBs that specifically contain a portion of stroma [36,37]; however, the vacuolar transport of plastid-associated ATI bodies is an autophagy-dependent process, as this body was not produced in the *atg5* mutants [76]. Therefore, ATI bodies represent a distinct form of autophagy vesicles that transport some stroma, thylakoid, and envelope components into the vacuole (Figure 2a). Plastid-associated ATI bodies are also observed inside the chloroplast, and ATI1 interacts with some thylakoid proteins in vivo [76]. It is thus proposed that plastid-associated ATI bodies form in chloroplasts and are then delivered into the vacuole via autophagosome-mediated transport (Figure 2a), although how such bodies are evacuated from chloroplasts remains unclear.

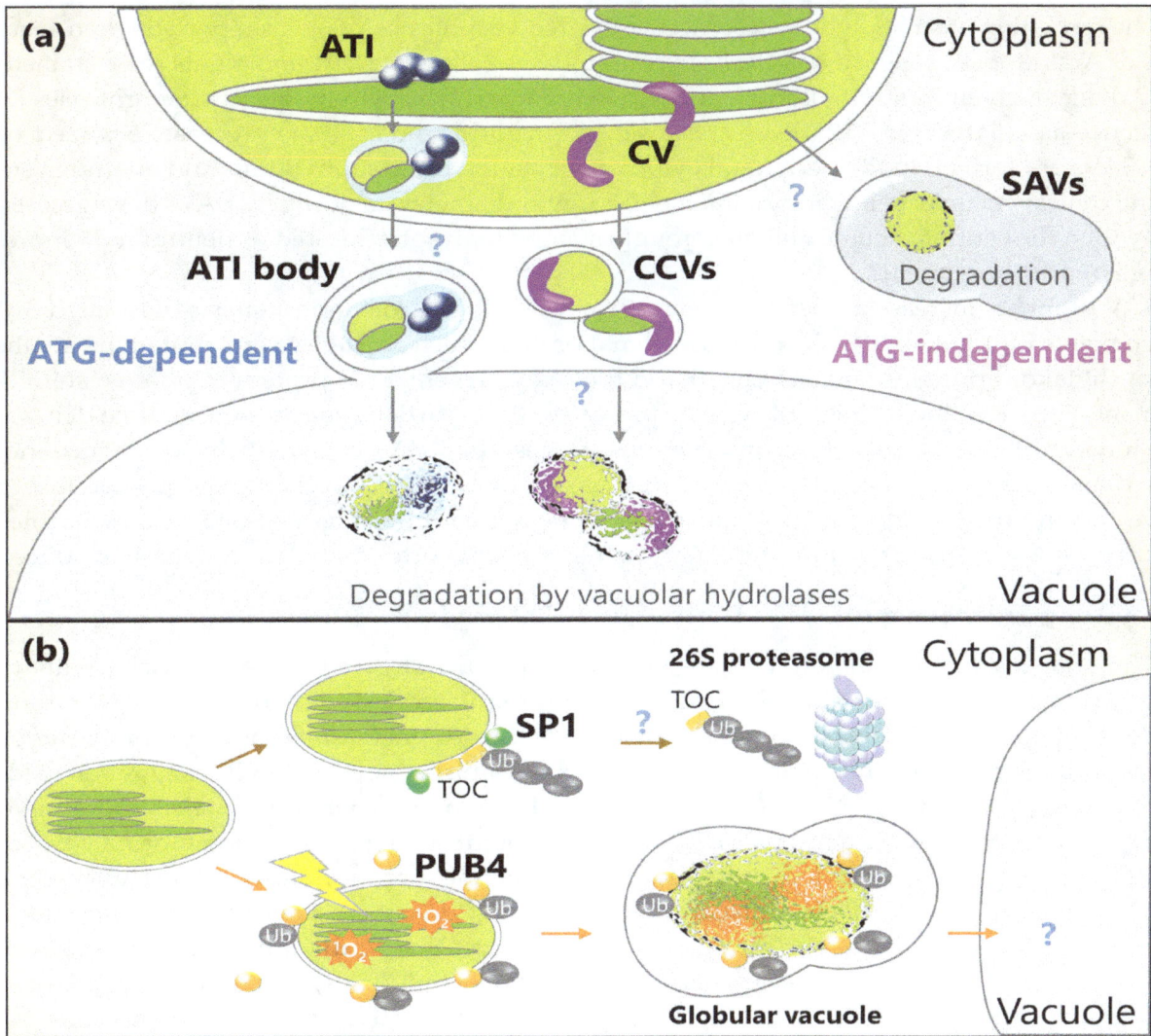

Figure 2. Schematic model for chloroplast protein turnover mediated by ATI bodies, CV-containing vesicles (CCVs), senescence-associated vacuoles (SAVs), or ubiquitination. (**a**) Plastid-associated ATI bodies are produced in chloroplasts and are then delivered into the central vacuole via an autophagy-dependent pathway. ATI bodies transport thylakoid, stroma, and envelope proteins. CV protein also interacts with thylakoid and stroma proteins, and then induces the production of CCVs that transport thylakoid, stroma, and envelope proteins into the central vacuole via an autophagy-independent pathway. SAVs are small lytic compartments that form in the cytoplasm. Stroma components are incorporated into the SAVs for digestion. (**b**) Chloroplast outer envelope-anchored E3 ligase, SP1, ubiquitinates TOC proteins and facilitates their degradation by 26S proteasome. Cytoplasmic E3 ligase PUB4 ubiquitinates oxidative chloroplasts accumulating $^1O_2$ for the digestion of such chloroplasts in their entirety.

The appearance of plastid-associated ATI bodies in energy-starved seedlings or senescing leaves suggests that ATI bodies also contribute to amino acid recycling during starvation or senescence as part of the autophagy process, although the link between the induction level of the ATI bodies and changes in free amino acid content has not been evaluated. Additionally, plastid-associated ATI bodies are produced under salt stress, and ATI-knockdown plants have reduced salt tolerance [76]. The activation of autophagosome production and the reduced tolerance of *atg* mutants to salt stress were also observed in Arabidopsis plants [50,77]. These findings suggest that ATI bodies are involved in salt stress-induced chloroplast protein turnover.

5. Senescence-Associated Vacuoles

The formation of small, lytic senescence-associated vacuoles (SAVs) was reported when senescing leaves of Arabidopsis, soybean (*Glycine max*), and tobacco (*Nicotiana tabacum*) plants were stained with R-6502 dye, which emits strong fluorescence upon the hydrolytic activity of cysteine proteases [78,79]. Senescence-associated gene 12 (SAG12) is a senescence-induced cysteine protease localized within SAVs. SAVs are formed in the peripheral cytoplasmic region of mesophyll cells and are much smaller than the central vacuole, being approximately 0.7 μm in diameter. In addition, SAVs have greater lytic activity than the central vacuole and are strongly stained by lysotracker red or neutral red, fluorescent markers of acidic organelles.

SAV numbers increase as leaf senescence progresses [80]. Proteomic analysis of isolated SAVs in tobacco plants indicated that SAVs contain stromal proteins such as Rubisco and glutamine synthetase, but not thylakoid proteins such as LHCII and the reaction center D1 protein in photosystem II [79]. Treatment with a specific inhibitor of cysteine proteases, E-64, partially suppressed Rubisco degradation in the tobacco leaf discs [80]. These observations suggest that SAVs contribute to senescence-induced Rubisco degradation, similar to RCBs; however, *atg7* mutants produced SAVs [79]. Therefore, SAVs may be an autophagy-independent, extra-chloroplastic route for the degradation of stromal proteins in senescent leaves (Figure 2a). How stromal proteins are transported into the SAVs remains uncertain.

6. Autophagy-Independent Vesicles Derived from Chloroplasts

The *chloroplast vesiculation* (*CV*) gene encodes a plastid-targeted protein in rice plants that is strongly upregulated under abiotic stress and downregulated by cytokinin [81]. In Arabidopsis, expression of the *CV-GFP* construct under the control of the dexamethasone-inducible promoter caused the formation of a type of chloroplast-derived vesicle exhibiting strong CV-GFP signal referred to as CV-containing vesicles (CCVs) [81]. CCVs are around 1 μm in diameter and contain stroma, envelope, and thylakoid proteins, as demonstrated by immunoblot analysis of some chloroplast proteins, co-immunoprecipitation assays of potential CV-interacting protein, and confocal microscopy of fluorescent marker proteins of chloroplast stroma [81]. CCVs do not associate with the autophagosome marker GFP-ATG8a, and the *atg5* mutation does not affect the production of CCVs. Additionally, CCVs do not associate with SAVs stained with lysotracker red. Thus, CCVs are part of a vacuolar degradation process for chloroplasts that is independent of autophagy and SAVs (Figure 2a).

Immuno-EM analysis showed that CV-GFP was associated with the thylakoid or envelope membranes before CCV production [81]. The interaction of CV with PsbO protein, a subunit in the thylakoid-bound photosystem II complex, was confirmed by co-immunoprecipitation detection and a bimolecular fluorescence complementation (BiFC) assay. These results indicate that CV interacts with some proteins inside the chloroplast before CCVs form. The C-terminal domain of CV, which is largely conserved among CV orthologs of various plant species, is required for CCV production [81]; however, how chloroplast-targeted CV induces the formation of CCVs and chloroplast destabilization has not been evaluated.

In Arabidopsis plants, endogenous *CV* was upregulated in senescing leaves and leaves subjected to oxidative stress or salt stress [81]. Consistent with this, transient expression of *CV* caused accelerated leaf senescence, and the suppression of *CV* transcript by miRNA led to increased leaf longevity under salt stress. Similarly, in rice plants, the RNAi silencing of *CV* expression led to delayed leaf senescence under water deficit stress, and the transient overexpression of *CV-GFP* under the control of the β-estradiol-inducible promoter accelerated leaf senescence symptoms [82]. Elevated *CV* transcript levels were observed in UVB-damaged Arabidopsis leaves [67]. *CV* may activate the destabilization and degradation of chloroplasts through the formation of CCVs during senescence, especially under stress conditions in Arabidopsis and rice plants.

7. Ubiquitin E3 Ligase-Associated Chloroplast Degradation

The ubiquitin proteasome system (UPS) is an evolutionarily conserved major protein degradation system in eukaryotic cells [83–85]. During UPS-mediated proteolysis, the polypeptide ubiquitin acts as a sorting signal for the degradation of specific proteins by the 26S proteasome in the ubiquitination cascade. Ubiquitin is activated by E1 proteins and then transferred to E2 ubiquitin conjugating enzymes. The transfer of ubiquitin from E2s to target proteins requires E3 ubiquitin ligases. The resulting ubiquitinated proteins are selectively incorporated into the 26S proteasome complex for breakdown. Eukaryotic genomes generally encode a large family of E3s and Arabidopsis plants can theoretically express more than 1500 of these proteins [86–88]. The ubiquitination of specific proteins by individual E3s allows for highly controlled, selective protein degradation by the UPS.

The UPS was shown to contribute to the degradation of chloroplast proteins in an experiment using suppressor of ppi1 locus 1 (SP1) isolated from Arabidopsis plants [89]. SP1 is a chloroplast outer envelope-anchored E3 ligase that ubiquitinates some proteins of the translocon on the outer chloroplast membrane (TOC) complex (Figure 2b). Most nucleus-encoded chloroplast proteins are imported into chloroplasts through the TOC and translocon on the inner chloroplast membrane (TIC) complexes [90]. During the greening of etiolated seedlings, etioplasts, which are a type of plastid present in non-green tissues, are converted to mature chloroplasts; therefore, large amount of photosynthetic proteins encoded in the nuclear genome are expressed and imported into the plastid via TIC-TOC complexes. *sp1* mutant plants exhibit delayed maturation of chloroplasts during the greening of etiolated seedlings [89]. Thus, SP1 likely serves as a control for protein import into chloroplasts via the turnover of the TOC complex when etioplasts develop into functional chloroplasts. *sp1* mutants also showed delayed leaf yellowing during dark-induced accelerated senescence; conversely, SP1-overexpressing plants showed an enhanced decline of photosynthetic efficiency [89]. SP1-mediated TOC turnover may further regulate protein import into chloroplasts when functional chloroplasts are actively degraded during senescence.

SP1 induces the degradation of TOC during oxidative stress caused by salt or osmotic stress, thereby attenuating protein import into the chloroplasts [91]. Under these stress conditions, accumulation of hydrogen peroxide (H_2O_2), a type of ROS, was enhanced in *sp1* mutants and was alleviated in SP1-overexpressing plants. SP1-mediated degradation of the TOC complex by UPS suppresses photosynthetic activity and thereby limits ROS production [91], since ROS are produced during photosynthesis. SP1-mediated TOC turnover may therefore control the chloroplast proteome and photosynthetic capacity in response to stress. It is still unclear how ubiquitinated proteins on outer-envelope proteins are solubilized to allow degradation by UPS localized in the cytoplasm.

A recent study reported that a cytosol-localized E3 ligase functions in the degradation of entire chloroplasts [92]. When dark-germinated, etiolated seedlings of the Arabidopsis mutant of plastid *ferrochelatase 2* (*fc2*) are transferred from darkness to light, their chloroplasts over-accumulate singlet oxygen ($^1O_2$), thereby leading to the death of photosynthetic cells and compromised greening of plants. A suppressor mutant (referred to as *pub4–6* [92]) of this inhibited greening phenomenon had an amino acid substitution in *plant u-box 4* (*PUB4*), which encodes a cytosolic ubiquitin E3 ligase. Although EM analysis indicated that entire chloroplasts were digested in the cytoplasm during compromised greening in *fc2* plants, this degradation of chloroplasts was lower in *fc2 pub4–6* plants, even though $^1O_2$ accumulation was not suppressed. Therefore, PUB4-related ubiquitination triggers the digestion of entire chloroplasts that are accumulating $^1O_2$ (Figure 2b). However, unlike the *pub4–6* mutation, the T-DNA insertional knockout mutations of *PUB4* (referred to as *pub4–1* and *pub4–2*) did not suppress the phenotype of the *fc2* mutant during greening [92]. Therefore, it is unclear how PUB4 is involved in the ubiquitination of $^1O_2$ accumulating chloroplasts and their subsequent degradation.

In mammalian cells, ubiquitination largely acts as a trigger of autophagic removal of dysfunctional organelles [93]. During mitophagy, depolarized mitochondria are ubiquitinated by the E3 ligase Parkin, allowing for the autophagic removal of targeted mitochondria into the lysosome [94–97]. During the greening of *fc2* mutants, some chloroplasts appeared to be degraded in the cytoplasm, and the interaction between degrading chloroplasts and the vacuole via a globule-like structure was observed [92]. Such observations were distinct from the vacuolar chloroplasts that result from chlorophagy in leaves exposed to strong visible light (1200–2000 $\mu mol \cdot m^{-2} \cdot s^{-1}$), where entire chloroplasts exhibiting thylakoid membranes are localized in the central vacuole in EM imaging [67]. Furthermore, the *pub4–6* and *atg10* mutants are phenotypically distinct, as *atg10* plants showed accelerated senescence during dark treatment compared to wild-type plants, but *pub4–6* plants did not [92]. Therefore, PUB4-related ubiquitination is unlikely a simple trigger of autophagy.

8. Future Perspectives

Our understanding of the diverse extraplastidic pathways mediating chloroplast protein degradation has progressed in the past decades. Table 1 compares their relationships to core autophagy machinery, plant species, induction stimuli, and degradation targets. It is clear that multiple pathways are induced during diverse stress conditions, such as sugar starvation, senescence, and oxidative stress (Table 1). Thus, new questions about chloroplast turnover arise, including why plants have multiple processes for chloroplast protein turnover, and how these processes are differentially utilized. Future research should examine how extraplastidic systems are coordinated with intraplastidic proteolysis. The RCB pathway is activated during the earlier stages of dark treatment and chlorophagy is induced during the later stages [52,67], suggesting that several pathways are induced at distinct time points during leaf senescence and stress responses. An important role of intrachloroplastic proteases in chloroplast protein turnover during photodamage was largely demonstrated [15,16]. Therefore, extraplastidic pathways that are induced during photooxidative stress might be triggered when intraplastidic proteolysis is insufficient for maintaining chloroplast functions.

Table 1. List of extraplastidic degradation pathways described.

Pathway	Relationship to Core Autophagy Machinery	Analyzed Species	Degradation Targets	Stimuli [b]	References
RCBs (Rubisco-containing bodies)	dependent	Arabidopsis, rice, wheat	stroma, envelope	sugar starvation, senescence	[36–38,41,42,44,52–54]
Chlorophagy	dependent	Arabidopsis	entire chloroplasts	photodamage, senescence	[52,67]
ATI bodies	dependent	Arabidopsis	stroma, thylakoid, envelope	sugar starvation, salt stress, senescence	[76]
SAVs (Senescence-associated vacuoles)	independent	Arabidopsis, soybean, tobacco	stroma	senescence	[78–80]
CCVs (Chloroplast vesiculation-containing vesicles)	independent	Arabidopsis, rice	stroma, thylakoid, envelope	senescence, salt stress, oxidative stress	[81,82]
E3 ligase SP1	– [a]	Arabidopsis	TOC proteins on outer envelope	senescence, greening, oxidative stress	[89,91]
E3 ligase PUB4	– [a]	Arabidopsis	entire chloroplasts	Oxidative stress ($^1O_2$)	[92]

[a] The link of the E3 ubiquitin ligases to autophagy has not been directly examined. [b] Stimuli inducing the respective pathways.

In Arabidopsis *atg* plants, both *CV* expression and proteasome activity are increased [43,98], suggesting a complementary relationship among some of the chloroplast-associated degradation systems. However, since senescence symptoms are largely accelerated in *atg* mutants due to the over-accumulation of salicylic acid [99], the increase in *CV* expression or proteasome activity in *atg* plants can also be interpreted as a result of accelerated senescence and cell death. To better understand the process of chloroplast protein turnover and to decipher the relationships among the diverse pathways mediating this process, the mechanisms regulating these pathways will need to be elucidated. It would be fascinating to determine whether the distinct pathways that mediate chloroplast degradation share a common upstream regulatory mechanism, or whether they are regulated independently. In addition, how small vesicles delivering portions of chloroplasts, including RCBs, ATI bodies, and CCVs, are derived from entire chloroplasts largely remains to be explained.

The extraplastidic routes for chloroplast protein turnover were largely identified using Arabidopsis plants (Table 1). This advance greatly expanded our understanding of chloroplast protein turnover in important cereals, such as rice and maize [38,54,56,82]. Chloroplast degradation is strongly linked to nitrogen remobilization and the changes of photosynthetic capacity that are important determinants of productivity in crop plants. Therefore, manipulating chloroplast protein turnover might be an effective strategy to improve the productivity of crops. In rice plants, RNAi-mediated silencing of *CV* led to an increase in grain yield under water deficit stress [74]. Studies showed that Arabidopsis plants overexpressing one of the core *ATGs* had an enhanced stress tolerance [100,101]. In addition, SP1-overexpressing Arabidopsis plants had improved tolerance to oxidative stress [91]. Therefore, elucidating the molecular basis of multiple processes for chloroplast protein turnover in Arabidopsis plants may suggest strategies to improve the productivity and stress tolerance of crop plants.

Acknowledgments: This work was supported, in part, by KAKENHI (Grant Numbers 17H05050, awarded to Masanori Izumi and 16J03408, awarded to Sakuya Nakamura), the JSPS Research Fellowship for Young Scientists (awarded to Sakuya Nakamura), Building of Consortia for the Development of Human Resources in Science and Technology (awarded to Masanori Izumi), JST PRESTO (Grant Number JPMJPR16Q1, awarded to Masanori Izumi), and the Program for Creation of Interdisciplinary Research at Frontier Research Institute for Interdisciplinary Sciences, Tohoku University, Japan (awarded to Masanori Izumi).

Author Contributions: Masanori Izumi conceived the topic of this review; Masanori Izumi and Sakuya Nakamura wrote the paper; Sakuya Nakamura designed the figures with the support of Masanori Izumi.

References

1. Makino, A.; Osmond, B. Effects of nitrogen nutrition on nitrogen partitioning between chloroplasts and mitochondria in pea and wheat. *Plant Physiol.* **1991**, *96*, 355–362. [CrossRef] [PubMed]

2. Makino, A.; Sakuma, H.; Sudo, E.; Mae, T. Differences between maize and rice in N-use efficiency for photosynthesis and protein allocation. *Plant Cell Physiol.* **2003**, *44*, 952–956. [CrossRef] [PubMed]

3. Evans, J.R. Photosynthesis and nitrogen relationships in leaves of C_3 plants. *Oecologia* **1989**, *78*, 9–19. [CrossRef] [PubMed]

4. Mae, T.; Ohira, K. The remobilization of nitrogen related to leaf growth and senescence in rice plants (*Oryza sativa* L.). *Plant Cell Physiol.* **1981**, *22*, 1067–1074. [CrossRef]

5. Masclaux-Daubresse, C.; Daniel-Vedele, F.; Dechorgnat, J.; Chardon, F.; Gaufichon, L.; Suzuki, A. Nitrogen uptake, assimilation and remobilization in plants: Challenges for sustainable and productive agriculture. *Ann. Bot.* **2010**, *105*, 1141–1157. [CrossRef] [PubMed]

6. Stitt, M.; Zeeman, S.C. Starch turnover: Pathways, regulation and role in growth. *Curr. Opin. Plant Biol.* **2012**, *15*, 282–292. [CrossRef] [PubMed]

7. Baena-González, E.; Sheen, J. Convergent energy and stress signaling. *Trends Plant Sci.* **2008**, *13*, 474–482. [CrossRef] [PubMed]

8. Chaves, M.M.; Flexas, J.; Pinheiro, C. Photosynthesis under drought and salt stress: Regulation mechanisms from whole plant to cell. *Ann. Bot.* **2009**, *103*, 551–560. [CrossRef] [PubMed]

9.   Araújo, W.L.; Tohge, T.; Ishizaki, K.; Leaver, C.J.; Fernie, A.R. Protein degradation—An alternative respiratory substrate for stressed plants. *Trends Plant Sci.* **2011**, *16*, 489–498. [CrossRef] [PubMed]

10.   Sonoike, K. Various aspects of inhibition of photosynthesis under light/chilling stress: "Photoinhibition at chilling temperatures" versus "Chilling damage in the light". *J. Plant Res.* **1998**, *111*, 121–129. [CrossRef]

11.   Li, Z.R.; Wakao, S.; Fischer, B.B.; Niyogi, K.K. Sensing and responding to excess light. *Annu. Rev. Plant Biol.* **2009**, *60*, 239–260. [CrossRef] [PubMed]

12.   Tikkanen, M.; Mekala, N.R.; Aro, E.M. Photosystem II photoinhibition-repair cycle protects Photosystem I from irreversible damage. *Biochim. Biophys. Acta* **2014**, *1837*, 210–215. [CrossRef] [PubMed]

13.   Takahashi, S.; Badger, M.R. Photoprotection in plants: A new light on photosystem II damage. *Trends Plant Sci.* **2011**, *16*, 53–60. [CrossRef] [PubMed]

14.   Kataria, S.; Jajoo, A.; Guruprasad, K.N. Impact of increasing Ultraviolet-B (UV-B) radiation on photosynthetic processes. *J. Photochem. Photobiol B* **2014**, *137*, 55–66. [CrossRef] [PubMed]

15.   Nishimura, K.; Kato, Y.; Sakamoto, W. Chloroplast proteases: Updates on proteolysis within and across suborganellar compartments. *Plant Physiol.* **2016**, *171*, 2280–2293. [CrossRef] [PubMed]

16.   Van Wijk, K.J. Protein maturation and proteolysis in plant plastids, mitochondria, and peroxisomes. *Annu Rev. Plant Biol.* **2015**, *66*, 75–111. [CrossRef] [PubMed]

17.   Nakatogawa, H.; Suzuki, K.; Kamada, Y.; Ohsumi, Y. Dynamics and diversity in autophagy mechanisms: Lessons from yeast. *Nat. Rev. Mol. Cell Biol.* **2009**, *10*, 458–467. [CrossRef] [PubMed]

18.   Mizushima, N.; Komatsu, M. Autophagy: Renovation of cells and tissues. *Cell* **2011**, *147*, 728–741. [CrossRef] [PubMed]

19.   Anding, A.L.; Baehrecke, E.H. Cleaning house: Selective autophagy of organelles. *Dev. Cell* **2017**, *41*, 10–22. [CrossRef] [PubMed]

20.   Tsukada, M.; Ohsumi, Y. Isolation and characterization of autophagy-defective mutants of *Saccharomyces erevisiae*. *FEBS Lett.* **1993**, *333*, 169–174. [CrossRef]

21.   Chung, T.; Suttangkakul, A.; Vierstra, R.D. The ATG autophagic conjugation system in maize: ATG transcripts and abundance of the ATG8-lipid adduct are regulated by development and nutrient availability. *Plant Physiol.* **2009**, *149*, 220–234. [CrossRef] [PubMed]

22.   Xia, K.F.; Liu, T.; Ouyang, J.; Wang, R.; Fan, T.; Zhang, M.Y. Genome-wide identification, classification, and expression analysis of autophagy-associated gene homologues in rice (*Oryza sativa* L.). *DNA Res.* **2011**, *18*, 363–377. [CrossRef] [PubMed]

23.   Meijer, W.H.; van der Klei, I.J.; Veenhuis, M.; Kiel, J.A.K.W. *ATG* genes involved in non-selective autophagy are conserved from yeast to man, but the selective Cvt and pexophagy pathways also require organism-specific genes. *Autophagy* **2007**, *3*, 106–116. [CrossRef] [PubMed]

24.   Yoshimoto, K.; Hanaoka, H.; Sato, S.; Kato, T.; Tabata, S.; Noda, T.; Ohsumi, Y. Processing of ATG8s, ubiquitin-like proteins, and their deconjugation by ATG4s are essential for plant autophagy. *Plant Cell* **2004**, *16*, 2967–2983. [CrossRef] [PubMed]

25.   Suzuki, N.N.; Yoshimoto, K.; Fujioka, Y.; Ohsumi, Y.; Inagaki, F. The crystal structure of plant ATG12 and its biological implication in autophagy. *Autophagy* **2005**, *1*, 119–126. [CrossRef] [PubMed]

26.   Xiong, Y.; Contento, A.L.; Bassham, D.C. AtATG18a is required for the formation of autophagosomes during nutrient stress and senescence in *Arabidopsis thaliana*. *Plant J.* **2005**, *42*, 535–546. [CrossRef] [PubMed]

27.   Doelling, J.H.; Walker, J.M.; Friedman, E.M.; Thompson, A.R.; Vierstra, R.D. The APG8/12-activating enzyme APG7 is required for proper nutrient recycling and senescence in *Arabidopsis thaliana*. *J. Biol. Chem.* **2002**, *277*, 33105–33114. [CrossRef] [PubMed]

28.   Phillips, A.R.; Suttangkakul, A.; Vierstra, R.D. The ATG12-conjugating enzyme ATG10 is essential for autophagic vesicle formation in *Arabidopsis thaliana*. *Genetics* **2008**, *178*, 1339–1353. [CrossRef] [PubMed]

29.   Suttangkakul, A.; Li, F.Q.; Chung, T.; Vierstra, R.D. The ATG1/ATG13 protein kinase complex Is both a regulator and a target of autophagic recycling in *Arabidopsis*. *Plant Cell* **2011**, *23*, 3761–3779. [CrossRef] [PubMed]

30.   Chung, T.; Phillips, A.R.; Vierstra, R.D. ATG8 lipidation and ATG8-mediated autophagy in Arabidopsis require ATG12 expressed from the differentially controlled *ATG12A* and *ATG12B* loci. *Plant J.* **2010**, *62*, 483–493. [CrossRef] [PubMed]

31. Thompson, A.R.; Doelling, J.H.; Suttangkakul, A.; Vierstra, R.D. Autophagic nutrient recycling in Arabidopsis directed by the ATG8 and ATG12 conjugation pathways. *Plant Physiol.* **2005**, *138*, 2097–2110. [CrossRef] [PubMed]

32. Li, F.; Chung, T.; Vierstra, R.D. AUTOPHAGY-RELATED11 plays a critical role in general autophagy- and senescence-induced mitophagy in Arabidopsis. *Plant Cell* **2014**, *26*, 788–807. [CrossRef] [PubMed]

33. Mae, T.; Kai, N.; Makino, A.; Ohira, K. Relation between ribulose bisphosphate carboxylase content and chloroplast number in naturally senescing primary leaves of wheat. *Plant Cell Physiol.* **1984**, *25*, 333–336. [CrossRef]

34. Ono, K.; Hashimoto, H.; Katoh, S. Changes in the number and size of chloroplasts during senescence of primary leaves of wheat grown under different conditions. *Plant Cell Physiol.* **1995**, *36*, 9–17. [CrossRef]

35. Martinoia, E.; Heck, U.; Dalling, M.J.; Matile, P. Changes in chloroplast number and chloroplast constituents in senescing barley leaves. *Biochem. Physiol. Pflanz.* **1983**, *178*, 147–155. [CrossRef]

36. Chiba, A.; Ishida, H.; Nishizawa, N.K.; Makino, A.; Mae, T. Exclusion of ribulose-1,5-bisphosphate carboxylase/oxygenase from chloroplasts by specific bodies in naturally senescing leaves of wheat. *Plant Cell Physiol.* **2003**, *44*, 914–921. [CrossRef] [PubMed]

37. Ishida, H.; Yoshimoto, K.; Izumi, M.; Reisen, D.; Yano, Y.; Makino, A.; Ohsumi, Y.; Hanson, M.R.; Mae, T. Mobilization of rubisco and stroma-localized fluorescent proteins of chloroplasts to the vacuole by an *ATG* gene-dependent autophagic process. *Plant Physiol.* **2008**, *148*, 142–155. [CrossRef] [PubMed]

38. Izumi, M.; Hidema, J.; Wada, S.; Kondo, E.; Kurusu, T.; Kuchitsu, K.; Makino, A.; Ishida, H. Establishment of monitoring methods for autophagy in rice reveals autophagic recycling of chloroplasts and root plastids during energy limitation. *Plant Physiol.* **2015**, *167*, 1307–1320. [CrossRef] [PubMed]

39. Gao, C.J.; Zhuang, X.H.; Shen, J.B.; Jiang, L.W. Plant ESCRT complexes: Moving beyond endosomal sorting. *Trends Plant Sci.* **2017**, *22*, 986–998. [CrossRef] [PubMed]

40. Spitzer, C.; Li, F.Q.; Buono, R.; Roschzttardtz, H.; Chung, T.J.; Zhang, M.; Osteryoung, K.W.; Vierstra, R.D.; Otegui, M.S. The endosomal protein CHARGED MULTIVESICULAR BODY PROTEIN1 regulates the autophagic turnover of plastids in Arabidopsis. *Plant Cell* **2015**, *27*, 391–402. [CrossRef] [PubMed]

41. Izumi, M.; Wada, S.; Makino, A.; Ishida, H. The autophagic degradation of chloroplasts via rubisco-containing bodies is specifically linked to leaf carbon status but not nitrogen status in Arabidopsis. *Plant Physiol.* **2010**, *154*, 1196–1209. [CrossRef] [PubMed]

42. Izumi, M.; Hidema, J.; Makino, A.; Ishida, H. Autophagy contributes to nighttime energy availability for growth in Arabidopsis. *Plant Physiol.* **2013**, *161*, 1682–1693. [CrossRef] [PubMed]

43. Barros, J.A.S.; Cavalcanti, J.H.F.; Medeiros, D.B.; Nunes-Nesi, A.; Avin-Wittenberg, T.; Fernie, A.R.; Araujo, W.L. Autophagy deficiency compromises alternative pathways of respiration following energy deprivation in *Arabidopsis thaliana*. *Plant Physiol.* **2017**, *175*, 62–76. [CrossRef] [PubMed]

44. Hirota, T.; Izumi, M.; Wada, S.; Makino, A.; Ishida, H. Vacuolar protein degradation via autophagy provides substrates to amino acid catabolic pathways as an adaptive response to sugar starvation in *Arabidopsis thaliana*. *Plant Cell Physiol.* **2018**. [CrossRef] [PubMed]

45. Hildebrandt, T.M.; Nesi, A.N.; Araujo, W.L.; Braun, H.P. Amino acid catabolism in plants. *Mol. Plant* **2015**, *8*, 1563–1579. [CrossRef] [PubMed]

46. Araújo, W.L.; Ishizaki, K.; Nunes-Nesi, A.; Larson, T.R.; Tohge, T.; Krahnert, I.; Witt, S.; Obata, T.; Schauer, N.; Graham, I.A.; et al. Identification of the 2-hydroxyglutarate and Isovaleryl-CoA dehydrogenases as alternative electron donors linking lysine catabolism to the electron transport chain of *Arabidopsis* mitochondria. *Plant Cell* **2010**, *22*, 1549–1563. [CrossRef] [PubMed]

47. Ishizaki, K.; Larson, T.R.; Schauer, N.; Fernie, A.R.; Graham, I.A.; Leaver, C.J. The critical role of *Arabidopsis* electron-transfer flavoprotein: Ubiquinone oxidoreductase during dark-induced starvation. *Plant Cell* **2005**, *17*, 2587–2600. [CrossRef] [PubMed]

48. Ishizaki, K.; Schauer, N.; Larson, T.R.; Graham, I.A.; Fernie, A.R.; Leaver, C.J. The mitochondrial electron transfer flavoprotein complex is essential for survival of Arabidopsis in extended darkness. *Plant J.* **2006**, *47*, 751–760. [CrossRef] [PubMed]

49. Peng, C.; Uygun, S.; Shiu, S.H.; Last, R.L. The impact of the branched-chain ketoacid dehydrogenase complex on amino acid homeostasis in Arabidopsis. *Plant Physiol.* **2015**, *169*, 1807–1820. [CrossRef] [PubMed]

50. Liu, Y.; Xiong, Y.; Bassham, D.C. Autophagy is required for tolerance of drought and salt stress in plants. *Autophagy* **2009**, *5*, 954–963. [CrossRef] [PubMed]

51. Chen, L.; Liao, B.; Qi, H.; Xie, L.J.; Huang, L.; Tan, W.J.; Zhai, N.; Yuan, L.B.; Zhou, Y.; Yu, L.J.; et al. Autophagy contributes to regulation of the hypoxia response during submergence in *Arabidopsis thaliana*. *Autophagy* **2015**, *11*, 2233–2246. [CrossRef] [PubMed]

52. Wada, S.; Ishida, H.; Izumi, M.; Yoshimoto, K.; Ohsumi, Y.; Mae, T.; Makino, A. Autophagy plays a role in chloroplast degradation during senescence in individually darkened leaves. *Plant Physiol.* **2009**, *149*, 885–893. [CrossRef] [PubMed]

53. Ono, Y.; Wada, S.; Izumi, M.; Makino, A.; Ishida, H. Evidence for contribution of autophagy to rubisco degradation during leaf senescence in *Arabidopsis thaliana*. *Plant Cell Environ.* **2013**, *36*, 1147–1159. [CrossRef] [PubMed]

54. Wada, S.; Hayashida, Y.; Izumi, M.; Kurusu, T.; Hanamata, S.; Kanno, K.; Kojima, S.; Yamaya, T.; Kuchitsu, K.; Makino, A.; et al. Autophagy supports biomass production and nitrogen use efficiency at the vegetative stage in rice. *Plant Physiol.* **2015**, *168*, 60–73. [CrossRef] [PubMed]

55. Guiboileau, A.; Yoshimoto, K.; Soulay, F.; Bataillé, M.P.; Avice, J.C.; Masclaux-Daubresse, C. Autophagy machinery controls nitrogen remobilization at the whole-plant level under both limiting and ample nitrate conditions in Arabidopsis. *New Phytol.* **2012**, *194*, 732–740. [CrossRef] [PubMed]

56. Li, F.Q.; Chung, T.; Pennington, J.G.; Federico, M.L.; Kaeppler, H.F.; Kaeppler, S.M.; Otegui, M.S.; Vierstra, R.D. Autophagic recycling plays a central role in maize nitrogen remobilization. *Plant Cell* **2015**, *27*, 1389–1408. [CrossRef] [PubMed]

57. Guiboileau, A.; Avila-Ospina, L.; Yoshimoto, K.; Soulay, F.; Azzopardi, M.; Marmagne, A.; Lothier, J.; Masclaux-Daubresse, C. Physiological and metabolic consequences of autophagy deficiency for the management of nitrogen and protein resources in Arabidopsis leaves depending on nitrate availability. *New Phytol.* **2013**, *199*, 683–694. [CrossRef] [PubMed]

58. Kurusu, T.; Koyano, T.; Hanamata, S.; Kubo, T.; Noguchi, Y.; Yagi, C.; Nagata, N.; Yamamoto, T.; Ohnishi, T.; Okazaki, Y.; et al. OsATG7 is required for autophagy-dependent lipid metabolism in rice postmeiotic anther development. *Autophagy* **2014**, *10*, 878–888. [CrossRef] [PubMed]

59. Ishida, H.; Izumi, M.; Wada, S.; Makino, A. Roles of autophagy in chloroplast recycling. *Biochim. Biophys. Acta* **2014**, *1837*, 512–521. [CrossRef] [PubMed]

60. Kim, J.; Lee, H.; Lee, H.N.; Kim, S.H.; Shin, K.D.; Chung, T. Autophagy-related proteins are required for degradation of peroxisomes in Arabidopsis hypocotyls during seedling growth. *Plant Cell* **2013**, *25*, 4956–4966. [CrossRef] [PubMed]

61. Shibata, M.; Oikawa, K.; Yoshimoto, K.; Kondo, M.; Mano, S.; Yamada, K.; Hayashi, M.; Sakamoto, W.; Ohsumi, Y.; Nishimura, M. Highly oxidized peroxisomes are selectively degraded via autophagy in Arabidopsis. *Plant Cell* **2013**, *25*, 4967–4983. [CrossRef] [PubMed]

62. Yoshimoto, K.; Shibata, M.; Kondo, M.; Oikawa, K.; Sato, M.; Toyooka, K.; Shirasu, K.; Nishimura, M.; Ohsumi, Y. Organ-specific quality control of plant peroxisomes is mediated by autophagy. *J. Cell Sci.* **2014**, *127*, 1161–1168. [CrossRef] [PubMed]

63. Goto-Yamada, S.; Mano, S.; Nakamori, C.; Kondo, M.; Yamawaki, R.; Kato, A.; Nishimura, M. Chaperone and protease functions of LON protease 2 modulate the peroxisomal transition and degradation with autophagy. *Plant Cell Physiol.* **2014**, *55*, 482–496. [CrossRef] [PubMed]

64. Farmer, L.M.; Rinaldi, M.A.; Young, P.G.; Danan, C.H.; Burkhart, S.E.; Bartel, B. Disrupting autophagy restores peroxisome function to an *Arabidopsis lon2* mutant and reveals a role for the LON2 protease in peroxisomal matrix protein degradation. *Plant Cell* **2013**, *25*, 4085–4100. [CrossRef] [PubMed]

65. Liu, Y.; Burgos, J.S.; Deng, Y.; Srivastava, R.; Howell, S.H.; Bassham, D.C. Degradation of the endoplasmic reticulum by autophagy during endoplasmic reticulum stress in Arabidopsis. *Plant Cell* **2012**, *24*, 4635–4651. [CrossRef] [PubMed]

66. Yang, X.C.; Srivastava, R.; Howell, S.H.; Bassham, D.C. Activation of autophagy by unfolded proteins during endoplasmic reticulum stress. *Plant J.* **2016**, *85*, 83–95. [CrossRef] [PubMed]

67. Izumi, M.; Ishida, H.; Nakamura, S.; Hidema, J. Entire photodamaged chloroplasts are transported to the central vacuole by Autophagy. *Plant Cell* **2017**, *29*, 377–394. [CrossRef] [PubMed]

68. Izumi, M.; Nakamura, S. Vacuolar digestion of entire damaged chloroplasts in *Arabidopsis thaliana* is accomplished by chlorophagy. *Autophagy* **2017**, *13*, 1239–1240. [CrossRef] [PubMed]

69. Izumi, M.; Nakamura, S. Partial or entire: Distinct responses of two types of chloroplast autophagy. *Plant Signal. Behav.* **2017**, *12*, e1393137. [CrossRef] [PubMed]

70. Ichimura, Y.; Kirisako, T.; Takao, T.; Satomi, Y.; Shimonishi, Y.; Ishihara, N.; Mizushima, N.; Tanida, I.; Kominami, E.; Ohsumi, M.; et al. A ubiquitin-like system mediates protein lipidation. *Nature* **2000**, *408*, 488–492. [CrossRef] [PubMed]

71. Noda, N.N.; Ohsumi, Y.; Inagaki, F. Atg8-family interacting motif crucial for selective autophagy. *FEBS Lett.* **2010**, *584*, 1379–1385. [CrossRef] [PubMed]

72. Kanki, T.; Wang, K.; Cao, Y.; Baba, M.; Klionsky, D.J. Atg32 is a mitochondrial protein that confers selectivity during mitophagy. *Dev. Cell* **2009**, *17*, 98–109. [CrossRef] [PubMed]

73. Okamoto, K.; Kondo-Okamoto, N.; Ohsumi, Y. Mitochondria-anchored receptor Atg32 mediates degradation of mitochondria via selective autophagy. *Dev. Cell* **2009**, *17*, 87–97. [CrossRef] [PubMed]

74. Mochida, K.; Oikawa, Y.; Kimura, Y.; Kirisako, H.; Hirano, H.; Ohsumi, Y.; Nakatogawa, H. Receptor-mediated selective autophagy degrades the endoplasmic reticulum and the nucleus. *Nature* **2015**, *522*, 359–362. [CrossRef] [PubMed]

75. Honig, A.; Avin-Wittenberg, T.; Ufaz, S.; Galili, G. A new type of compartment, defined by plant-specific atg8-interacting proteins, is induced upon exposure of Arabidopsis plants to carbon starvation. *Plant Cell* **2012**, *24*, 288–303. [CrossRef] [PubMed]

76. Michaeli, S.; Honig, A.; Levanony, H.; Peled-Zehavi, H.; Galili, G. Arabidopsis ATG8-INTERACTING PROTEIN1 is involved in autophagy-dependent vesicular trafficking of plastid proteins to the vacuole. *Plant Cell* **2014**, *26*, 4084–4101. [CrossRef] [PubMed]

77. Luo, L.M.; Zhang, P.P.; Zhu, R.H.; Fu, J.; Su, J.; Zheng, J.; Wang, Z.Y.; Wang, D.; Gong, Q.Q. Autophagy is rapidly induced by salt stress and is required for salt tolerance in Arabidopsis. *Front. Plant Sci.* **2017**, *8*, 1459. [CrossRef] [PubMed]

78. Otegui, M.S.; Noh, Y.S.; Martinez, D.E.; Vila Petroff, M.G.; Andrew Staehelin, L.; Amasino, R.M.; Guiamet, J.J. Senescence-associated vacuoles with intense proteolytic activity develop in leaves of Arabidopsis and soybean. *Plant J.* **2005**, *41*, 831–844. [CrossRef] [PubMed]

79. Martinez, D.E.; Costa, M.L.; Gomez, F.M.; Otegui, M.S.; Guiamet, J.J. 'Senescence-associated vacuoles' are involved in the degradation of chloroplast proteins in tobacco leaves. *Plant J.* **2008**, *56*, 196–206. [CrossRef] [PubMed]

80. Carrion, C.A.; Costa, M.L.; Martinez, D.E.; Mohr, C.; Humbeck, K.; Guiamet, J.J. In vivo inhibition of cysteine proteases provides evidence for the involvement of 'senescence-associated vacuoles' in chloroplast protein degradation during dark-induced senescence of tobacco leaves. *J. Exp. Bot.* **2013**, *64*, 4967–4980. [CrossRef] [PubMed]

81. Wang, S.H.; Blumwald, E. Stress-induced chloroplast degradation in Arabidopsis is regulated via a process independent of autophagy and senescence-associated vacuoles. *Plant Cell* **2014**, *26*, 4875–4888. [CrossRef] [PubMed]

82. Sade, N.; Umnajkitikorn, K.; Rubio Wilhelmi, M.D.M.; Wright, M.; Wang, S.; Blumwald, E. Delaying chloroplast turnover increases water-deficit stress tolerance through the enhancement of nitrogen assimilation in rice. *J. Exp. Bot.* **2017**, *69*, 867–878. [CrossRef] [PubMed]

83. Komander, D.; Rape, M. The ubiquitin code. *Annu. Rev. Biochem.* **2012**, *81*, 203–229. [CrossRef] [PubMed]

84. Vierstra, R.D. The expanding universe of ubiquitin and ubiquitin-like modifiers. *Plant Physiol.* **2012**, *160*, 2–14. [CrossRef] [PubMed]

85. Shu, K.; Yang, W.Y. E3 ubiquitin ligases: Ubiquitous actors in plant development and abiotic stress responses. *Plant Cell Physiol.* **2017**, *58*, 1461–1476. [CrossRef] [PubMed]

86. Hua, Z.H.; Vierstra, R.D. The cullin-ring ubiquitin-protein ligases. *Annu. Rev. Plant Biol.* **2011**, *62*, 299–334. [CrossRef] [PubMed]

87. Kraft, E.; Stone, S.L.; Ma, L.G.; Su, N.; Gao, Y.; Lau, O.S.; Deng, X.W.; Callis, J. Genome analysis and functional characterization of the E2 and RING-type E3 ligase ubiquitination enzymes of Arabidopsis. *Plant Physiol.* **2005**, *139*, 1597–1611. [CrossRef] [PubMed]

88. Stone, S.L.; Hauksdottir, H.; Troy, A.; Herschleb, J.; Kraft, E.; Callis, J. Functional analysis of the RING-type ubiquitin ligase family of Arabidopsis. *Plant Physiol.* **2005**, *137*, 13–30. [CrossRef] [PubMed]

89. Ling, Q.H.; Huang, W.H.; Baldwin, A.; Jarvis, P. Chloroplast biogenesis is regulated by direct action of the ubiquitin-proteasome system. *Science* **2012**, *338*, 655–659. [CrossRef] [PubMed]

90. Jarvis, P.; López-Juez, E. Biogenesis and homeostasis of chloroplasts and other plastids. *Nat. Rev. Mol. Cell Biol.* **2013**, *14*, 787–802. [CrossRef] [PubMed]

91. Ling, Q.H.; Jarvis, P. Regulation of chloroplast protein import by the ubiquitin E3 Ligase SP1 is important for stress tolerance in plants. *Curr. Biol.* **2015**, *25*, 2527–2534. [CrossRef] [PubMed]

92. Woodson, J.D.; Joens, M.S.; Sinson, A.B.; Gilkerson, J.; Salom, P.A.; Weigel, D.; Fitzpatrick, J.A.; Chory, J. Ubiquitin facilitates a quality-control pathway that removes damaged chloroplasts. *Science* **2015**, *350*, 450–454. [CrossRef] [PubMed]

93. Kraft, C.; Peter, M.; Hofmann, K. Selective autophagy: Ubiquitin-mediated recognition and beyond. *Nat. Cell Biol.* **2010**, *12*, 836–841. [CrossRef] [PubMed]

94. Matsuda, N.; Sato, S.; Shiba, K.; Okatsu, K.; Saisho, K.; Gautier, C.A.; Sou, Y.S.; Saiki, S.; Kawajiri, S.; Sato, F.; et al. PINK1 stabilized by mitochondrial depolarization recruits Parkin to damaged mitochondria and activates latent Parkin for mitophagy. *J. Cell Biol.* **2010**, *189*, 211–221. [CrossRef] [PubMed]

95. Narendra, D.; Tanaka, A.; Suen, D.F.; Youle, R.J. Parkin is recruited selectively to impaired mitochondria and promotes their autophagy. *J Cell Biol.* **2008**, *183*, 795–803. [CrossRef] [PubMed]

96. Narendra, D.P.; Jin, S.M.; Tanaka, A.; Suen, D.F.; Gautier, C.A.; Shen, J.; Cookson, M.R.; Youle, R.J. PINK1 is selectively stabilized on impaired mitochondria to activate Parkin. *PLoS Biol.* **2010**, *8*, e1000298. [CrossRef] [PubMed]

97. Vives-Bauza, C.; Zhou, C.; Huang, Y.; Cui, M.; de Vries, R.L.; Kim, J.; May, J.; Tocilescu, M.A.; Liu, W.; Ko, H.S.; et al. PINK1-dependent recruitment of Parkin to mitochondria in mitophagy. *Proc. Natl. Acad. Sci. USA* **2010**, *107*, 378–383. [CrossRef] [PubMed]

98. Have, M.; Balliau, T.; Cottyn-Boitte, B.; Derond, E.; Cueff, G.; Soulay, F.; Lornac, A.; Reichman, P.; Dissmeyer, N.; Avice, J.C.; et al. Increase of proteasome and papain-like cysteine protease activities in autophagy mutants: Backup compensatory effect or pro cell-death effect? *J. Exp. Bot.* **2017**. [CrossRef]

99. Yoshimoto, K.; Jikumaru, Y.; Kamiya, Y.; Kusano, M.; Consonni, C.; Panstruga, R.; Ohsumi, Y.; Shirasu, K. Autophagy negatively regulates cell death by controlling NPR1-dependent salicylic acid signaling during senescence and the innate immune response in *Arabidopsis*. *Plant Cell* **2009**, *21*, 2914–2927. [CrossRef] [PubMed]

100. Wang, P.; Sun, X.; Jia, X.; Ma, F. Apple autophagy-related protein MdATG3s afford tolerance to multiple abiotic stresses. *Plant Sci.* **2017**, *256*, 53–64. [CrossRef] [PubMed]

101. Xia, T.M.; Xiao, D.; Liu, D.; Chai, W.T.; Gong, Q.Q.; Wang, N.N. Heterologous expression of ATG8c from soybean confers tolerance to nitrogen deficiency and increases yield in Arabidopsis. *PLoS ONE* **2012**, *7*, e37217. [CrossRef] [PubMed]

14

Complete Chloroplast Genome Sequence and Phylogenetic Analysis of *Quercus acutissima*

Xuan Li [1], Yongfu Li [1], Mingyue Zang [1], Mingzhi Li [2] and Yanming Fang [1,*]

[1] Co-Innovation Center for Sustainable Forestry in Southern China, College of Biology and the Environment, Key Laboratory of State Forestry Administration on Subtropical Forest Biodiversity Conservation, Nanjing Forestry University, 159 Longpan Road, Nanjing 210037, China; xuanli18851128817@163.com (X.L.); liyongfu199417@gmail.com (Y.L.); sanskritm@163.com (M.Z.)

[2] Genepioneer Biotechnologies Co. Ltd., Nanjing 210014, China; limzhi87@foxmail.com

* Correspondence: jwu4@njfu.edu.cn

Abstract: *Quercus acutissima*, an important endemic and ecological plant of the *Quercus* genus, is widely distributed throughout China. However, there have been few studies on its chloroplast genome. In this study, the complete chloroplast (cp) genome of *Q. acutissima* was sequenced, analyzed, and compared to four species in the Fagaceae family. The size of the *Q. acutissima* chloroplast genome is 161,124 bp, including one large single copy (LSC) region of 90,423 bp and one small single copy (SSC) region of 19,068 bp, separated by two inverted repeat (IR) regions of 51,632 bp. The GC content of the whole genome is 36.08%, while those of LSC, SSC, and IR are 34.62%, 30.84%, and 42.78%, respectively. The *Q. acutissima* chloroplast genome encodes 136 genes, including 88 protein-coding genes, four ribosomal RNA genes, and 40 transfer RNA genes. In the repeat structure analysis, 31 forward and 22 inverted long repeats and 65 simple-sequence repeat loci were detected in the *Q. acutissima* cp genome. The existence of abundant simple-sequence repeat loci in the genome suggests the potential for future population genetic work. The genome comparison revealed that the LSC region is more divergent than the SSC and IR regions, and there is higher divergence in noncoding regions than in coding regions. The phylogenetic relationships of 25 species inferred that members of the *Quercus* genus do not form a clade and that *Q. acutissima* is closely related to *Q. variabilis*. This study identified the unique characteristics of the *Q. acutissima* cp genome, which will provide a theoretical basis for species identification and biological research.

Keywords: *Quercus*; chloroplast genome; phylogenetic relationship

1. Introduction

Oak trees provide humans with materials used in food, clothing, and houses, while oak forests supply living organisms and animals with comfortable habitats, good air, and sufficient and pure moisture. Oak trees are linked to Chinese culture, and are also often called eucalyptus or pecking trees. In China, eucalyptus is regarded as a mysterious tree, growing silently, watching its ancestors forge ahead, and passing through generation to generation. Many countries regard oaks as sacred trees, and consider them to be magical and a symbol of longevity, strength, and pride.

The genus *Quercus* L. (Oak) contains more than 400 species that are widespread in the northern hemisphere [1]. These species play important roles in China's forest ecosystem. *Quercus* L. (Oak)'s taxonomy, genetic structure, and breeding is complicated because of its wide variety of species, diverse forms, complex habitat conditions, and gene exchanges between species. Many studies have used nuclear simple sequence repeat (SSR) chloroplast DNA makers to study phylogeny and population variation [2,3]. Previously, studies found a conflict (inconsistency) between the phylogeny of plastid data and nuclear data in Senecioneae and Neotropical Catasetinae [4,5]. Therefore, it is not sufficient to study *Quercus* simply by using plastid regions. With the rapid development of next-generation sequencing, genome acquisition is now cheaper and faster than traditional Sanger sequencing. Complete chloroplast (cp) genome size data will be necessarily used to infer the phylogenetic relationship of *Quercus* or Fagaceae in future studies.

The genus is characterized by a high variability of morphological and ecological traits, the occurrence of mixed stands, the presence of large population sizes, and high levels of gene flow within the *Quercus* complex [6–11]. A new classification of *Quercus* L. was proposed by Denk with eight sections: *Cyclobalanopsis, Cerris, Ilex, Lobatae, Quercus, Ponticae, Protobalanus*, and *Virentes* [12]. In China, *Quercus* is divided into five morphology-based sections: *Quercus, Aegilops, Heterobalanus, Engleriana*, and *Echinolepides* [13–15]. Due to incomplete sampling and the use of markers with insufficient phylogenetic signals and complex evolutionary problems, the relationships among *Quercus* species are not fully understood.

Q. acutissima is an ecological and economic tree species in deciduous broad-leaved forests in the temperate zone of East Asia, widely distributed on the Hu Huanyong line or in Southeast China (latitude from 18° to 41° N and longitude from 91° to 123° E) [16]. This line from Heilongjiang Province to Tengchong, Yunnan Province, is roughly inclined in a 45° straight line. The development, origin, and reproduction of China are linked with *Q. acutissima*. Therefore, we need to protect, cultivate, and utilize *Q. acutissima*, and this has received substantial attention in phylogeny and biogeography studies. Most previous studies have focused on its population structure [17], breeding [18], forest management [19], and physiology [20]. Studies on the genetic variation of *Q. acutissima* using simple sequence repeat (SSR) and cpDNA makers have been carried out in China and South Korea [16,21]. According to this research, the distribution of *Q. acutissima* often overlaps with other oak trees, i.e., *Q. variabilis* and *Q. chenii* [22]. There is often a variety of species found in the population, although this has usually been determined from a comparison of morphology, rather than at a molecular level. Therefore, an analysis of the complete cp genome of *Q. acutissima* will help to identify the species further.

In the present study, we constructed the whole chloroplast genome of *Q. acutissima* by using next-generation sequencing and applying a combination of de novo and reference-guided assembly. Here, we describe the whole chloroplast genome sequence of *Q. acutissima* and the characterization of long repeats and simple sequence repeats (SSRs). We compare and analyze the chloroplast genome of *Q. acutissima* and the chloroplast genome of other members of Fagaceae. It is expected that the results will provide a theoretical basis for the determination of phylogenetic status and future scientific research.

2. Results and Discussion

2.1. Features of Q. Acutissima cpDNA

A total number of 63 million pair-end reads were produced with 9.82 Gb of clean data. Data from all of the reads were deposited in the NCBI Sequence Read Archive (SRA) under accession number MH607377. The size of the complete cp genome is 161,124 bp (Figure 1). The cp genome displayed a typical quadripartite structure, including a pair of IR (25,816 bp) separated by the large single copy (LSC; 90,423 bp) and small single copy (SSC; 19,069 bp) regions (Figure 1 and Table 1). The DNA G + C contents of the LSC, SSC, and IR regions, and the whole genome are 34.62, 30.84, 42.78, and 36.08 mol %, respectively, which is also similar to the chloroplast genomes of other *Quercus* species (Figure A1; Table 2). The DNA G + C content is a very important indicator of species affinity [23]. It is obvious that the DNA G + C content of the IR region is higher than that of other regions (LSC, SSC). This phenomenon is very common in other plants [23,24]. GC skewness has been shown to be an indicator of DNA lead chains, lag chains, replication origin, and replication terminals [25–27].

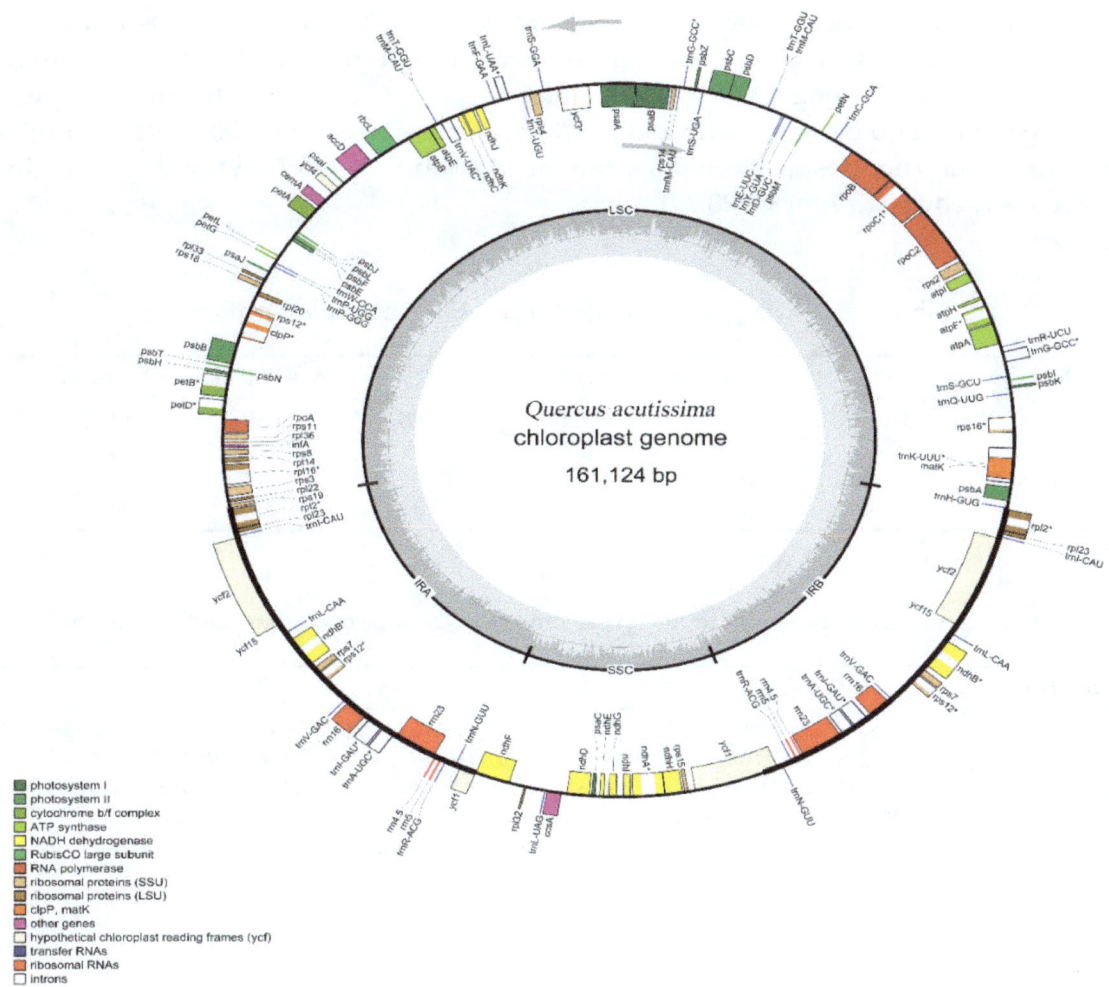

Figure 1. Chloroplast genome map of *Q. acutissima*. Genes inside the circle are transcribed clockwise, and those outside are transcribed counterclockwise. Genes of different functions are color-coded. The darker gray in the inner circle shows the GC content, while the lighter gray shows the AT content.

Table 1. Summary of five *Quercus* chloroplast genome features.

Genome Features	Q. acutissima	Q. variabilis	Q. dolicholepis	C. mollissima	L. balansae	F. engleriana
Genome size (bp)	161,124	161,077	161,237	160,799	161,020	158,346
LSC length (bp)	90,423	90,387	90,461	90,432	90,596	87,667
SSC length (bp)	19,068	19,056	19,048	18,995	19,160	18,895
IR length (bp)	51,632	51,634	51,728	51,372	51,264	51,784
Number of genes	136	134	134	130	134	131
Number of protein–coding genes	88	86	86	83	87	83
Number of tRNA genes	40	40	40	37	39	40
Number of rRNA genes	8	8	8	8	8	8

Plant chloroplast genomes may have 63–209 genes, but most are concentrated between 110 and 130, with a highly conserved composition and arrangement, including photosynthetic genes, chloroplast transcriptional expression-related genes, and some other protein-coding genes [28]. In the *Q. acutissima* chloroplast genome, 136 functional genes were predicted and divided into six groups, including eight rRNA genes, 40 tRNA genes, and 88 protein-coding genes (Tables 1 and 3). In addition, 14 tRNA genes, eight rRNA genes, and 15 protein-coding genes are duplicated in the IR regions (Figure 1). The LSC region includes 62 protein-coding and 25 tRNA genes, while the SSC region includes 13 protein-coding genes (Table A1).

Based on the protein-coding sequences and tRNA genes, the frequency of codon usage was estimated for the *Q. acutissima* cp genome and is summarized in Table A2. In total, all genes are encoded by 6311 codons. Among these, leucine, with 2824 (44.4%) codons, is the most frequent amino acid in the cp genome, and cysteine, with 293 (1.1%), is the least frequent (Table 3). A- and U-ending codons are common. The most preferred synonymous codons (relative synonymous codon usage values (RSCU) > 1) end with A or U [23,29].

Table 2. Base composition of the *Q. acutissima* chloroplast genome.

Region	A (%)	T (U) (%)	C (%)	G (%)	A + T (%)	G + C (%)
LSC	31.99	33.4	17.74	16.88	65.39	34.62
SSC	34.46	34.71	16.24	14.6	69.17	30.84
IR	28.61	28.61	21.39	21.39	57.22	42.78
Total	31.69	32.24	18.46	17.62	63.93	36.08

Table 3. List of genes annotated in the cp genomes of *Q. acutissima* sequenced in this study.

Function	Genes
RNAs, transfer	*trnH-GUG, trnK-UUU, trnQ-UUG, trnS-GCU, trnG-GCC, trnR-UCU, trnC-GCA, trnD-GUC, trnY-GUA, trnE-UUC, trnT-GGU, trnM-CAU, trnS-UGA, trnG-GCC, trnfM-CAU, trnS-GGA, trnT-UGU, trnL-UAA, trnF-GAA, trnV-UAC, trnM-CAU, trnT-GGU, trnW-CCA, trnP-UGG, trnP-GGG, trnI *-CAU, trnL-CAA *, trnV-GAC, trnI-GAU *, trnA-UGC, trnR-ACG, trnN-GUU, trnL-UAG, trnN-GUU, trnR-ACG, trnA-UGC, trnV-GAC*
RNAs, ribosomal	*rrn23 *, rrn16 *, rrn5 *, rrn4.5 **
Transcription and splicing	*rpoC1 *, rpoC2, rpoA, rpoB*
Translation, ribosomal proteins	
Small subunit	*rps2, rps3, rps4, rps7, rps8, rps11, rps12 **, rps14, rps15, rps16 *, rps18, rps19*
Large subunit	*rpl2 *, rpl14, rpl16 *, rpl20, rpl22, rpl23, rpl32, rpl33, rpl36*
Photosynthesis	
ATP synthase	*atpE, atpB, atpA, atpF *, atpH, atpI*
Photosystem I	*psaI, psaB, psaA, psaC, psaJ, ycf3 *, ycf4*
Photosystem II	*psbD, psbC, psbZ, psbT, psbH, psbK, psbI, psbJ, psbF, psbE, psbM, psbN, psbL, psbA, psbB*
Calvin cycle	*rbcL*
Cytochrome complex	*petN, petA, petL, petG, petB *, petD **
NADH dehydrogenase	*ndhB *, ndhI, ndhK, ndhC, ndhF, ndhD, ndhG, ndhE, ndhA, ndhH, ndhJ*
Others	*inFA, ycf15 *, ycf1 *, ycf2 *, accD, cemA, ccsA, clpP ***

* Genes containing one intron; ** genes containing two introns.

In total, we found 23 intron-containing genes, including 15 protein-coding genes, and eight tRNA genes (Table 4). 21 genes (13 protein-coding and eight tRNA genes) contain one intron, and two genes (*ycf3* and *clpP*) contain two introns. The *trnK-UUU* has the largest intron (2505 bp), and the *trnL-UAA* has the smallest intron (483bp). Studies have shown that *ycf3* is required for stable accumulation of photosystem I complexes [30]. Therefore, we speculate that the *ycf3* intron gain of *Q. acutissima* may be helpful for further study of the mechanism of photosynthesis evolution.

Table 4. The lengths of exons and introns in genes with introns in the *Q. acutissima* chloroplast genome.

Gene	Location	Exon I (bp)	Intron I (bp)	Exon II (bp)	Intron II (bp)	Exon III (bp)
rps16	LSC	42	898	195		
atpF	LSC	144	780	411		
rpoC1	LSC	432	827	1626		
ycf3	LSC	127	718	228	778	155
clpP	LSC	69	844	294	649	228
petB	LSC	6	841	642		
petD	LSC	9	640	474		
rpl16	LSC	9	1102	399		
rpl2	RepeatA	390	628	471		
ndhB	RepeatA	777	680	756		
rps12	RepeatA	10	537	231		
ndhA	SSC	551	1040	541		
rps12	RepeatB			232	536	26
ndhB	RepeatB	777	680	756		
rpl2	RepeatB	390	628	471		
trnG-GCC	LSC	23	734	37		
trnK-UUU	LSC	37	2505	35		
trnL-UAA	LSC	35	483	50		
trnV-UAC	LSC	36	630	37		
trnI-GAU	RepeatA	42	950	35		
trnA-UGC	RepeatA	38	800	35		
TRNA-UGC	RepeatB	38	800	35		
trnI-GAU	RepeatB	42	950	35		

2.2. Comparative Analysis of Genomic Structure

The chloroplast sequence are often used to measure the genetic diversity within a species, the gene flow between species, and the size of ancestral populations of separated sister species [31]. Thus, it is necessary to understand the chloroplast differences between species. The complete cp genome sequence of *Q. acutissima* was compared to those of *Q. variabilis*, *Q. dolicholepis*, *Castanea mollissima*, *Lithocarpus balansae*, and *Fagus engleriana*. *F. engleriana* has the smallest cp genome with the largest IR region (51,784 bp), and *Q. dolicholepis* has the largest cp genome (Table 1). We assumed that the different lengths of the SSC and IR regions is the main reason for variety in sequence lengths. To verify the possibility of genome divergence, sequence identity was calculated for six species' chloroplast DNA using the program mVISTA with *Q. variabilis* as a reference (Figure 2). The results of this comparison revealed that LSC regions are more divergent than SSC and IR regions and that higher divergence is found in noncoding than in coding regions. The complete cp genome sequence of *F. engleriana* is quite different from the five other plants. There was no significant difference between the chloroplast genome sequences of evergreen and deciduous trees. At the same time, the results of the sliding window indicated that the location of the variation in the cp genome among the six species occurred in the LSC and SSC regions (Figure A2). Significant variation was found in coding regions of some genes, including *psbI*, *rpl33*, *petB*, *rpl2*, *rps16*, *rpoC2*, *ndhK*, *ycf2*, *ycf1*, and *ndhI*. The highest divergence in noncoding regions was found in the intergenic regions of *trnK-rps16*, *rps 16-trnQ*, *psbK-psbI*, *trnS-trnG*, *atpH-atpI*, *atpI-rps2*, *rpoB-trnC*, *trnC-petN*, *psbM-trnD*, *trnD-trnY*, *trnE-trnM*, *trnT-petD*, *psbZ-trnG*, *trnT-trnL*, *trnF-ndhJ*, *rbcL-accD*, *psaI-ycf4*, *ycf4-cemA*, *petA-psbL*, *psaJ-rpl33*, *clpP-psbB*, *rpl14-rpl16*, *ndhF-rpl32*, *ccsA-ndhD*, *ndhD-psaC*, and *rps15-ycf1*.

The contraction and expansion of the IR region at the borders play important roles in evolution. They are common evolutionary events and a major cause of changes in the size of the chloroplast genome. They may also cause variation in the length of angiosperm plastid genome [32–34]. Detailed comparisons of the IR–SSC and IR–LSC boundaries among the cp genomes of the above six Fagaceae species were presented in Figure 3. The IR regions are relatively highly conserved in the *Quercus* genus—the *rpl2* gene in the *Quercus* cp genome is shifted by 62 bp from IRb to LSC at the LSC/IRb border, and by 62 bp from IRa to LSC at the IRa/LSC border. Compared to other species in the genus, the range of the IRa/SSC regions changes greatly. Compared with evergreen and deciduous species, we found significant differences in IRb/SSC. Some reports showed that *ycf1* is necessary for plant viability and encodes *Tic214*, an important component of the *Arabidopsis TIC* complex [35,36]. The *ycf1* gene crossed the SSC/IRb region, with 1041bp of *ycf1_like* within IRb (incompletely duplicated in IRb). The SSC/IRa junction is located in the *ycf1* region in all Fagaceae species chloroplast genomes and extends into the SSC region by different lengths depending on the genome (*Q. acutissima*, 4619 bp; *Q. variabilis*, 4620 bp; *Q. dolicholepis*, 4611 bp; *C. mollissima*, 4623 bp; *L. balansae*, 4626 bp; *F. engleriana*, 4633 bp); the IRa region includes 1041, 1041, 1068, 1059, 828, and 1049 bp of the *ycf1* gene.

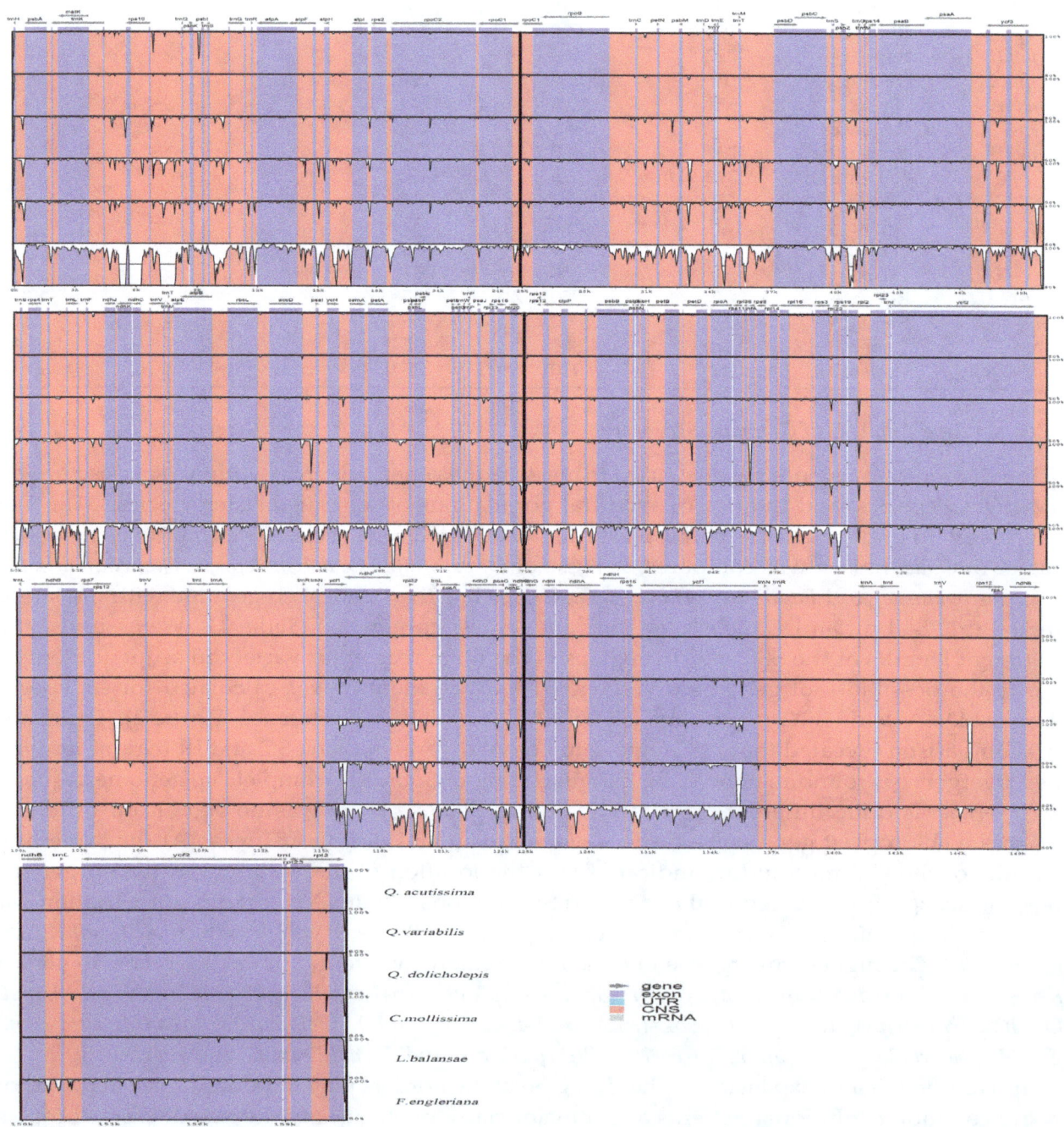

Figure 2. Complete chloroplast genome comparison of six species using the chloroplast genome of *Q. variabilis* as a reference. The grey arrows and thick black lines above the alignment indicate the genes' orientations. The Y-axis represents the identity from 50% to 100%.

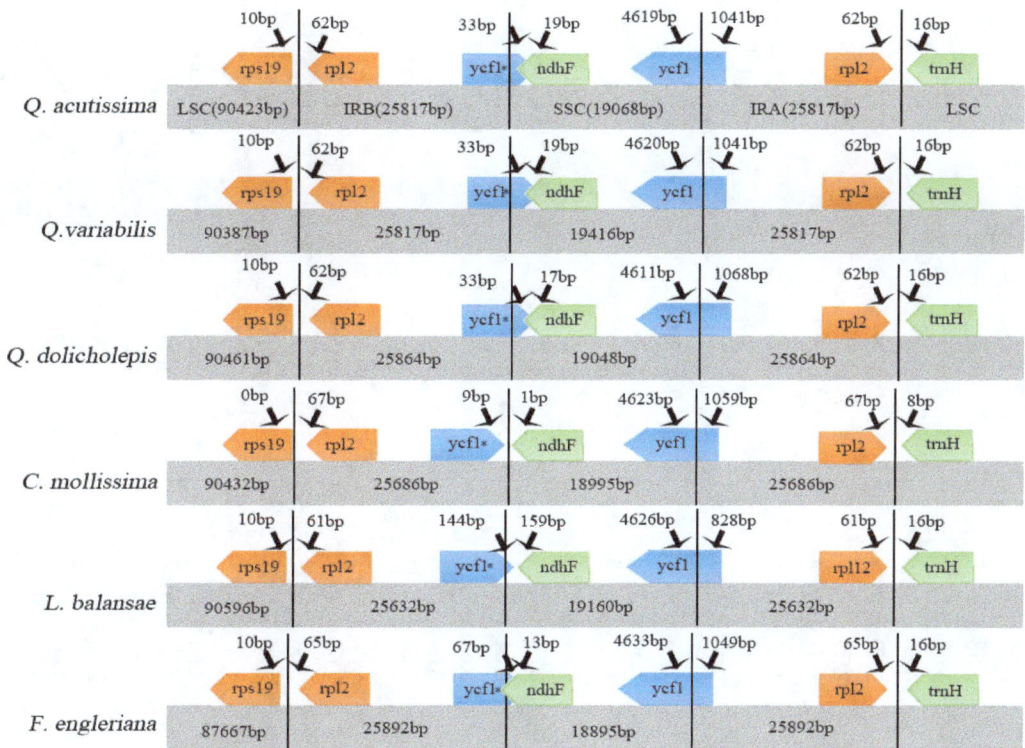

Figure 3. Comparison of the large single copy (LSC), small single copy (SSC), and inverted repeat (IR) regions in chloroplast genomes of four species. Genes are denoted by colored boxes. The gaps between the genes and the boundaries are indicated by the base lengths (bp). Extensions of the genes are indicated above the boxes.

2.3. Long-Repeat and SSR Analysis

For the repeat structure analysis (Table 5), 31 forward and 22 inverted repeats were detected in the *Q. acutissima* cp genome. Most of these repeats are between 19 and 46 bp. The longest forward repeat is 46 bp in length and is located in the LSC region. A total of 35, 18, and eight repeats were found in the LSC, SSC, IR regions, respectively. Seven forward repeats were located in IR, including one repeat associated with *ycf1* genes and one repeat related to the *trnV-UAC* and *trnA-UGC* genes. Most repeats in the intergenic spacers are distributed in the LSC region. Ten repeats are distributed in the SSC region, and only four of them are in the intergenic spacers.

As chloroplast-specific SSRs are uniparentally inherited and are inclined to undergo slipped-strand mispairing, they are often used in population genetics, species identification, and evolutionary process research of wild plants [37,38]. In addition, chloroplast genome sequences are highly conserved, and the SSR primer for chloroplast genomes can be transferred across species and genera. Yoko et al. used six maternally inherited chloroplast (cpDNA) simple sequence repeat (SSR) markers to study the genetic variation in *Q. acutissima* [39]. In this study, a total of 65 SSRs were found in *Q. acutissima*, most of them distributed in LSC and SSC and partly distributed in IR. These included 61 mononucleotide SSRs (93.85%) and four dinucleotide SSRs (6.15%) (Table 6). Compared with other *Quercus* species, fewer types of SSRs were identified in *Q. acutissima* [40]. Among them, two SSRs belonged to the C type, and the others all belonged to the A/T types. These results are consistent with the hypothesis that cpSSRs are generally composed of short polyadenine (polyA) or polythymine (polyT) repeats and rarely contain tandem guanine (G) or cytosine (C) repeats [41]. We also found that 12 SSRs were located in genes, and the remaining were all located in intergenic regions. These cpSSR markers could be used to examine the genetic structure, diversity, differentiation, and maternity in *Q. acutissima* and its relative species in future studies.

Table 5. Long repeat sequence in the *Q. acutissima* chloroplast genome.

ID	Repeat Start I	Type	Size (bp)	Repeat Start 2	Mismatch (bp)	E-Value	Gene	Region
1	6831	F	46	6853	0	1.47×10^{-18}	IGS	LSC
2	11,847	R	31	11,847	0	1.58×10^{-9}	IGS	LSC
3	6818	R	26	6818	0	1.62×10^{-6}	*rps16*	LSC
4	47,242	F	25	47,264	0	6.49×10^{-6}	IGS	LSC
5	6831	F	24	6875	0	2.59×10^{-5}	IGS	LSC
6	115,801	F	24	135,722	0	2.59×10^{-5}	*ycf1*	IRA; IRB
7	113,545	F	23	113,576	0	1.04×10^{-4}	IGS	IRA
8	118,844	R	23	118,844	0	1.04×10^{-4}	IGS	IRA
9	137,948	F	23	137,979	0	1.04×10^{-4}	IGS	IRB
10	11,371	F	22	41,193	0	4.15×10^{-4}	*trnG-GCC* (exon), *trnG-GCC*	LSC
11	9536	F	21	39,849	0	1.66×10^{-3}	*trnS-UGA, trnS-GCU*	LSC
12	10,319	F	21	18,682	0	1.66×10^{-3}	IGS	LSC
13	117,049	R	21	117,049	0	1.66×10^{-3}	*ndhF*	SSC
14	36,478	F	20	53,719	0	6.64×10^{-3}	IGS	LSC
15	53,720	F	20	130,481	0	6.64×10^{-3}	IGS	LSC; SSC
16	55,907	R	20	55,907	0	6.64×10^{-3}	*atpB*	LSC
17	57,271	F	20	142,064	0	6.64×10^{-3}	*trnV-UAC, trnA-UGC*	LSC; IRB
18	105,331	F	20	105,349	0	6.64×10^{-3}	IGS	IRA
19	146,178	F	20	146,196	0	6.64×10^{-3}	IGS	IRB
20	4930	F	19	36,476	0	2.66×10^{-2}	IGS	LSC
21	8915	R	19	8915	0	2.66×10^{-2}	IGS	LSC
22	13,541	R	19	76,642	0	2.66×10^{-2}	*atpA*	LSC
23	18,685	R	19	118,842	0	2.66×10^{-2}	*clpP*	LSC; SSC
24	21,297	R	19	54,183	0	2.66×10^{-2}	*rpoC2*	LSC
25	36,479	F	19	130,481	0	2.66×10^{-2}	IGS	LSC; SSC
26	39,957	R	19	39,957	0	2.66×10^{-2}	IGS	LSC
27	62,040	R	19	62,040	0	2.66×10^{-2}	IGS	LSC
28	64,751	R	19	64,751	0	2.66×10^{-2}	IGS	LSC
29	69,026	R	19	69,026	0	2.66×10^{-2}	IGS	LSC
30	71,277	R	19	71,277	0	2.66×10^{-2}	IGS	LSC
31	72,561	R	19	72,561	0	2.66×10^{-2}	IGS	LSC
32	4430	R	18	4430	0	1.06×10^{-1}	IGS	LSC
33	4437	F	18	24,828	0	1.06×10^{-1}	*rpoC1* (intron)	SSC

Table 5. *Cont.*

ID	Repeat Start 1	Type	Size (bp)	Repeat Start 2	Mismatch (bp)	E-Value	Gene	Region
34	4935	F	18	52,105	0	1.06×10^{-1}	IGS	LSC
35	4938	F	18	118,695	0	1.06×10^{-1}	IGS	LSC
36	6813	F	18	6847	0	1.06×10^{-1}	IGS	LSC
37	6813	F	18	6869	0	1.06×10^{-1}	IGS	LSC
38	6817	F	18	127,945	0	1.06×10^{-1}	*ndhA* (intron)	LSC
39	7369	F	18	7387	0	1.06×10^{-1}	IGS	LSC; SSC
40	7465	R	18	7465	0	1.06×10^{-1}	IGS	LSC; SSC
41	8589	R	18	34,768	0	1.06×10^{-1}	IGS	LSC; SSC
42	9996	R	18	9996	0	1.06×10^{-1}	IGS	LSC
43	10,283	F	18	31,730	0	1.06×10^{-1}	IGS	LSC
44	10,322	R	18	118,843	0	1.06×10^{-1}	IGS	LSC; IRA
45	10,548	F	18	133,365	0	1.06×10^{-1}	*ycf1*	LSC
46	31,728	F	18	125,951	0	1.06×10^{-1}	IGS	LSC
47	39,812	F	18	40,698	0	1.06×10^{-1}	*trnS-UGA*	LSC; SSC
48	40,022	R	18	69,093	0	1.06×10^{-1}	IGS	LSC
49	40,700	F	18	123,827	0	1.06×10^{-1}	IGS	LSC
50	43,446	F	18	45,670	0	1.06×10^{-1}	*psaB*	SSC
51	40,022	R	18	69,093	0	1.06×10^{-1}	IGS	LSC
52	40,700	F	18	123,827	0	1.06×10^{-1}	IGS	LSC
53	43,446	F	18	45,670	0	1.06×10^{-1}	*psaB, psaA*	LSC

F: forward; I: inverted; IGS: intergenic space.

Table 6. Simple sequence repeats (SSRs) in the *Q. acutissima* chloroplast genome.

ID	Repeat Motif	Length (bp)	Start	End	Region	Gene
1	(A)10	9	1809	1818	LSC	
2	(C)14	13	4433	4446	LSC	
3	(T)11	10	4697	4707	LSC	
4	(A)10	9	4939	4948	LSC	*trnK-UUU*
5	(T)11	10	7001	7011	LSC	
6	(T)10	9	7746	7755	LSC	
7	(A)10	9	8174	8183	LSC	
8	(A)12	11	8590	8601	LSC	
9	(A)11	10	8920	8930	LSC	
10	(A)10	9	9465	9474	LSC	
11	(A)10	9	10,161	10,170	LSC	
12	(A)11	10	13,547	13,557	LSC	
13	(T)12	11	15,345	15,356	LSC	
14	(T)10	9	16,160	16,169	LSC	
15	(A)12	11	18,692	18,703	LSC	*rpoC2*
16	(T)12	11	21,295	21,306	LSC	*rpoC2*
17	(T)14	13	25,299	25,312	LSC	
18	(T)10	9	28,563	28,572	LSC	
19	(T)10	9	29,651	29,660	LSC	
20	(T)11	10	30,275	30,285	LSC	
21	(C)14	13	30,428	30,441	LSC	
22	(T)11	10	31,731	31,741	LSC	
23	(A)10	9	32,094	32,103	LSC	
24	(A)10	9	33,986	33,995	LSC	
25	(A)13	12	34,775	34,787	LSC	
26	(A)10	9	34,955	34,964	LSC	
27	(A)10	9	36,485	36,494	LSC	
28	(AT)6	11	39,819	39,830	LSC	
29	(T)10	9	41,238	41,247	LSC	*trnfM-CAU*
30	(T)11	10	53,217	53,227	LSC	
31	(A)10	9	53,726	53,735	LSC	
32	(T)15	14	54,110	54,124	LSC	
33	(A)11	10	54,990	55,000	LSC	
34	(T)10	9	55,713	55,722	LSC	
35	(T)10	9	59,591	59,600	LSC	
36	(T)10	9	60,063	60,072	LSC	
37	(T)10	9	64,092	64,101	LSC	*accD*
38	(A)11	10	64,266	64,276	LSC	
39	(AT)7	13	64,570	64,583	LSC	
40	(T)14	13	64,945	64,958	LSC	
41	(T)13	12	66,170	66,182	LSC	
42	(T)11	10	68,616	68,626	LSC	*petA*
43	(T)11	10	70,730	70,740	LSC	
44	(T)11	10	71,398	71,408	LSC	
45	(T)11	10	73,389	73,399	LSC	
46	(AT)6	11	77,274	77,285	LSC	*clpP*
47	(TA)7	13	82,928	82,941	LSC	*petD*
48	(A)11	10	85,781	85,791	LSC	
49	(T)10	9	86,100	86,109	LSC	
50	(T)10	9	88,820	88,829	LSC	
51	(T)11	10	114,070	114,080	IRA	
52	(T)12	11	118,582	118,593	SSC	
53	(A)11	10	118,695	118,705	SSC	
54	(T)11	10	119,000	119,010	SSC	
55	(A)10	9	119,794	119,803	SSC	
56	(T)11	10	122,199	122,209	SSC	*ndhD*
57	(A)10	9	122,546	122,555	SSC	
58	(AT)8	15	123,832	123,847	SSC	
59	(T)11	10	125,812	125,822	SSC	
60	(T)11	10	125,954	125,964	SSC	
61	(T)11	10	130,262	130,272	SSC	
62	(A)10	9	130,487	130,496	SSC	
63	(T)10	9	133,465	133,474	SSC	*ycf1*
64	(T)13	12	134,042	134,054	SSC	*ycf1*
65	(A)11	10	137,468	137,478	SSC	

2.4. Phylogenetic Analysis

Phylogenetic analysis was completed on an alignment of concatenated nucleotide sequences of all chloroplast genomes from 25 angiosperm species (Figure 4). We used the Bayesian inference (BI) method based on RAxML to build a phylogenetic tree, and *Malus prunifolia* and *Ulmus gaussenii* were used as the outgroup. Support is generally high for almost all relationships inferred from all chloroplast genome data based on BI methods (the support values have a range of 0.8956 to 1). It is noteworthy that the species in genus *Quercus* do not form a clade. Several evergreen tree species gather together to form one clade. *Q. acutissima* and *Q. variabilis* are sister species and are frequently mixed in Chinese endemic species; the second clade splits into two subclades. *F. engleriana* is in the top position, while *Q. acutissima* appears to be more closely related to *Q. variabilis*, *Q. dolicholepis*, and *Q. baronii*. In general, the topologies of the other branches (genus *Fagus*, *Trigonobalanus*, *Lithocarpus*, and *Castanopsis*) are almost the same based on two nuclear loci (ITS and CRC) [3].

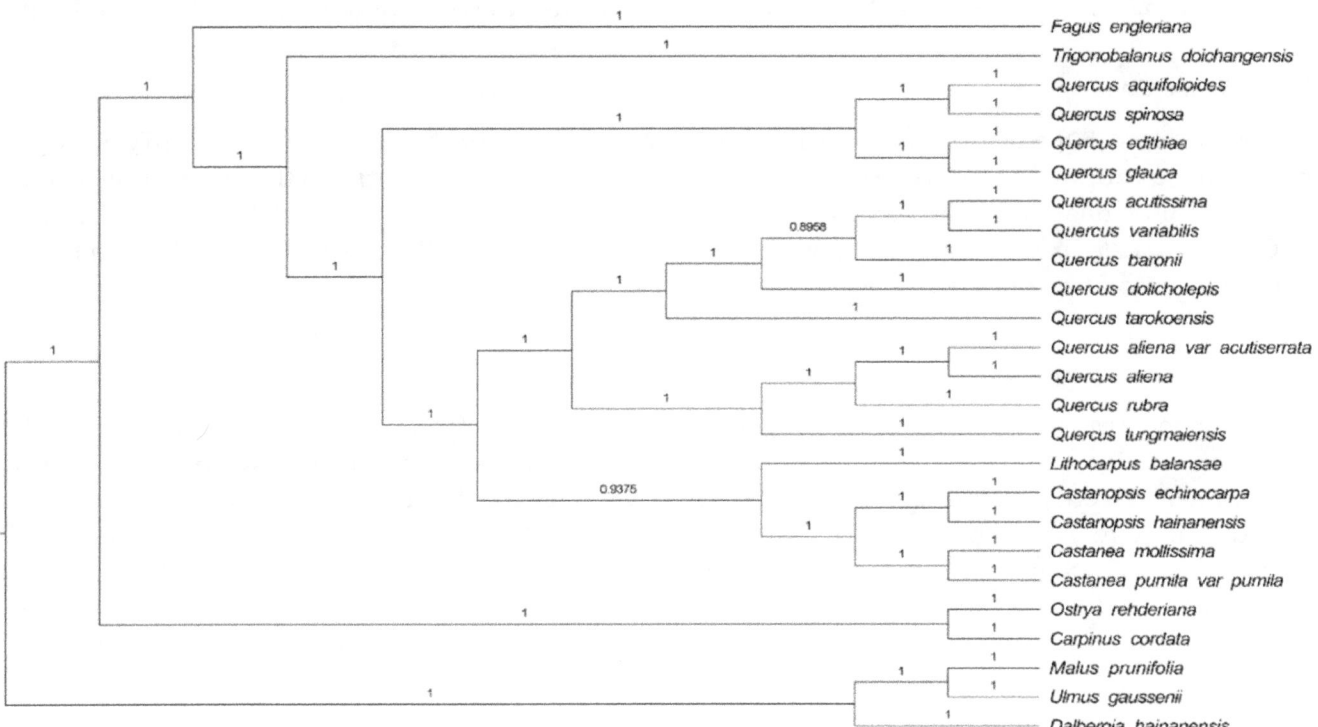

Figure 4. Bayesian inference (BI) phylogenetic tree reconstruction including 25 species based on all chloroplast genomes. *Malus prunifolia* and *Ulmus gaussenii* were used as the outgroup.

3. Materials and Methods

3.1. Sampling, DNA Extraction, Sequencing, and Assembly

Q. acutissima was planted in Nanjing Forestry University and Zijin Mountain in Nanjing, China (32°04′ N, 118°48′ E; 32°04′ N, 118°50′ E), respectively. Fresh leaves were collected and wrapped in ice and immediately stored at −80 °C until analysis. Genomic DNA was isolated by the modified method CTAB [42]. Agarose gel electrophoresis and one drop spectrophotometer (OD-1000, Shanghai Cytoeasy Biotech Co., Ltd., Shanghai, China) were used to detect DNA integrity and quality. Shotgun libraries (250 bp) were constructed using pure DNA according to the manufacturer's instructions. Sequencing was performed with an Illumina Hiseq 2500 platform (Nanjing, China), yielding at least 9.82 GB of clean data for *Q. acutissima*. Firstly, all of the raw reads were trimmed by Fastqc. Next, we performed a BLAST analysis between trimmed reads and references (*Q. variabilis* and *Q. dolicholepis*) to extract cp-like reads. Finally, we used the chloroplast-like reads to assemble sequences using NOVOPlasty [43]. NOVOPlasty assembled part reads and stretched as far as possible until a circular genome formed. When the assembly result was within the expected range, the overlap was larger than 200 bp, and the assembly formed a ring.

3.2. Annotation and Analysis of the cpDNA Sequences

CpGAVAS was used to annotate the sequences; DOGMA (http://dogma.ccbb.utexas.edu/) and BLAST were used to check the results of the annotation [44,45]. tRNAscanSE was used to identify the tRNAs [46]. The circular gene maps of the species of *Q. acutissima* were drawn using the OGDRAWv1.2 program [47] (http://ogdraw.mpimp-golm.mpg.de/). An analysis of variation in synonymous codon usage, relative synonymous codon usage values (RSCU), codon usage, and the GC content of the complete plastid genomes and commonly analyzed CDS was conducted. MISA(available online: http://pgrc.ipk-gatersleben.de/misa/misa.html) [48] and REPuter (available online: https://bibiserv. cebitec.uni-bielefeld.de/reputer/) [49] was used to visualize the SSRs and long repeats, respectively.

3.3. Genome Comparison

MUMmer [50] was used for pairing sequence alignment of the cp genome. The mVISTA [51] program was applied to compare the complete cp genome of *Q. acutissima* to the other published cp genomes of its related species, i.e., *Q. variabilis* (KU240009), *Q. dolicholepis* (KU240010), *C. mollissima* (HQ336406), *L. balansae* (KP299291), and *F. engleriana* (KX852398) with the shuffle-LAGAN mode [52], using the annotation of *Q. variabilis* as a reference.

3.4. Phylogenetic Analysis

Phylogenies were constructed by Bayesian inference (BI) analysis using the 25 cp genome of the Fagaceae species sequences from the NCBI Organelle Genome and Nucleotide Resources database. The sequences were initially aligned using MAFFT [53]. Then, the visualization and manual adjustment of multiple sequence alignment were conducted in BioEdit [54]. An IQ-tree was used to select the best-fitting evaluation of models of nucleotide sequences [55]. TVM + F + R4 and GTR + G were selected as the best substitution models for the BI analyses. BI analyses were conducted using Mrbayes [56]. *Malus prunifolia* (NC_031163), and the *Ulmus gaussenii* (NC_037840) were used as the outgroups.

4. Conclusions

In this study, we reported and analyzed the complete cp genome of *Q. acutissima*, an endemic and ecological tree species in China. The chloroplast genome was shown to be more conservative with similar characteristics to other genus *Quercus* species. Compared to the cp genomes of five other oak species, its LSC were shown to be more divergent among the four regions, and noncoding regions showed higher divergence. An analysis of the phylogenetic relationships among six species found *Q. acutissima* to be closely related to *Q. variabilis*. The developmental position of the tree in the Fagaceae family is consistent with previous studies. The results of this study provide an assembly of a whole chloroplast genome of *Q. acutissima* which might facilitate genetics, breeding, and biological discoveries in the future.

Author Contributions: X.L. performed most of the experiments, data analysis, and the writing of the manuscript; Y.L. participated in the data analysis; M.Z. and M.L. participated in the preprocessing of data; and Y.F. supervised the project and provided suggestions for the manuscript.

Acknowledgments: This research was supported by the National Natural Science Foundation of China (31770699, 31370666), the Priority Academic Program Development of Jiangsu Higher Education Institutions (PAPD), and the Nanjing Forestry University Excellent Doctoral Thesis Fund.

Abbreviations

LSC	Large single copy
SSC	Small single copy
IR	Inverted repeat
Cp	Chloroplast
BI	Bayesian inference
A	Adenine
T	Thymine
G	Guanine
C	Cytosine

Appendix A

Table A1. The number of genes in the *Q. acutissima* cp genome.

Region	Number of CDS	Number of tRNA	Number of rRNA	Total
LSC region	62	25	0	87
SSC region	13	1	0	14
IRA region	6	7	4	17
IRB region	7	7	4	18

Table A2. Codon-anticodon recognition patterns and codon usage of the *Q. acutissima* chloroplast genome.

Amino Acid	Codon	No.	RSCU	tRNA	Amino Acid	Codon	No.	RSCU	tRNA
Ala	GCG	164	0.47		Pro	CCA	313	1.13	*trnP-TGG*
Ala	GCC	224	0.64		Pro	CCC	226	0.82	
Ala	GCU	630	1.79		Pro	CCU	409	1.48	
Ala	GCA	388	1.1		Pro	CCG	161	0.58	
Cys	UGU	221	1.44		Gln	CAG	215	0.45	
Cys	UGC	86	0.56	*trnC-GCA*	Gln	CAA	731	1.55	*trnQ-TTG*
Asp	GAC	209	0.39	*trnD-GTC*	Arg	CGU	337	1.26	*trnR-ACG*
Asp	GAU	870	1.61		Arg	AGA	500	1.87	*trnR-TCT*
Glu	GAA	1064	1.5	*trnE-TTC*	Arg	CGA	358	1.34	
Glu	GAG	357	0.5		Arg	AGG	183	0.68	
Phe	UUU	983	1.3		Arg	CGG	118	0.44	
Phe	UUC	535	0.7	*trnF-GAA*	Arg	CGC	109	0.41	
Gly	GGU	580	1.27		Ser	AGC	125	0.37	*trnS-GCT*
Gly	GGG	330	0.72		Ser	UCU	557	1.66	
Gly	GGA	706	1.55		Ser	UCA	397	1.18	*trnS-TGA*
Gly	GGC	206	0.45	*trnG-GCC*	Ser	UCC	349	1.04	*trnS-GGA*
His	CAU	486	1.54		Ser	AGU	391	1.17	
His	CAC	145	0.46	*trnH-GTG*	Ser	UCG	193	0.58	
Ile	AUC	458	0.58		Thr	ACU	538	1.6	
Ile	AUA	758	0.97		Thr	ACG	160	0.48	
Ile	AUU	1139	1.45		Thr	ACC	247	0.73	*trnT-GGT*
Lys	AAG	379	0.5		Thr	ACA	402	1.19	*trnT-TGT*
Lys	AAA	1062	1.4		Val	GUU	508	1.41	
Leu	UUG	572	1.22	*trnL-CAA*	Val	GUC	181	0.5	*trnV-GAC*
Leu	UUA	894	1.9		Val	GUA	547	1.52	
Leu	CUU	583	1.24		Val	GUG	207	0.57	
Leu	CUA	373	0.79	*trnL-TAG*	Trp	UGG	462	1	*trnW-CCA*
Leu	CUC	204	0.43		Tyr	UAC	212	0.42	*trnY-GTA*
Leu	CUG	198	0.42		Tyr	UAU	792	1.58	
Met	AUG	620	1	*trnI-CAT*	Stop	UAA	47	1.6	
Asn	AAU	1004	1.5		Stop	UAG	22	0.75	
Asn	AAC	304	0.46		Stop	UGA	19	0.65	

RSCU: Relative Synonymous Codon Usage.

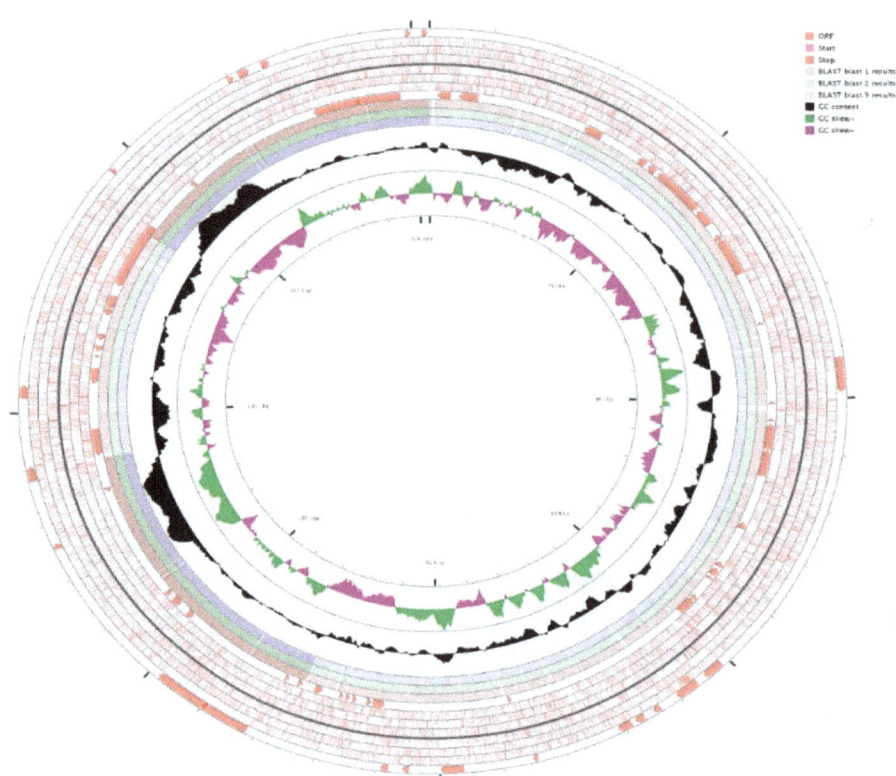

Figure A1. BLAST result of the chloroplast genome and the GC stew of *Q. acutissima*. BlAST 1 represents *L. balansae*; BlAST 2 represents *Q. variabilis*; BlAST 3 represents *Q. dolicholepis*.

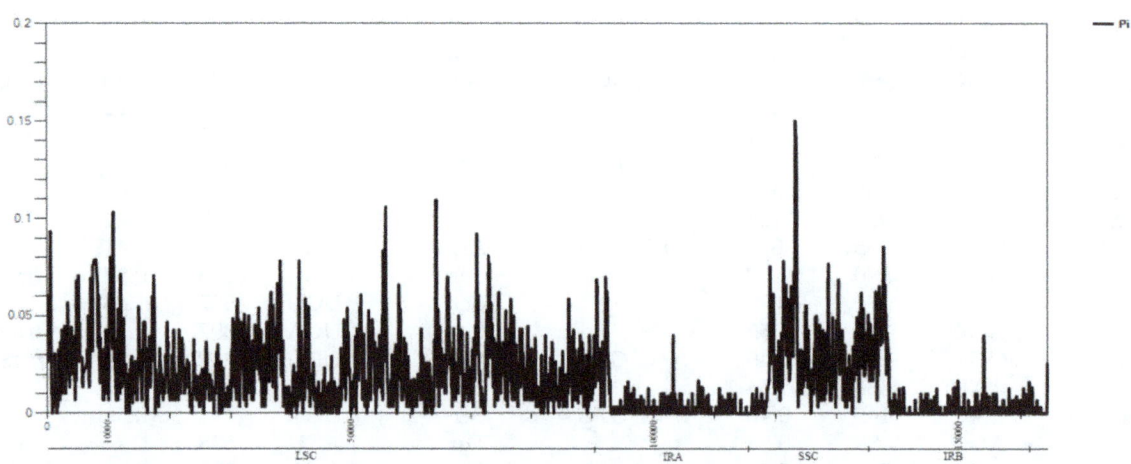

Figure A2. Percentage of variation in the complete cp genomes of the six species. The regions are oriented according to their locations in the genome.

References

1. Aldrich, P.R.; Cavender-Bares, J. Quercus. *Wild Crop Relat. Genom. Breed. Resour.* **2011**, 89–129. [CrossRef]
2. Manos, P.S.; Cannon, C.H.; Oh, S.H. Phylogenetic relationships and taxonomic status of the paleoendemic Fagaceae of western North America: Recognition of a new genus, *Notholithocarpus*. *Madroño* **2008**, *55*, 181–190. [CrossRef]
3. Oh, S.H.; Manos, P.S. Molecular phylogenetics and cupule evolution in Fagaceae as inferred from nuclear crabs claw sequences. *Taxon* **2008**, *57*, 434–451.

4.   Pelser, P.B.; Kennedy, A.H.; Tepe, E.J.; Shidler, J.B.; Nordenstam, B.; Kadereit, J.W.; Watson, L.E. Patterns and causes of incongruence between plastid and nuclear *Senecioneae* (Asteraceae) phylogenies. *Am. J. Bot.* **2010**, *97*, 856–873. [CrossRef] [PubMed]

5.   Pérezescobar, O.A.; Balbuena, J.A.; Gottschling, M. Rumbling orchids: How to assess divergent evolution between chloroplast endosymbionts and the nuclear host. *Syst. Biol.* **2016**, *65*, 51. [CrossRef] [PubMed]

6.   Curtu, A.L.; Gailing, O.; Leinemann, L.; Finkeldey, R. Genetic variation and differentiation within a natural community of five oak species (*Quercus* spp.). *Plant Biol.* **2006**, *9*, 116–126. [CrossRef] [PubMed]

7.   Kleinschmit, J.; Kleinschmit, J.G.R.; Vukelic, J.; Anic, I. *Quercus robur-Quercus petraea*: A critical review of the species concept. *Glasnik Za Šumske Pokuse* **2000**, *37*, 441–452.

8.   Denk, T.; Grimm, G.W. The oaks of western Eurasia: Traditional classifications and evidence from two nuclear markers. *Taxon* **2010**, *59*, 351–366.

9.   Kremer, A.; Abbott, A.G.; Carlson, J.E.; Manos, P.S.; Plomion, C.; Sisco, P.; Staton, M.E.; Ueno, S.; Vendramin, G.G. Genomics of Fagaceae. *Tree Genet. Genomes* **2012**, *8*, 583–610. [CrossRef]

10. Simeone, M.C.; Piredda, R.; Papini, A.; Vessella, F.; Schirone, B. Application of plastid and nuclear markers to DNA barcoding of Euro-Mediterranean oaks (*Quercus*, Fagaceae): Problems, prospects and phylogenetic implications. *Bot. J. Linn. Soc.* **2013**, *172*, 478–499. [CrossRef]

11. Hipp, A.L. Should hybridization make us skeptical of the oak phylogeny? *Int. Oaks* **2015**, *26*, 9–17.

12. Denk, T.; Grimm, G.W.; Manos, P.S.; Deng, M.; Hipp, A.L. An updated infrageneric classification of the oaks: Review of previous taxonomic schemes and synthesis of evolutionary patterns. In *Oaks Physiological Ecology. Exploring the Functional Diversity of Genus Quercus* L.; Springer: Cham, Switzerland, 2017; pp. 13–38.

13. Zhou, Z.; Wilkinson, H.; Wu, Z. Taxonomical and evolutionary implications of the leaf anatomy and architecture of *Quercus* L. Subgenus *Quercus* from China. *Cathaya* **1995**, *7*, 1–34.

14. Pu, C.; Zhou, Z.; Luo, Y. A cladistic analysis of *Quercus* (Fagaceae) in China based on leaf epidermic and architecture. *Acta Bot. Yunnanica* **2002**, *24*, 689–698.

15. Peng, Y.S.; Chen, L.; Li, J.Q. Study on Numerical Taxonomy of *Quercus* L. (Fagaceae) in China. *J. Plant Sci.* **2007**, *25*, 149–157.

16. Zhang, X.; Yao, L.I.; Fang, Y. Geographical distribution and prediction of potential ranges of *Quercus acutissima* in China. *Acta Bot. Boreali-Occident. Sin.* **2014**, *34*, 1685–1692.

17. Zhang, X.; Li, Y.; Liu, C.; Xia, T.; Zhang, Q.; Fang, Y. Phylogeography of the temperate tree species *Quercus acutissima* in China: Inferences from chloroplast DNA variations. *Biochem. Syst. Ecol.* **2015**, *63*, 190–197. [CrossRef]

18. Hui, L.; Xie, H.; Jiang, Z.; Li, C.; Zhang, G. Photosynthetic response of potted *Quercus acutissima* Carruth seedlings under different soil moisture conditions. *Sci. Soil Water Conserv.* **2013**, *11*, 93–97.

19. Fang, S.; Liu, Z.; Cao, Y.; Liu, D.; Yu, M.; Tang, L. Sprout development, biomass accumulation and fuelwood characteristics from coppiced plantations of *Quercus acutissima*. *Biomass Bioenergy* **2011**, *35*, 3104–3114. [CrossRef]

20. Wu, T.; Wang, G.G.; Wu, Q.; Cheng, X.; Yu, M.; Wang, W.; Yu, X. Patterns of leaf nitrogen and phosphorus stoichiometry among *Quercus acutissima* provenances across China. *Ecol Complex.* **2014**, *17*, 32–39. [CrossRef]

21. Choi, H.S.; Kim, Y.Y.; Hong, K.N.; Hong, Y.P.; Hyun, J.O. Genetic structure of a population of *Quercus acutissima* in Korea revealed by microsatellite markers. *Korean J. Genet.* **2005**, *27*, 267–271.

22. Huang, L.; Xiao, L.I.; Yan, J. Studies on Introduction of North American Oaks. China Forestry Science and Technology. 2005. Available online: http://xueshu.baidu.com/s?wd=paperuri%3A%2866d7b49f4975cf2de13aa699e48387b1%29&filter=sc_long_sign&tn=SE_xueshusource_2kduw22v&sc_vurl=http%3A%2F%2Fen.cnki.com.cn%2FArticle_en%2FCJFDTOTAL-LKKF200501009.htm&ie=utf-8&sc_us=11198188077522908127 (accessed on 16 August 2018).

23. Shen, X.; Wu, M.; Liao, B.; Liu, Z.; Bai, R.; Xiao, S.; Li, X.; Zhang, B.; Xu, J.; Chen, S. Complete chloroplast genome sequence and phylogenetic analysis of the medicinal plant *Artemisia annua*. *Molecules* **2017**, *22*, 1330. [CrossRef] [PubMed]

24. Guo, S.; Guo, L.; Zhao, W.; Xu, J.; Li, Y.; Zhang, X.; Shen, X.; Wu, M.; Hou, X. Complete chloroplast genome sequence and phylogenetic analysis of *Paeonia ostii*. *Molecules* **2018**, *23*, 246. [CrossRef] [PubMed]

25. Lobry, J.R. Asymmetric substitution patterns in the two DNA strands of bacteria. *Mol. Biol. Evol.* **1996**, *13*, 660–665. [CrossRef] [PubMed]

26. Necsulea, A.; Lobry, J. A new method for assessing the effect of replication on DNA base composition asymmetry. *Mol. Biol. Evol.* **2007**, *24*, 2169–2179. [CrossRef] [PubMed]

27. Tillier, E.R.; Collins, R.A. The contributions of replication orientation, gene direction, and signal sequences to base-composition asymmetries in bacterial genomes. *J. Mol. Evol.* **2000**, *50*, 249–257. [CrossRef] [PubMed]

28. Jansen, R.K.; Raubeson, L.A.; Boore, J.L.; Depamphilis, C.W.; Chumley, T.W.; Haberle, R.C.; Wyman, S.K.; Alverson, A.J.; Peery, R.; Herman, S.J. Methods for obtaining and analyzing whole chloroplast genome sequences. *Method Enzymol.* **2005**, *395*, 348.

29. Shetty, S.M.; Md Shah, M.U.; Makale, K.; Mohd-Yusuf, Y.; Khalid, N.; Othman, R.Y. Complete chloroplast genome sequence of *Musa balbisiana* corroborates structural heterogeneity of inverted repeats in wild progenitors of cultivated bananas and plantains. *Plant Genome* **2016**, *9*. [CrossRef] [PubMed]

30. Boudreau, E.; Takahashi, Y.; Lemieux, C.; Turmel, M.; Rochaix, J.D. The chloroplast *ycf3* and *ycf4* open reading frames of Chlamydomonas reinhardtii are required for the accumulation of the photosystem I complex. *Embo J.* **1997**, *16*, 6095–6104. [CrossRef] [PubMed]

31. Cavender Bares, J.; González Rodríguez, A.; Eaton, D.A.R.; Hipp, A.A.L.; Beulke, A.; Manos, P.S. Phylogeny and biogeography of the American live oaks (*Quercus* subsection *Virentes*): A genomic and population genetics approach. *Mol Ecol.* **2015**, *24*, 3668–3687. [CrossRef] [PubMed]

32. Kode, V.; Mudd, E.A.; Iamtham, S.; Day, A. The tobacco plastid *accD* gene is essential and is required for leaf development. *Plant J.* **2005**, *44*, 237–244. [CrossRef] [PubMed]

33. Raubeson, L.A.; Peery, R.; Chumley, T.W.; Dziubek, C.; Fourcade, H.M.; Boore, J.L.; Jansen, R.K. Comparative chloroplast genomics: Analyses including new sequences from the angiosperms *Nuphar advena* and *Ranunculus macranthus*. *BMC Genom.* **2007**, *8*, 174. [CrossRef] [PubMed]

34. Yao, X.; Tang, P.; Li, Z.; Li, D.; Liu, Y.; Huang, H. The first complete chloroplast genome sequences in *Actinidiaceae*: Genome structure and comparative analysis. *PLoS ONE* **2015**, *10*, e129347. [CrossRef] [PubMed]

35. Dong, W.; Xu, C.; Li, C.; Sun, J.; Zuo, Y.; Shi, S.; Cheng, T.; Guo, J.; Zhou, S. *Ycf1*, the most promising plastid DNA barcode of land plants. *Sci. Rep.* **2015**, *5*, 8348. [CrossRef] [PubMed]

36. Kikuchi, S.; Bédard, J.; Hirano, M.; Hirabayashi, Y.; Oishi, M.; Imai, M.; Takase, M.; Ide, T.; Nakai, M. Uncovering the protein translocon at the chloroplast inner envelope membrane. *Science* **2013**, *339*, 571. [CrossRef] [PubMed]

37. Provan, J. Novel chloroplast microsatellites reveal cytoplasmic variation in *Arabidopsis thaliana*. *Mol. Ecol.* **2000**, *9*, 2183–2185. [CrossRef] [PubMed]

38. Flannery, M.L.; Mitchell, F.J.; Coyne, S.; Kavanagh, T.A.; Burke, J.I.; Salamin, N.; Dowding, P.; Hodkinson, T.R. Plastid genome characterisation in *Brassica* and Brassicaceae using a new set of nine SSRs. *Theor. Appl. Genet.* **2006**, *113*, 1221–1231. [CrossRef] [PubMed]

39. Saito, Y.; Tsuda, Y.; Uchiyama, K.; Saito, Y.; Tsuda, Y.; Uchiyama, K.; Fukuda, T.; Seto, Y.; Kim, P.G.; Shen, H.L.; et al. Genetic Variation in *Quercus acutissima* Carruth., in Traditional Japanese Rural Forests and Agricultural Landscapes, Revealed by Chloroplast Microsatellite Markers. *Forests* **2017**, *8*, 451. [CrossRef]

40. Yang, Y.; Zhu, J.; Feng, L.; Zhou, T.; Bai, G.; Yang, J.; Zhao, G. Plastid genome comparative and phylogenetic analyses of the key genera in Fagaceae: Highlighting the effect of codon composition bias in phylogenetic inference. *Front. Plant Sci.* **2018**, *9*, 82. [CrossRef] [PubMed]

41. Wang, L.; Wuyun, T.N.; Du, H.; Wang, D.; Cao, D. Complete chloroplast genome sequences of *Eucommia ulmoides*: Genome structure and evolution. *Tree Genet. Genomes* **2016**, *12*, 12. [CrossRef]

42. Doyle, J.J. A rapid DNA isolation procedure for small quantities of fresh leaf tissue. *Phytochem. Bull.* **1987**, *19*, 11–15.

43. Dierckxsens, N.; Mardulyn, P.; Smits, G. Novoplasty: *De novo* assembly of organelle genomes from whole genome DNA. *Nucleic Acids Res.* **2017**, *45*, e18. [PubMed]

44. Chang, L.; Shi, L.; Zhu, Y.; Chen, H.; Zhang, J.; Lin, X.; Guan, X. CpGAVAS, an integrated web server for the annotation, visualization, analysis, and GenBank submission of completely sequenced chloroplast genome sequences. *BMC Genom.* **2012**, *13*, 715.

45. Wyman, S.K.; Jansen, R.K.; Boore, J.L. Automatic annotation of organellar genomes with DOGMA. *Bioinformatics* **2004**, *20*, 3252–3255. [CrossRef] [PubMed]

46. Schattner, P.; Brooks, A.N.; Lowe, T.M. The tRNAscan-SE, snoscan and snoGPS web servers for the detection of tRNAs and snoRNAs. *Nucleic Acids Res.* **2005**, *33*, W686. [CrossRef] [PubMed]

47. Lohse, M.; Drechsel, O.; Bock, R. Organellar Genome DRAW (OGDRAW): A tool for the easy generation of high-quality custom graphical maps of plastid and mitochondrial genomes. *Curr. Genet.* **2007**, *52*, 267–274. [CrossRef] [PubMed]

48. Mudunuri, S.B.; Nagarajaram, H.A. IMEx: Imperfect Microsatellite Extractor. *Bioinformatics* **2007**, *23*, 1181–1187. [CrossRef] [PubMed]

49. Kurtz, S.; Choudhuri, J.V.; Ohlebusch, E.; Schleiermacher, C.; Stoye, J.; Giegerich, R. REPuter: The manifold applications of repeat analysis on a genomic scale. *Nucleic Acids Res.* **2001**, *29*, 4633–4642. [CrossRef] [PubMed]

50. Kurtz, S.; Phillippy, A.; Delcher, A.L.; Smoot, M.; Shumway, M.; Antonescu, C.; Salzberg, S.L. Versatile and open software for comparing large genomes. *Genome Biol.* **2004**, *5*, R12. [CrossRef] [PubMed]

51. Mayor, C.; Brudno, M.; Schwartz, J.R.; Poliakov, A.; Rubin, E.M.; Frazer, K.A.; Pachter, L.S.; Dubchak, I. VISTA: Visualizing global DNA sequence alignments of arbitrary length. *Bioinformatics* **2000**, *16*, 1046–1047. [CrossRef] [PubMed]

52. Frazer, K.A.; Pachter, L.; Poliakov, A.; Rubin, E.M.; Dubchak, I. VISTA: Computational tools for comparative genomics. *Nucleic Acids Res.* **2004**, *32*, W273. [CrossRef] [PubMed]

53. Katoh, K.; Kuma, K.; Toh, H.; Miyata, T. MAFFT version 5: Improvement in accuracy of multiple sequence alignment. *Nucleic Acids Res.* **2005**, *33*, 511–518. [CrossRef] [PubMed]

54. Hall, T.A. BioEdit: A user-friendly biological sequence alignment editor and analysis program for windows 95/98/NT. *Nucleic Acids Symp. Ser.* **1999**, *41*, 95–98.

55. Lam-Tung, N.; Schmidt, H.A.; Arndt, V.H.; Quang, M.B. IQ-TREE: A fast and effective stochastic algorithm for estimating maximum-likelihood phylogenies. *Mol. Biol. Evol.* **2015**, *32*, 268–274.

56. Huelsenbeck, J.P.; Ronquist, F. MRBAYES: Bayesian inference of phylogenetic trees. *Bioinformatics* **2001**, *17*, 754–755. [CrossRef] [PubMed]

Permissions

List of Contributors

Cai-Yun Zhang and Xiao-Lu Mo
Guangdong Food and Drug Vocational College, Guangzhou 510520, China

Tong-Jian Liu and Jian-Rong Li
Key Laboratory of Plant Resources Conservation and Sustainable Utilization, South China Botanical Garden, Chinese Academy of Sciences, Guangzhou 510650, China

Xue-Jun Ge and Hui-Run Huang
Key Laboratory of Plant Resources Conservation and Sustainable Utilization, South China Botanical Garden, Chinese Academy of Sciences, Guangzhou 510650, China
Center of Conservation Biology, Core Botanical Gardens, Chinese Academy of Sciences, Guangzhou 510650, China

Gang Yao
South China Limestone Plants Research Centre, College of Forestry and Landscape Architecture, South China Agricultural University, Guangzhou 510642, China

Hai-Fei Yan
Key Laboratory of Plant Resources Conservation and Sustainable Utilization, South China Botanical Garden, Chinese Academy of Sciences, Guangzhou 510650, China
Center of Plant Ecology, Core Botanical Gardens, Chinese Academy of Sciences, Guangzhou 510650, China

Geoffrey E. Burrows and Celia Connor
School of Agricultural and Wine Sciences, Charles Sturt University, Wagga Wagga, NSW 2678, Australia

Mei Jiang, Haimei Chen, Shuaibing He, Liqiang Wang, Amanda Juan Chen and Chang Liu
Key Laboratory of Bioactive Substances and Resource Utilization of Chinese Herbal Medicine from Ministry of Education, Institute of Medicinal Plant Development, Chinese Academy of Medical Sciences, Peking Union Medical College, Beijing 100193, China

Mira Park
Unit of Polar Genomics, Korea Polar Research Institute, Incheon 21990, Korea
Department of Life Science, Sogang University, Seoul 04107, Korea

Byeong-ha Lee
Department of Life Science, Sogang University, Seoul 04107, Korea

Hyun Park, Hyoungseok Lee and Jungeun Lee
Unit of Polar Genomics, Korea Polar Research Institute, Incheon 21990, Korea
Polar Science, University of Science & Technology, Daejeon 34113, Korea

Huiyu Wu, Narong Shi, Xuyao An, Cong Liu, Li Cao, Yi Feng, Daojie Sun and Lingli Zhang
College of Agronomy, Northwest A&F University, Yangling 712100, China

Hongfei Fu
College of Food Science and Engineering, Northwest A&F University, Yangling 712100, China

Hiroki Irieda
Academic Assembly, Institute of Agriculture, Shinshu University, Nagano 399-4598, Japan

Daisuke Shiomi
Department of Life Science, College of Science, Rikkyo University, Tokyo 171-8501, Japan

Wencai Wang
Institute of Clinical Pharmacology, Guangzhou University of Chinese Medicine, Guangzhou 510000, China

Siyun Chen
Germplasm Bank of Wild Species, Kunming Institute of Botany, Chinese Academy of Sciences, Kunming 650201, China

Xianzhi Zhang
College of Forestry, Northwest A&F University, Yangling 712100, China

Malte Mader, Birte Pakull, Céline Blanc-Jolivet, Maike Paulini-Drewes, Zoéwindé Henri-Noël Bouda, Bernd Degen and Birgit Kersten
Thünen Institute of Forest Genetics, Sieker Landstrasse 2, D-22927 Grosshansdorf, Germany

Ian Small
Australian Research Centre of Excellence in Plant Energy Biology, School of Molecular Sciences, The University of Western Australia, 35 Stirling Highway, Crawley, WA 6009, Australia

Zhitao Niu, Qingyun Xue, Hui Wang, Xuezhu Xie, Shuying Zhu, Wei Liu and Xiaoyu Ding
College of Life Sciences, Nanjing Normal University, Nanjing 210023, China

Shaoyu Zheng
School of Landscape and Architecture, Zhejiang Agriculture and Forestry University, Hangzhou 311300, China

Cuihua Gu
School of Landscape and Architecture, Zhejiang Agriculture and Forestry University, Hangzhou 311300, China
Department of Biology, Colorado State University, Fort Collins, CO 80523, USA

Luke R. Tembrock
Department of Biology, Colorado State University, Fort Collins, CO 80523, USA

Zhiqiang Wu
Department of Ecology, Evolution, and Organismal Biology, Ames, IA 50011, USA

Lea Vojta, Andrea Čuletić and Hrvoje Fulgosi
Laboratory for Plant Molecular Biology and Biotechnology, Division of Molecular Biology, Ruđer Bošković Institute, Bijenička cesta 54, 10000 Zagreb, Croatia

Deng-Feng Xie, Yan Yu, Yi-Qi Deng, Juan Li, Hai-Ying Liu, Song-Dong Zhou and Xing-Jin He
Key Laboratory of Bio-Resources and Eco-Environment of Ministry of Education, College of Life Sciences, Sichuan University, Chengdu 610065, Sichuan, China

Masanori Izumi
Frontier Research Institute for Interdisciplinary Sciences, Tohoku University, Sendai 980-8578, Japan
Department of Environmental Life Sciences, Graduate School of Life Sciences, Tohoku University, Sendai 980-8577, Japan
Precursory Research for Embryonic Science and Technology (PRESTO), Japan Science and Technology Agency, Kawaguchi 332-0012, Japan

Sakuya Nakamura
Department of Environmental Life Sciences, Graduate School of Life Sciences, Tohoku University, Sendai 980-8577, Japan

Xuan Li, Yongfu Li, Mingyue Zang and Yanming Fang
Co-Innovation Center for Sustainable Forestry in Southern China, College of Biology and the Environment, Key Laboratory of State Forestry Administration on Subtropical Forest Biodiversity Conservation, Nanjing Forestry University, 159 Longpan Road, Nanjing 210037, China

Mingzhi Li
Genepioneer Biotechnologies Co. Ltd., Nanjing 210014, China

Index